Life and Motion of Socio-Economic Units

Also in the GISDATA Series

1 *GIS and generalization: methodology and practice*
edited by J. C. Müller, J.-P. Lagrange and R. Weibel

2 *Geographic objects with indeterminate boundaries*
edited by P. A. Burrough and A. U. Frank

3 *GIS diffusion: the adoption and use of geographical information systems in local government in Europe*
edited by I. Masser, H. Campbell and M. Craglia

4 *Spatial analytical perspective on GIS*
edited by M. Fischer, H. J. Scholten and D. Unwin

5 *European geographic information infrastructures: opportunities and pitfalls*
edited by P. Burrough and I. Masser

6 *GIS and health*
edited by A. Gatrell and M. Löytönen

7 *Spatial models and GIS*
edited by A. S. Fotheringham and M. Wegener

8 *Life and motion of socio-economic units*
edited by A. U. Frank, J. Raper and J.-P. Cheylan

9 *Remote sensing and urban analysis*
edited by J.-P. Donnay, M. Barnsley and P. Longley

Series Editors

I. Masser and F. Salgé

Life and Motion of Socio-Economic Units

EDITORS

ANDREW U. FRANK, JONATHAN RAPER
AND
JEAN-PAUL CHEYLAN

GISDATA 8

SERIES EDITORS

I. MASSER and F. SALGÉ

First published 2001 by
Taylor & Francis
11 New Fetter Lane, London EC4P 4EE

Simultaneously published in the USA and Canada
by Taylor & Francis
29 West 35th Street, New York, NY 1001

Taylor & Francis is an imprint of the Taylor & Francis Group

Printed and bound in Great Britain by
TJ International Ltd, Padstow, Cornwall

Publisher's Note

This book has been prepared from camera-ready copy provided by the authors.

British Library Cataloguing in Publication Data

A catalogue record for this book is available from the British Library

Library of Congress Cataloging in Publication Data

Life and motion of socio-economic units / Andrew Frank, Jonathan Raper and Jean-Paul Cheylan [editors].
p. cm. – (GISDATA; 8)
Includes bibliographical references and index.
1. Geographical informations systems. 2. Geographical perception.
I. Frank, Andrew U. II. Raper, Jonathan. III. Cheylan, Jean-Paul.
IV Series
G70.212.L52 2000
910′.285—dc 21 99-045691

ISBN 0-7484-0845-2

TABLE OF CONTENTS

The GIS Data Series

Series Editors' Preface

Over the last few years there have been many signs that a European GIS community is coming into existence. This is particularly evident in the launch of the first of the European GIS (EGIS) conferences in Amsterdam in April 1990, the publication of the first issue of a GIS journal devoted to European issues (*GIS Europe*) in February 1992, the creation of a multipurpose European ground-related information network (MEGRIN) in June 1993, and the establishment of a European organisation for geographic information (EUROGI) in October 1993. Set in the context of increasing pressures towards greater European integration, these developments can be seen as a clear indication of the need to exploit the potential of a technology that can transcend national boundaries to deal with a wide range of social and environmental problems that are also increasingly seen as transcending the national boundaries within Europe.

The GISDATA scientific programme is very much part of such developments. Its origins go back to January 1991, when the European Science Foundation funded a small workshop at Davos in Switzerland to explore the need for a European level GIS research programme. Given the tendencies noted above it is not surprising that participants of this workshop felt very strongly that a programme of this kind was urgently needed to overcome the fragmentation of existing research efforts within Europe. They also argued that such a programme should concentrate on fundamental research and it should have a strong technology transfer component to facilitate the exchange of ideas and experience at a crucial stage in the development of an important new research field. Following this meeting a small coordinating group was set up to prepare more detailed proposals for a GIS scientific programme during 1992. A central element of these proposals was a research agenda of priority issues groups together under the headings of geographic databases, geographic databases, geographic data integration and social and environmental applications.

The GISDATA scientific programme was launched in January 1993. It is a four-year scientific programme of the Standing Committee of Social Sciences of the European Science Foundation. By the end of the programme more than 300 scientists from 20 European countries will have directly participated in GISDATA activities and many others will have utilised the networks built up as a result of them. Its objectives are:

- to enhance existing national research efforts and promote collaborative ventures overcoming European-wide limitations in geographic data integration, database design and social and environmental applications;
- to increase awareness of the political, cultural, organisational, technical and informational barriers to the increased utilisation and inter-operability of GIS in Europe;
- to promote the ethical use of integrated information systems, including GIS, which handle socio-economic data by respecting the legal restrictions on data privacy at the national and European levels;
- to facilitate the development of appropriate methodologies for GIS research at the European level;
- to produce output of high scientific value; and
- to build up a European network of researchers with particular emphasis on young researchers in the GIS field.

A key feature of the GISDATA programme is the series of specialist meetings that has been organised to discuss each of the issues outlined in the research agenda. The organisation of each of these meetings is in the hands of a small task force of leading European experts in the field. The aim of these meetings is to stimulate research networking at the European level on the issues involved and also to produce high quality output in the form of books, special issues of major journals and other materials.

With these considerations in mind, and in collaboration with Taylor & Francis, the GISDATA series has been established to provide a showcase for this work. It will present the products of selected specialist meetings in the form of edited volumes of specially commissioned studies. The basic objectives of the GISDATA series is to make the findings of these meetings accessible to as wide an audience as possible to facilitate the development of the GIS field as a whole.

For these reasons the work described in the series is likely to be of considerable importance in the context of the growing European GIS community. However, given that GIS is essentially a global technology most of the issues discussed in these volumes have their counterparts in research in other parts of the world. In fact there is already a strong UK dimension to the GISDATA programme as a result of the collaborative links that have been established with the National Center for Geographic Information and Analysis through the United States National Science Foundation. As a result it is felt that the subject matter contained in these volumes will make a significant contribution to global debates on geographic information systems research.

Ian Masser
François Salgé

General Introduction

The world in which we live is permanently changing and these changes vary widely: some changes are extremely slow, nearly unnoticeable for us, for example, changes in climate or geology; others are rapid, like the movement of cars or wind. The objects changing are sometimes very large, sometimes very small (Gibson, 1979).

Nevertheless, most of the world is stable and with respect to this stable framework, we observe some elements change. In general, we seem to assume that we and most of our environment remain stable and few objects around us change. This is sometimes an illusion, as perceived changes are the effect of changes in our perception, in the classification or the observation of phenomena.

The interaction between objects have spatial and temporal ranges (Fraser, 1981): people interact with their environment centred around their places of living for 'three scores and ten' years. Glaciers interact with the region around them for thousands of years. A grass plant interacts with the plants and soil within a foot for a single year. Each science concentrates on some cluster of interaction, with a particular range of space or time in which the interaction of interests occurs (Morrison, 1982). For faster or slower interaction the larger or smaller processes create an environment which is relative to the process. It becomes then necessary to aggregate phenomena for a smaller or shorter interaction scale to bring them to the scale of interaction of the larger or smaller process. GIS—and to some degree geography in general—cross these scale boundaries and have to deal with the effects of aggregation, both in time and space.

People's interest in change in their environment is high. Many decisions are determined by change; often change is the more important signal than a static situation. The media report on change. Typically, politicians react to change with changes in the socio-administrative system. Planning observes changes in the environment and tries to stir them in a desired direction. Geography has evolved from a mostly descriptive attitude and concentrates on studying processes in space and time, i.e., changes, their causes and effects. The field of spatio-temporal processes in geographic space is a broad and rich field for scientific studies. Accurate data are necessary to back theories with actual observations. Even more data are necessary to describe changes.

Geographic Information Systems were designed to accurately collect, manage and render a description of the world around us (FIG Fédération Internationale des Géomètres, 1981). The current design of commercial GIS assumes the world as static and ignores processes that change the world. The GIS contains only the resulting, currently valid data—previous, historical data is lost. The changes in the world are captured in an update process. Updating is a practical problem; it is an administrative process built around the GIS, more as an afterthought.

The variable time is not included in current GIS, which store static snapshots of the world, following Sinton's general scheme for cartography: Time is fixed, Location is controlled and Theme is observed (Sinton, 1978). Often the time of the snapshot is not even well defined. This triad, which is imposed by technical limitations upon cartography, has been carried forward to the current commercial GIS.

This restriction of GIS to cartographic snapshots has reduced the technical problems in their design and in the challenge they pose to their users. But it has also reduced the potential of GIS to contribute to the solution of actual administrative, planning and scientific questions. It has also increased the difficulties to integrate GIS as a tool in administrative processes. GIS are not maps and are not restricted by the physical constraints of cartographic presentation. GIS could be used to collect time varying data, exploit it with respect to time and show the results graphically, only current commercially available systems are very much restricted. Several of the contributions develop this theme and May Yuan proposes a view of a GIS which has theme, space and time as equal partners (Chapter 15). We present here a continuous chain of arguments, from Jonathan Raper (Chapter 1), Jean-Paul Cheylan (Chapter 3) to Stuart Aitken (Chapter 13), which point to the impossibility to capture the social reality of a territory in the straightjacket of static Euclidean geometry, which can be linked to this remaining restriction inherited from cartography. To lift this restriction is possible today.

The field of spatio-temporal GIS is vast and the number of applications is enormous. Despite the importance ascribed to this field and the research efforts made in the past years, progress has been slim (Barrera *et al.*, 1991; NCGIA 1993; Egenhofer and Golledge, 1998). A first approach to introduce time into the GIS is to add to the two spatial dimensions current GIS handle, a third for the height and a fourth for time. Understanding time and space as a four-dimensional isotropic continuum has been very effective for physics and leads to substantial insight into the structure of matter and the smallest particles of which it consists. It is doubtful if a similar approach is usable in geography. Geographic space is not isotropic and movement in the plane is much different and experienced differently from a movement in the vertical. Unlike spatial dimensions, the arrow for time is fixed and processes progress from the past to the future and we cannot reverse this at will (Franck, Chapter 8); (Couclelis and Gale, 1986).

To make progress in research, it is often necessary to concentrate on a small part of a large problem. In this volume the focus is on socio-economic units, as a particular type of geographic object and their change in time. It is well understood that this is only one of several important research issues on the way to create a widely useful temporal geographic information system, but it is hoped that this focus will result in progress in particular questions.

The interaction of people with space can first be seen as a very direct one: people move around in space. This can be seen abstractly as the movement of points against a fixed frame.

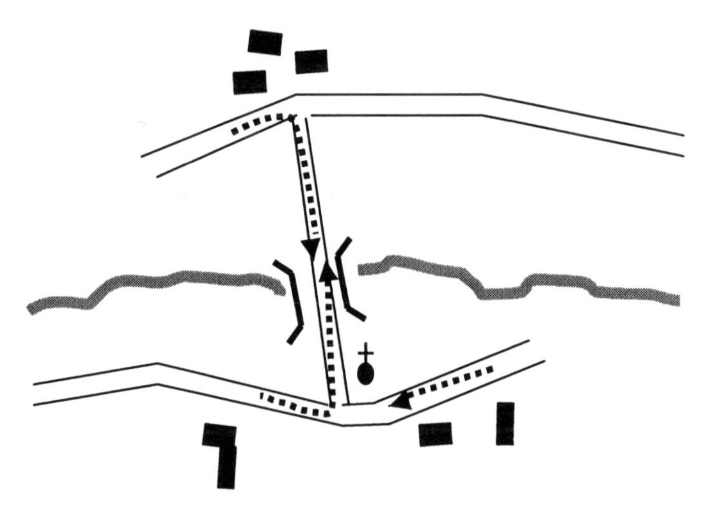

Figure 1 Movement of points (vehicle)

But people also create by social processes larger areal units in space, for example, by the division of land into countries, which serve as fixed frames of reference for point movements. These areal units themselves change if sufficiently large time periods are considered.

Figure 2 Movement of area (country)

We have selected this second topic in the hope of

- opening a new topic of research, avoiding areas where already several competent research efforts are underway;
- defining the area well enough that productive detailed research can follow; and
- demonstrating that research results in this subfield have important practical applications.

Torsten Hägerstrand pioneered research in time geography addressing the first question. He traced the movement of individuals in space and time. He described the path of a person from home to work place in the morning and back to home in the evening and depicted it graphically as a path in three dimensions (the x–y plane of urban space and the time dimension as z). Generalised, time geography investigates the movement of one (or few) mobile elements in a space that is much larger than the elements, and the only thing that changes is the position of the elements in this space—everything else remains fixed in time. The town does not change (noticeably) during our travel, nor are changes in the elements moving considered. The conceptual framework is rich and widely applicable for studies in transportation and application of GIS to transportation problems. A currently important application of 'point movement' in geography and GIS is traffic, from systems to observe traffic volume, flow of individual cars to systems to advise drivers about optimal routes (Golledge, 1998).

This book concentrates on the second aspect of the human dimension of space and time in geography: the change of socially created large geographic objects, used to subdivide space for administrative and economic purposes. Social processes, mostly political, subdivide geographic space in delimited areas, some as large as a continent, some as small as a parcel of land owned by a person. These areas we call here spatial Socio-Economic Units. This social spatial framework is considered fixed for most of our activities, it is part of the fixed spatial framework against which we observe change—units like France, Washington D.C., and even parcels often remain unchanged for hundreds of years. Nevertheless, by careful observation, we find that these socio-economic units are occasionally changing and this framework is not as stable as we generally believe. Municipal boundaries change as the result of administrative reform; war leads to the appearance or disappearance of countries, etc.

The term 'spatial socio-economic' was selected to describe in most general terms spatial units that are the result of some social, cultural, economic, behavioural, etc. processes. There are many such processes and they follow their specific logic and produce different spatial units. The spatial socio-economic units most often considered are the reporting units for the national statistics and the political subdivisions, from communes, districts to countries. Many spatial socio-economic units consist of smaller units and are hierarchically aggregated—but particulars of the hierarchy problem are not within the scope of the effort reported here (for recent work on spatial hierarchies see Timpf, 1998). The prototypical spatial socio-economic unit has sharply defined and precisely known boundaries, but not all spatial socio-economic units follow this model: Aggregates of postal addresses for a common postal code are important, widely used spatial socio-economic units, but not in all countries do they represent delimited areas (Raper *et al.*, 1992); (Reis, Chapter 21). Other examples of socio-economic units with vague boundaries are the service areas—the area serviced by a particular school or hospital is often not precisely fixed and changes over time. Some socio-economic units seem to be more real and appear as objectively, often even physically existing: spatial socio-economic units are sometimes clearly marked in the world (national boundaries as an extreme case), influence other spatial processes (for example, the effects of urban planning zones) and have or become 'physical reality'. Others remain purely administrative and barely noticed by the general public, and some exist only in the perception of the local residents (for example, territories of gangs or sacred places). Despite their conceptual stability, such units slowly change.

This book tries to systematise this specific sub-area of spatio-temporal GIS: what are the elements that describe change to the socio-economic units in time. This requires asking general questions about the spatial socio-economic units themselves: what are their properties, how are they created, how do they change and disappear. The contributions in this volume are grouped in 5 parts, each having a particular focus.

Part 1 sets the stage and introduces the problem in abstract terms. It provides the intellectual framework to start an abstract discussion about the 'life and motion' of spatial socio-economic units, concentrating on their generic aspects and abstracting away the particulars of each special case. On this abstract level of discussion, Jonathan Raper initiates the discussion: spatial socio-economic units are human artifacts (Searle, 1984); (Berger and Luckmann, 1996), used for the collection of data with a life span, from creation to deletion. Spatial socio-economic units can also move in space over time. The second chapter by Andrew Frank differentiates types of change, separating changes in attribute properties from motion of the spatial socio-economic unit and catastrophic events ('life'), which create new spatial socio-economic units or make them disappear. Jean-Paul Cheylan further analyses the interaction between these changes and the final chapter by Marinos Kavouras links the abstract study of spatial socio-economic units with their use in administration and in particular planning.

Part 2 inquires in the philosophical background. Spatial socio-economic units are not physical objects and their properties are thus not as simple to comprehend as the red or green colour of an apple. It starts with an analysis by Carola Eschenbach of a seemingly simple query "How did the population of European capitals change during the last 100 years?", revealing the possible ontological commitments one could make and their effects on the results of the query. The chapter by Barry Smith reviews the metaphysical categories of Aristotle and concludes that spatial socio-economic units are not physical objects and share only few properties of these. From the investigation by Roberto Casati follows that spatial socio-economic units are more similar to shadows, which differentiate different parts of a region, without having a physical existence. But time is also not uniform and the chapter by Georg Franck points out four different uses of time—from a reversible physical time, to the irreversible biological clock, the experienced time with the exceptional point 'present' and two different measures for time in economics, which link space and time.

Part 3 reviews formal properties of a spatio-temporal GIS, starting with a review of the logical foundation by Michael Worboys and stressing three different views on space, time and theme—revisiting a fundamental law of cartography (Sinton, 1978) and applying it to the temporal GIS. Damir Medak contributes an analysis on a highly formalised level separating different 'lifestyles' for spatial socio-economic units. The remaining chapters consider the spatio-temporal database perspective: Therese Libourel investigates changes in the database schema, a point often forgotten, but of eminent importance for databases which should be used for long periods of time. Emmanuel Stephanakis and Timos Sellis link the abstract requirements to database implementation issues and query processing—the perspective on time in temporal databases can be abstracted to a sequence of transactions on which queries are possible.

Part 4 considers geographic applications. The chapter by Stuart Aitken is situated in the urban planning context and critically assesses the spatial socio-economic units, asking how they are created, by whom and to what end. The following chapter by Denis Gautier presents a complex application of a spatio-temporal GIS and the corresponding analysis to understand the evolution of a forest under different economic pressures. May Yuan further increases the links between space, theme and time with an example of service areas of supermarkets. Mauro Salvemini stresses how planning, especially city planning, implies a complex structure of the world where spatial objects of different nature interact with very different time scales. The evolution of land ownership and the legal rules, which guide a land registration system—a key example for a temporal GIS, is the subject of a highly formal treatment by Khaled Al-Taha.

The last part investigates the definition of spatial socio-economic units: Erik Stubkjær's chapter analyses in detail the interaction between changing the definitions of spatial socio-economic units and their collection—in a way reflecting the perspective of Libourel's review of the database schema. Stan Openshaw and Seraphim Alvanides then discuss the effects various aggregations of spatial socio-economic units have on the presentation of socio-economic data and then assess various methods for an optimal design of spatial socio-economic units. Mike Coombes and Stan Openshaw report on a practical project concerning a specific set of SSEUs in Britain, the so-called Travel-to-Work Areas, revealing changes in the size of spatial socio-economic units associated with longer journeys to work over the last decades. The chapter by Rui Pereira Reis deals with the actual decisions made in the creation of Portugal's new postcodes. The last chapter by Jostein Ryssevik reports on a pragmatic solution for a database of socio-economic data from the Norwegian census, which allows the approximate transformation of collected data to a selected set of spatial socio-economic units—adjusting for changes in their boundaries.

In conclusion, the volume presses for a stronger interaction between the social processes that create and change spatial socio-economic units, and the management and use of the data collected for them. The simplifying fiction that spatial socio-economic units are fixed, well defined areal units, which the static implementation of GIS today imply, considerably limits the use of GIS for the analysis of complex social interaction as it is necessary for, for example, urban planning. This view is an artifact from the cartographic tradition of GIS and must be replaced by a balanced triad of theme, space and time (Yuan, Chapter 15). The future GIS must provide data types for time varying objects, for example, spatial socio-economic units or moving objects (Erwig *et al.*, 1997), and relate them to forces which make them change and to abstract representations of change. This volume hopefully attempts to justify this need from a fundamental geographic analysis, but also from the practical needs—mostly felt in urban planning—and relates it to the formal mathematical treatment necessary for its realisation in future GIS software. The chapters here also indicate the relation of this research effort with the proper treatment of hierarchies and similar structures and scale in general—a currently active research area in GIS.

The theoretical analysis presented here should justify the development of GIS products by demonstrating their rich applicability and provide a theory for their implementation.

ACKNOWLEDGEMENTS

This interdisciplinary study is the part of the GISDATA project, which was initiated by Ian Masser and François Salgé, and is financed by the European Science Foundation. Max Craglia administrated the project and pushed it forward along its slow and wearisome path from the initial idea to the printed book. Marinos Kavouras acted as the gracious host for the meeting in Nafplion, where most of the chapters in this volume were presented initially. Authors then revised their manuscript based on the discussion and we invited a small number of additional contributions to form this hopefully comprehensive volume. Last, but not least, Mag. Roswitha Markwart managed the preparation of the book manuscript at the Technical University in Vienna. We appreciate the numerous and various contributions of them all.

Andrew Frank
Jonathan Raper
Jean-Paul Cheylan

REFERENCES

Barrera, R., Frank, A.U. and Al-Taha, K., 1991, Temporal relations in Geographic Information Systems: A workshop at the University of Maine. *SIGMOD Record*, **20** (3), pp. 85–91.

Berger, P.L. and Luckmann, T., 1996, *The Social Construction of Reality*, (New York: Doubleday).

Couclelis, H. and Gale, N., 1986, Space and spaces. *Geografiske Annaler* 68B, pp. 1–12.

Egenhofer, M.J. and Golledge, R.G., Eds, 1998, *Spatial and Temporal Reasoning in Geographic Information Systems*, (New York: Oxford University Press).

Gibson, J.J., 1979, *The Ecological Approach to Visual Perception*, (Boston: Houghton-Mifflin).

Erwig, M., Güting, R.H., Schneider, M. *et al.*, 1997, *Spatio-Temporal Data Types: An Approach to Modeling and Querying Moving Objects in Databases*. Report No. 224. (Hagen: FernUniversität).

FIG Fédération Internationale des Géomètres, Ed, 1981, *XVIe Congres International des Géomètres*. (Montreux, Switzerland).

Fraser, J.T., Ed., 1981, *The Voices of Time*, (Amherst: The University of Massachusetts Press).

Golledge, R.G., 1998, The relationship between GIS and disaggregate behavioral travel modeling. *Geographical Systems*, **5** (1–2), pp. 9–17.

Morrison, P., 1982, *Powers of Ten*, (San Francisco, CA: W.H. Freeman).

NCGIA, 1993, *Time in Geographic Space*. Report on I–10 Specialist Meeting, NCGIA.

Raper, J.F., Rhind, D. and Sheperd, J., 1992, *Postcodes—The New Geography*, (London: Longman).

Searle, J.R., 1984, *Minds, Brains and Science*, (Cambridge, MA: Harvard University Press).

Sinton, D., 1978, The inherent structure of information as a constraint to analysis: Mapped thematic data as a case study. In *Harvard Papers on Geographic Information Systems*, edited by Dutton, G. (Reading, MA: Addison-Wesley). Vol. 6.

Timpf, S., 1998, *Hierarchical Structures in Map Series*. Ph.D. thesis, Department of Geoinformation. (Vienna: Technical University Vienna).

PART ONE

Setting the Stage

INTRODUCTION

The notion 'spatial socio-economic' was selected to describe in the most general terms spatial units in geographic space that are the result of some social, cultural, economic, or behavioural process. There are many such processes, each following their own specific logic and producing different spatial units. In this first part of the book, an overview of the issues is provided covering the social processes that create the spatial socio-economic units, the framework used to identify particular aspects of the topic and the relevant application areas. The units being discussed in this book are explicitly spatial and are distinguished from socio-economic units in general.

Socio-economic units are part of the socially constructed reality and thus the product of social processes. Sociology has studied the social environment and the mechanism leading to the construction of apparently real objects (Berger and Luckmann, 1996) and there are very extensive debates in philosophy (Searle, 1984) questioning their ontological status, the level and particulars of the 'realness' of social artefacts.

The spatial socio-economic units most often considered are the reporting units of the national statistics and the political subdivisions, both of them having different levels of aggregations: the commune, the district, the province, and the nation for the political subdivision, and similarly for the spatial units used for reporting the results of statistical surveys. There are also jurisdictional spatial socio-economic units identifying national sovereignty, which sometimes extend beyond the physical boundaries of the state such as marine limits or 'security zones'. However, there are also non-state spatial socio-economic units maintained by commercial companies and utility providers, which sometimes define territories of provision and entitlement.

The first chapter by Jonathan Raper addresses these fundamental questions to clarify what spatial socio-economic units are in general and how they are constructed in a social process. It reviews the naive motivation of the positivists for the creation of spatial socio-economic units, often found in early geography texts, and reflects the more recent critical comments up to a post-modern radical questioning of reality and the need for spatial socio-economic units. Notwithstanding these spirited discussions, spatial socio-economic units are directly experienced in various ways. The extent and construction of spatial socio-economic units changes with technology, for example, when the distance one can easily and quickly travel is increased or through the use of telecommunication, but nevertheless, spatial socio-economic units are necessary for human understanding of space and its administration, despite their very shaky ontological foundation.

The analyses of the social forces behind this construction are mostly static, thus leading to a perception of spatial socio-economic units as stable. The analysis by Raper can be extended to consider the changes in the social process that lead to the formation of spatial socio-economic units and which then hints at the reasons for changes in the spatial socio-economic units. The results of the analysis of the use of spatial socio-economic units for territorial control applies *mutatis mutandis* to other situations such as areal units created by other processes, for example, watersheds (see Franck, Chapter 8; Aitken, Chapter 13; Openshaw, Chapter 19).

The second chapter by Andrew Frank approaches the subject of spatial socio-economic units analytically. It first separates the change of properties of spatial objects like spatial socio-economic units from changes in the spatial socio-economic units proper. Hence, the changing number of inhabitants of a country is a change in the properties of the spatial socio-economic unit, but not in its essence. It further differentiates between changes that an individual spatial socio-economic unit endures without changing its identity: a nation state can grow or shrink, it can even move in space, without losing its identity (the best known example is probably Poland). But spatial socio-economic units can also change drastically, when two units are merged (e.g., the 'Anschluss' of Austria into the German Reich) or when from a single SSEU two new ones are created (for example, the splitting of Czechoslovakia into the Czech Republic and Slovakia). These 'life changes' create new spatial socio-economic units or terminate existing units, and are more complex than mere movements. This viewpoint abstracts from the scale of resolution and the particulars of the spatial socio-economic unit. The chapter reviews the nine temporal constructs proposed by Al-Taha and Barrera (1994) for 'life changes', and gives a list of the different kinds of movements or other changes that can be considered as movements. This is by analogy with the way that ecologists talk about the growth of forest, or about the movement of a biotope slowly up-hill under pressure from civilisation, both the aggregation effects from individual plants emerging and disappearing in a particular pattern.

Socio-economic units are justified as spatial units for which data can be collected, for administrative processes or for scientific analysis. Jean-Paul Cheylan in his chapter creates in generic terms a framework that links the data collection methods to the temporal databases and query languages. This identifies alternatives and therefore design decisions to be made for the construction of databases for data related to spatial socio-economic units. The chapter links the analysis from the spatio-temporal database community with geographic thought and points out the critical needs of GIS applications. It points out the central role the concept of a permanent identity of an object in time plays. This is typically represented in a database by a stable identifier for each object. It then extends the framework presented in the previous chapter by Frank with the concept of a genealogy, covering the cases where one object spawns new objects, or objects are merged, from simple life changes, like creation or destruction.

Urban planning is the quintessential spatio-temporal effort—planning is not possible if we cannot assume that there is an evolution, or a change, in time. Planning goes beyond science, as it applies the results of science to arrive at a prediction of a possible future and the selection of a course of action that leads to a desirable future. The last chapter by Marinos Kavouras points to the fundamental view of space and time used in planning theory and shows how different application areas pose entirely different demands on the temporal GIS. He gives details of the requirements that planning applications have—a topic that is later reconsidered by Aitken (Chapter 13) and Salvemini (Chapter 16). The methods to form objects and the spatial reasoning required separate planning from administrative uses or scientific analysis. Kavouras points out the interaction between scale, resolution and generalisation—concepts that are currently analysed in the cartographic literature (for a review see Timpf, 1998)—and extends the discussion to temporal scale, resolution and generalisation. It becomes clear that a single answer, a single set of functionality for a GIS, is not possible—but possible multiple methods can be fitted within a unifying concept, thus allowing integration of multiple viewpoints in a system. At the end, it points to the very difficult task of building a temporal GIS to record uncertain information, as can be found in historical texts, and to link these to the known locations of historic sites and the like. This seems to be a small, very particular demand by historians and archaeologists using GIS (Allen *et al.*, 1990), but the extension of GIS to manage spatial and temporal information without exact position or date is in fact a very general demand, as most spatial information used by humans has a relative or indirect relation to space.

REFERENCES

Allen, K.M.S., Green, S.W., *et al.*, 1990, *Interpreting Space: GIS and Archaeology*, (London: Taylor & Francis).

Al-Taha, K. and Barrera, R., 1994, Identities through time. In *Proceedings of the International Workshop on Requirements for Integrated Geographic Information Systems*, New Orleans, Louisiana.

Berger, P.L. and Luckmann, T., 1996, *The Social Construction of Reality*, (New York: Doubleday).

Searle, J.R., 1984, *Minds, Brains and Science*, (Cambridge, MA: Harvard University Press).

Timpf, S., 1998, Hierarchical Structures in Map Series. Ph.D. thesis, Department of Geoinformation, (Vienna: Technical University Vienna).

CHAPTER ONE

Defining Spatial Socio-Economic Units: Retrospective and Prospective

Jonathan Raper

1.1 INTRODUCTION

It is a characteristic of all human societies that concepts of geographic space play a part in social behaviour on both an individual and collective level. Geographic space, defined by human experience at scales between the table-top and the planetary, is simultaneously a resource, a medium for interaction and an ordering and is, therefore, a pervasive part of human experience. As a consequence individuals, cultures and governance create and sustain representations of geographic space in a very wide range of forms, which are communicated and reproduced through language. While research on geographic space has given rise to a wide range of studies from many theoretical perspectives in recent decades (see Raper, 1996), this essay is concerned with the kind of spatial representations concerning the organisation and governance of society that must be made explicit so that they can be communicated unambiguously. In this sense spatial representations are seen as part of the process described by Hacking as 'representing and intervening' (1983, p. 31).

Weak forms of such spatial representations would be, for example, the spatial configuration of a set of dwellings establishing the group of citizens entitled to service from a particular hospital. A strong form of such a spatial representation would be the boundary of a legal jurisdiction. Such units have become known as spatial socio-economic units (SSEUs). The use of SSEUs in developed societies has progressed from statutory and collective usage in governance to *ad hoc* and private usage in commerce for service delivery and the marketing of products. Yet there have been few attempts to study the formalisation of these entities or to monitor their application, despite the growing influence of practices such as geodemographic analysis employing geographic information systems (GIS). This essay aims to review the work done on the formalisation of SSEUs and to make a case for their potential use in the exploration of social processes (Raper *et al.*, 1992).

1.2 DEFINITIONS

The creation of explicit representations of space such as SSEUs must depend ultimately on ontological foundations. There have been several recent studies of the ontological properties of entities defined in geographic space. Casati (this volume) has been concerned with entities which are abstract in their defining 'condition' stating that they can be causally sustained by process even if they lack physical reality, and gives the nation as an example. By contrast Smith (1995) was concerned to distinguish two kinds of explicitly geographic entities which have the 'identity of existence'. Firstly, he defined 'bone fide' objects as 'discontinuities in physical reality', whereas, secondly, he defined all other objects as 'fiat' objects which are created by human whim. Eschenbach (this volume) discusses the ontology of space employed by geographic entities noting that SSEUs require absolute geographic referencing in order to define their shape and location. She points out ways in which the identity and mereology of SSEUs is frequently ambiguous and complex as in the case of the changing definitions of Berlin and 'Germany's capital' during the 20th century.

Hence, the creation of SSEUs requires the use of a basic assumption, i.e., that some defining 'condition' or 'identity of existence' can be held constant over geographic space and over some span of time. This assumption is in itself highly problematic as many such defining conditions can interpenetrate spatially and temporally, yet such units are universally used in developed societies. Perhaps two broad cases of a defining condition can be identified: in one case the condition can be a restriction on access or activity within a geographic area defined physically with reference to the world of entities. In the other case the condition refers to the characteristics of the people or property defining the spatial unit—in other words the characteristics of an aggregation. The 'holding constant' of the condition is usually defined by reference to scales of social, economic or political activity significant to a particular society.

In the light of this definition it is proposed here that SSEUs can be classified according to their characteristics on three scales:

1. purposes, ranging from the symbolic (for example, religious) to the instrumental (for example, governmental);
2. lifespans, ranging from the transient (short lived) to the permanent (long lived); and
3. spatial identities, ranging from the diffuse (vaguely defined) to the concrete (sharply defined).

At one end of a spectrum of possible spatial units communities and social groups tend to informally create symbolic spatial units which are transient and diffuse in nature, for example, neighbourhoods (defined by identity), favourite vistas (defined by concepts of landscape) or perceptions of 'dangerous places-to-be' (defined by social behaviour). By contrast, at the other end of the spectrum, governments and commerce tend to define instrumental spatial units using structured approaches that are largely permanent and concrete in nature (though not necessarily completely unchanging). Such spatial units are made, for example, to control access to space (defined by ownership), to limit the uses of space (defined by jurisdiction), to distribute resources efficiently (defined by supply and demand locations) or to characterise localities (defined by a socio-economic classification).

The definition and the contesting of these SSEUs is an important social and political process and involves communities, government and commerce through such means as governance, spatial planning, service delivery or property trading. Fundamental to the process of making and remaking SSEUs is the need to communicate the status of the units widely, and hence, the need to describe and publish their nature and extent. In order to debate and realise 'spatial units' (especially where governance is concerned) some form of representation of space is desirable: while symbolic, transient and diffuse units may employ oral descriptions referenced to landmarks or street furniture, the instrumental, permanent and concrete units have traditionally been defined in terms of boundaries which are attached to physical objects identifiable on the ground and subject to a wide consensus. Such boundaries may be defined on 'official' maps in some jurisdictions.

1.3 THE POSITIVIST ORIGINS AND MOTIVATION FOR SPATIAL SOCIO-ECONOMIC UNITS

The origins of the case for the definition of SSEUs can be found in what has been termed 'positivist theories of space' (Johnston, 1991). Many of the studies by geographers carried out before 1970 were based solely on the principles of positivism, that observable facts are the only possible forms of knowledge. Positivism is generally considered to have developed, firstly, from Comte's (1842) view that the methods of the physical sciences could be applied to human behaviour and organisation; and, secondly, from Mach's (1883) view that observational confirmation is the key characteristic of scientific statements and that metaphysical statements have no meaning (Ray, 1991). Logical positivist philosophers such as Carnap and Ayer developed the methods of positivism in the early decades of this century emphasising empiricist methods based on verifiability from a realist perspective on perception. Geographers of the 1950s and 1960s largely accepted the claims of positivism, i.e., that spatial processes and 'objects' (areas) of interest could be observed and explanatory laws could be formulated about them.

Three distinct groups of theories have been developed along positivistic lines. Firstly, there are theories that aim to explain spatial organisation of land use. For example, in the early years of this century Christaller's Central Place Theory proposed that settlements were located at the centres of zones of influence which themselves were spatially and hierarchically ordered by settlement interactions; Von Thunen's theory predicted spatial zonation of land use in and around cities according to the cost of land and transportation of products to markets; and Weber's theories suggested that spatial variations of resources governed spatial zonation of land use. These rather 'pure' theories were elaborated in the 1950s and 1960s by adding into the studies 'realistic' conditions such as the spatial distribution of demand, 'economy of scale' effects, cumulative development, and agglomeration processes in order to explain observed spatial patterns (Lloyd and Dicken, 1977). In a later stage of development in the 1970s, these theories were modified by adding 'uncertain' or probabilistic decision-making as a factor in so-called behaviourist studies.

Secondly, 'positivist' theories of space were also developed to explain the evolution of regions. Notable examples include the 'geographical matrix' of Berry (1964) in which geography is represented as a matrix of places and characteristics at a series of specific times; the study of the 'spatial structure' generated by interaction between regions in developed societies developed by Haggett (1965); and, the use of system theory to model the behaviour of a regional economy (Chorley and Kennedy, 1971) and agricultural economics (Chapman, 1977).

Thirdly, a number of attempts were made since the 1970s to formally define 'zone design' methodologies. Many of these were based on 'Broadbent's rule' (Broadbent, 1970), i.e., the relation between trip length and zone radius governs the number of zones that are required to represent a spatial system. This rule is based on the axiom that an 'interaction' takes place in a spatial system if a trip crosses a zone boundary. As a 'rule-of-thumb' for the design of such zone systems, 90% of interactions should cross a zonal boundary (Masser and Brown, 1978). Trips were usually defined as straight-line distances, although distances through road networks were sometimes used. The problem with these approaches is that they only defined the appropriate size of zones or ways in which they could be aggregated hierarchically. Reis and Raper (1994) examined the possible approaches to the larger problem of actually constructing zones from basic geometry since many parts of the world still have no SSEUs defined. Wood *et al.* (1999) suggested that SSEUs could be derived from the second derivatives of a smooth function representing population density and illustrated their approach using London as an example.

However, spatial organisation, regional evolution and zone formation theories share certain 'positivist' assumptions: that space is treated as isotropic physical space; that time can be held constant over space; that 'places' such as communities, neighbourhoods or cities could be delineated with spatial precision; that the characteristics of places could be represented using quantitative indicators; that spatial processes such as commuting or diffusion of disease can be represented objectively; that 'geographic individuals' (whether persons, communities or socio-economic classes) behave rationally and reproducibly; and that the creators of SSEUs hold 'universally' supported values. Positivist theories of space have also normally adopted the methodology of critical rationalism (from Popper), which involved falsification rather than verification as an approach to theorising. However, the many falsifications of the positivist theories of space have usually been met with extensions to the theories or by exclusion of any 'falsifying' circumstances from the scope of the theory.

1.4 REACTIONS TO POSITIVIST THEORIES OF SOCIETY AND SPACE

In the period after 1970 there have been a wide range of reactions to the use of positivistic methodologies in the social sciences in general. In the social sciences one critique of positivism was provided by Foucault (1980) who questioned whether science could be equated with truth, and suggesting instead that truth is defined by the systems of power which create and sustain it. Foucault (1972) took a historical approach arguing that truth is always situated within the 'world-views' (or 'epistemes') that are created by a society in a particular period. By contrast, Habermas (1978) sought to show that human 'interests' structure knowledge and that different epistemologies are appropriate for each. Hence, positivism supplies a methodology for the study of the 'technical' interest expressed through work, hermeneutics (the study of meanings) provides a methodology for the 'practical' interest expressed through language and communication, and 'critical' theory provides a methodology for the 'emancipatory' interest expressed through power. The development of critical social theory is based upon the belief that positivism (focused on explanation of appearances) and hermeneutics (focused on understanding meanings) fail to provide any framework for the individual to develop the knowledge of how to act given contemporary structures of power in society. In fact, Habermas argues that the identification of knowledge with science is false and that positivistic science can only function through the 'technical' interest, which may not reflect a concern with values or impressions (Unwin, 1993).

One reaction to positivism in geography was the critique of the general assumptions of positivist theories of space and attempts to improve them through 'behavioural approaches'. Examples include the attempt to aggregate the 'geographic perceptions' of individual decision makers by creating 'mental maps' rather than assuming perfect knowledge of an isotropic physical space (Gould and White, 1974), and the experimental study of human mobility by Golledge (1980). Behaviourists, however, do not specifically account for the nature of the surrounding society in which individuals act (Smith, 1979). The post-positivist critique generally does not accept that behaviourist modifications of positivistic theories are adequate.

Reactions to positivistic theories of the role of space in human social behaviour also took the form of new humanistic theories of space. Hence, the application of idealism to geographical decision-making focused on the thinking and actions of individuals over space, for example, in route planning. Since in idealism no 'real world' is considered to exist, such decisions must be determined by individuals' own models of the real world (Guelke, 1974). Tuan (1977) sought to develop a new approach to the study of 'place' through the methods of phenomenology. In this fashion the sense of place defined by a 'landscape' or a 'neighbourhood' was defined by the aggregate of human appraisals of such environments—a method rarely if ever used to define SSEUs.

However, idealism has been criticised from several perspectives. Later social theory argued that individuals do not have complete freedom to act through simple self-awareness and are actually socially constrained (Johnston, 1991), implying that some form of 'external' social structure may exist. By contrast, from a cognitive science perspective Lakoff (1987) argued that it is the mind that creates the structures through which the world is understood: in a metaphysical sense, no 'mind-free' world is possible. The salient features of a mind-constructed world-view include the role of the body and human culture in defining organising concepts (called 'experientialism' by Lakoff). It is argued in experientialism that the world is perceived through 'schemas' that interface between the sensory apparatus and the mind. Frank and Mark (1991) have noted that many of these schemas are spatial (for example, container, blockage, path, surface, link, near–far, contact, centre–periphery and scale) and that others such as 'object' and 'part–whole' can be used in making spatial representations.

A variety of more radical alternatives to positivistic theories in geography and social science were proposed that were ultimately based on structuralist approaches derived from the work of Lévi-Strauss (1963). Structuralism argues that appearances and the underlying are not the same in a social context and that three levels of experience can be identified: the level of appearances (the 'superstructure'), processes (the 'infra-structure') and imperatives ('deep structure'). Gregory (1978) argued that, as far as the role of space in human behaviour is concerned, the superstructure could be equated with external spatial patterns (such as SSEUs), while the underlying infrastructure was responsible for spatial structured processes. Marxist analyses of society and space, for example, concerning the commodification of land in capitalism (Harvey 1982), are also developed using structuralism. Structuralists argue that the 'spatial' superstructure can only be understood through the processes of the underlying 'spatial' infrastructure: from a Marxist perspective the superstructure is largely irrelevant and analyses of it simply perpetuate the social relations which it reflects (Harvey, 1989). From this perspective SSEUs are artifacts which do not reflect underlying behaviour: in other words the imposition of SSEUs has more to do with control than understanding. However, the structuralist view that the socio-spatial processes of the infrastructure are deterministic has also been criticised since it leaves little room for the autonomy of individuals (Duncan and Ley, 1982).

Giddens (1979, 1981) developed the theory of structuration to integrate social structures and human agency in an explicitly spatial formulation. In structuration, social structures link the interaction amongst autonomous individuals with the reproduction of social systems across time and space. Hence, social systems are place and time-bound: Giddens (1984) defined a 'locale' as a 'setting for interactions' which can be an institution with spatial and/or temporal identity. Gregory and Urry (1985) argued that a spatial structure is a medium though which social relations can be produced and reproduced; hence, the spatial organisation of a society reflects a continuous socio-spatial dialectic referred to as the process of 'spatiality' by Soja (1985). Pred (1986) summarised this process in the maxim "the spatial becomes the social, and the social becomes the spatial" (p. 198). From this perspective it can be seen that SSEUs are far too static and sharply bounded by comparison with the social processes that drive general social diversification.

Another form of positivist critique is the development of 'transcendental realism' (Bhaskar, 1978), which seeks to re-examine a basic ontological question: what are the proper 'objects of knowledge'? Bhaskar argues that real structures exist outside human knowledge and experience, which can generate phenomena. In realism it is the interaction of such generative structures that should be studied as causal mechanisms for events and 'objects' (such as SSEUs). Sayer (1984) characterises realist research programmes as 'intensive' since they attempt to discover the causes of events as perceived on a case by case basis. Positivist empirical research programmes by contrast are 'extensive' involving descriptive generalisations, which cancel out individualistic behaviour. However, according to Sayer neither the causal mechanisms themselves nor the relationships between the mechanisms and the conditions are invariant in social theory. This implies that social science is an open system and, therefore, that the results of research are not generalisable. Realists, therefore, believe it is a mistake to develop a method to discriminate zones since the generative structures in social processes have ontological primacy over objects such as SSEUs.

In the 1990s these critiques of positivism have themselves been critically examined through the (re)thinking of the postmodernist movement. Postmodernism rejects theory and order and aims to examine the heterogeneity and diversity of social spaces at scales ranging from the individual to the settlement. For example, Soja (1989) argues that the very concept of a region or zone is now dead since there are often greater intra-regional differences than there are inter-regional differences. Highly heterogeneous cityscapes are simply reproduced everywhere in the same way. Friedland and Boden (1994) argue that the technologies of 'distanciation' (ability to act at a distance) such as telephone, fax and television has destroyed 'place' as a meaningful concept. Distanciation also means that capital (in the form of money) has a huge 'spatial elasticity', whereas labour cannot move as fast since it must physically move. Although postmodernist thought can be rejected as merely deconstructionist, it is implicit that SSEUs are rejected as 'modern' (ordered, regular) and not 'post-modern' (diverse, irregular).

1.5 THE CASE FOR HAVING REPRESENTATIONS OF HUMAN SPACES

Despite this powerful critique it can still be argued that spatial representations of social processes and human domains are valid and relevant. There seem to be two chief grounds on which to argue this point; viz. firstly, that spatial and temporal constraints in societies are central to human action (Friedland and Boden, 1994) and that such constraints can frequently be made explicit; and secondly, that SSEUs are *de facto* in use as policy instruments and as such their use must be monitored (Bennett, 1981).

That spatial and temporal constraints on human action exist in societies is rarely contested and has increasingly become a focus of attention, for example, gendered spaces (Massey, 1992). However, many of the studies of these constraints have focused on the discourse using intensive studies and not on the properties of space and time that give rise to them—the generative structures of Bhaskar. Friedland and Boden (1994) ask "how are spatio-temporal and cognitive structures co-constitutive of modernity?" To address this question a GIS holding street networks could be used to store information on the location and timing of perceived constraints ('no-go' areas to certain people) and how they were situated with reference to the world of entities. This could lead to the definition of new kinds of more socially relevant SSEUs that highlight problematic social phenomena such as drug dealing or prostitution. Such a study may prompt new insights into the generation of zones such as 'no-go' areas. In a GIS context the problem can be seen as a call for much more sophisticated spatio-temporal data structures than those currently available.

By contrast to the need to find ways to make constraints on action explicit and so amenable to spatial representation, concrete, spatially represented SSEUs are the definitive basis on which geodemographic analyses and political campaigning are now carried out in many developed nations. Yet the ontological basis for the aggregation of mobile populations into SSEUs and the dangers of the ecological fallacy in generalising about behaviour from SSEU demographic data seem to be substantially under-researched. In this case, GIS can be used to demonstrate spatial 'null hypotheses', i.e., that other SSEUs may exist which falsify the propositions that are advanced for any particular set of SSEUs.

That SSEUs are in use for a wide range of purposes is also undeniable. However, perhaps the 'positivistic' concepts behind SSEUs need to be updated. SSEUs need to be capable of temporal variation or of spatial disaggregation, i.e., to be defined over a functional unit such as a car that moves through space and time. The challenge for GIS in this is to provide the representational tools within which SSEUs can be reinvented.

REFERENCES

Bennett, R.J., 1981, Quantitative and theoretical geography in Western Europe. In *European Progress in Spatial Analysis*, edited by Bennett, R.J., pp. 1–33.

Berry, B.J.L., 1964, Approaches to regional analysis: a synthesis. *Annals of the Association of American Geographers*, **54**, pp. 2–11.

Bhaskar, R., 1978, *A Realist Theory of Science*, (Brighton: Harvester).

Broadbent, T.A., 1970, Notes on the design of operational models. *Environment and Planning*, **2**, pp. 469–476.

Chapman, G.P., 1977, *Human and Environmental Systems*, (London).

Chorley, R.J. and Kennedy, B.A., 1971, *Physical Geography: A Systems Approach*, (London: Prentice-Hall).

Comte, A., 1842, *Cours de philosophie positive*, (Paris).

Duncan, J. and Ley, D., 1982, Structural Marxism and human geography: a critical assessment. *Annals of the Association of American Geographers*, **72**, pp. 30–59.

Foucault, M., 1972, *The Archaeology of Knowledge*, (London: Tavistock).

Foucault, M., 1980, *Power/Knowledge. Selected Interviews and Other Writings 1972–77*, (Brighton: Harvester Press).

Frank, A.U. and Mark, D., 1991, Language issues for GIS. In *Geographic Information Systems: Principles and Applications*, edited by Maguire, D., Goodchild, M. and Rhind, D. (Longman: Harlow).

Friedland, R. and Boden, D., Eds, 1994, *NowHere: Space, Time and Modernity*, (Berkeley: University of California Press).

Giddens, A., 1979, *Central Problems in Social Theory: Action, Structure and Contradiction in Social Analysis*, (London: Macmillan).

Giddens, A., 1981, *A Contemporary Critique of Historical Materialism*. Vol. 1: Power, Property and the State, (London: Macmillan).

Giddens, A., 1984, *The Constitution of Society*, (Oxford: Polity Press).

Golledge, R., 1980, A behavioural view of mobility and migration research. *The Professional Geographer*, **32**, pp. 14–21.

Gould, P.R. and White, R., 1974, *Mental Maps*, (Harmondsworth: Penguin).

Gregory, D., 1978, *Ideology, Science and Human Geography*, (London: Hutchinson).

Gregory, D. and Urry, J., 1985, Introduction. In *Social Relations and Spatial Structures*, edited by Gregory, D. and Urry, J. (London: Macmillan), pp. 1–8.

Guelke, L., 1974, An idealist alternative in human geography. *Annals of the Association of American Geographers*, **14**, pp. 193–202.

Habermas, J., 1978, *Knowledge and Human Interests*, (London: Heinemann).

Hacking, I., 1983, *Representing and Intervening*, (Cambridge: Cambridge University Press).

Haggett, P., 1965, *Locational Analysis in Human Geography*, (London: Edward Arnold).

Harvey, D., 1982, *The Limits to Capital*, (Oxford: Blackwell).

Harvey, D., 1989, *The Urban Experience*, (Oxford: Basil Blackwell).
Johnston, R.J., 1991, *Geography and Geographers: Anglo-American Human Geography Since 1945*, (London: Edward Arnold).
Lakoff, G., 1987, *Women, Fire and Dangerous Things*, (Chicago: University of Chicago Press).
Lévi-Strauss, C., 1963, *Structuralist Anthropology*, (New York: Basic Books).
Lloyd, P.E. and Dicken, P., 1977, *Location in Space*, (London: Harper and Row).
Mach, E., 1883, *The Science of Mechanics*, (Berlin).
Masser, I. and Brown, P., 1978, *Spatial Interaction and Social Process*, (Leiden: Martinus Nijhoff).
Massey, D., 1992, *Space, Place and Gender*, (London: Polity Press).
Pred, A., 1986, *Place, Practice and Structure: Social and Spatial Transformation in S. Sweden 1750–1850*, (Cambridge: Polity Press).
Raper, J.F., 1996, Unsolved problems of spatial representation. In *Advances in GIS Research, Proceedings of 7th International Symposium on Spatial Data Handling*, Delft, Netherlands, edited by Kraak, M.-J. and Molenaar, M. (IGU), pp. 14.1–11.
Raper, J.F., Rhind, D.W. and Shepherd, J., 1992, *Postcodes: The New Geography*, (London: Longman).
Ray, C., 1991, *Time, Space and Philosophy*, (London: Routledge).
Reis, R.M.P. and Raper, J.F., 1994, Methodologies for the design and automated generation of postcodes from digital spatial data. In *Proceedings of European GIS Conference '94*, Paris, pp. 844–851.
Sayer, A., 1984, *Method in Social Science: a Realist Approach*, (London: Hutchinson).
Smith, B., 1995, On Drawing Lines on a Map. In *Spatial Information Theory—A Theoretical Basis for GIS, International Conference COSIT '95*, Lecture Notes in Computer Science 988, edited by Frank, A.U. and Kuhn, W. (Berlin: Springer-Verlag), pp. 475–484.
Smith, N., 1979, Geography, science and post-positivist modes of explanation. In *Progress in Human Geography*, **3**, pp. 365–383.
Soja, E.W., 1985, The spatiality of social life: towards a transformative retheorisation. In *Social Relations and Spatial Structures*, edited by Gregory, D. and Urry, J. (London: Macmillan), pp. 90–127.
Soja, E.W., 1989, *Postmodern Geographies: the Reassertion of Space in Critical Social Theory*, (London: Verso).
Tuan, Y.F., 1977, *Space and Place: the Perspective of Experience*, (Minneapolis: University of Minneapolis Press).
Unwin, T., 1993, *The Place of Geography*, (London: Longman).
Wood, J.D., Unwin, D.J., Stynes, K.S., Fisher, P.F. and Dykes, J.A., 1999, The use of the landscape metaphor in population mapping. *Environment and Planning B: Planning Design*, **26**, pp. 281–295.

CHAPTER TWO

Socio-Economic Units: Their Life and Motion

Andrew U. Frank

2.1 INTRODUCTION

Current Geographic Information Systems (GIS) are modelling static spatial situations. Using Sinton's terminology (Sinton, 1978), they hold time constant and then either vary theme or location—depending if they are vector (object) or raster (field) oriented (Frank, 1990; Goodchild, 1990). The world in which we are interested is in constant change; nothing is ever stable. This has been observed by the Greek philosophers and identified as one of the chief difficulties in understanding the world. We need to identify objects that we can see as (relatively) stable and against which we can compare others.

This is particularly important for socio-economic objects in geographic space, which are typically non-physical. By socio-economic objects we understand areal units that are used to describe social or economic phenomena. They encompass census tracts for statistical data collection, political subdivisions (borough, town, county, nation, etc.), urban and rural zones, areas delimited for demographic descriptions (ethnicity, religion, language, etc.). These socio-economic units are conceptual constructions (see Raper, this volume) and do not correspond directly to physical reality. Change and movement does, therefore, not have exactly the same meaning as with physical bodies.

Our conceptualisation of change is influenced by the perception of change of physical objects, mainly the moment of rigid physical objects. The experience with physical objects, primarily rigid bodies, but also with liquids, melting ice, etc. determines the cognitive categories we use to describe other objects and their movements, using metaphorical transformation (Lakoff and Johnson, 1980). It is thus recommended that the modes of change of the relevant physical objects are studied and formalised in order to transform the results to the socio-economic units and the special properties identified.

Socio-economic units often appear with an ontology similar to 'shadows' (Casati and Varzi, 1994). They are non-physical properties of an area, which can be moved without any movement of material. Shadows and holes are ontological categories that have only recently been systematically studied. The socio-economic units, as abstract objects, are perceived and understood in terms of such ontologies. Clarifying these will be an enormous first step.

2.2 CHANGE IN THE GIS

2.2.1 Importance of Change for GIS

There are very few applications where only the current situation is important. Most uses of GIS technology are interested in change. This is true for scientific and administrative uses of GIS.

Sciences in general, and geography can serve here as an example, are interested in processes that transform things. Processes are the general rules, the descriptive data is the particular, and thus of lesser standing. Scientists collect data in order to understand general rules, which are mostly seen as processes. The collection of data, the descriptive part of scientific activities, is necessary to have the foundation on which deductive work is building up. The need for support for time-related data was voiced by anthropologists, architects, atmosphere scientists, marine biologists, planners, urbanists, wild life specialists, etc. The topic has become of even more interest lately with the global change efforts (Mounsey and Tomlinson, 1988), where the object of the study is change itself, as it occurs on a global scale.

Administrative uses of GIS sometimes work properly with the current status only: tax mapping, facilities management for public utilities, forest management can do. But wherever legal aspects of liability, due process or the use of proper administrative procedures are involved, previous states are stored to assure that changes can be reconstructed.

2.2.2 Representation of Change

Support for temporal data in GIS software, so-called temporal GIS, has been asked for a long time (Smith, Boyle *et al.*, 1983; Allen, Green *et al.*, 1990). The lack of support for time related data in GIS may be one of the reasons why GIS is often seen only as a tool to help with the descriptive part of scientific work (Goodchild, 1990). A tool for geographers, helping with the descriptive part of science, but not helping with the heart of science, where processes are in the focus.

It is indeed surprising to see how poorly a current commercial GIS deals with data representing objects that change in time. The best we can currently do is to show a snapshot of the data and even this may represent data collected over a period of time (see Cheylan, this volume). Data quality standards demand that this is properly documented—but again, most data collections are lacking in this respect. Neither the software nor the data help with understanding changes.

To simulate change, snapshots of the situation at a specific time are accumulated. What would be required is a representation of change and the events that limit gradual change with the time they occur. To deduce the same information from the differences between snapshots is difficult and limited by the temporal resolution of the snapshots. This is a similar phenomenon as the one discussed in (Burrough and Frank, 1995), where the difference between the capabilities of current GIS software and the requirements was compared. The social sciences need tools to represent and analyse change.

2.2.3 Life and Motion as Two Types of Changes

Change comes in two forms: change of the objects of interest and change in the position or geometric form of these objects. For the first we use the heading *life* of objects: objects may appear and disappear (for example, a forest or a residential zone), two objects may merge (for example, two parcels or two towns), an object may split. For the second we use the heading *motion*: objects may move or may appear to move, with or without changing their form at the same time.

The life and motion of physical bodies, especially of human beings in space, are the prototypical notions dominating our experience. The motion of physical bodies in the natural sciences is sometimes complex, but the direct experience with the phenomena and the possibility to analyse the physical movements of particles constituting the bodies gives an approach to their understanding.

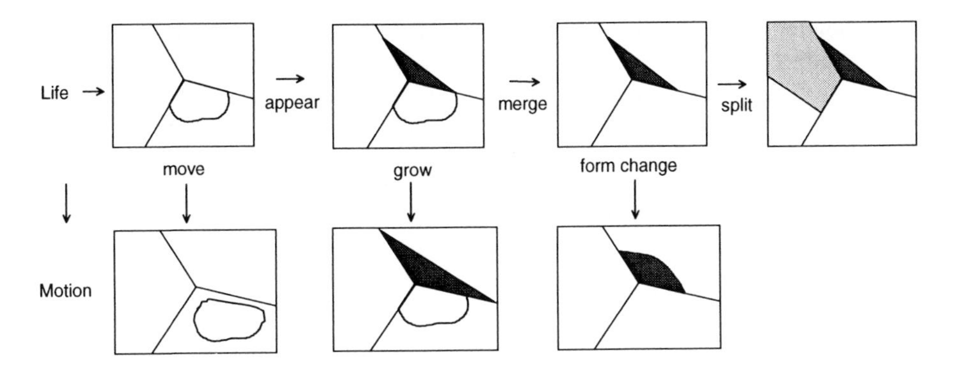

Figure 2.1 Objects' life and motion

More unusual are the *life and motion* of spatial units in economic and social sciences or administration. Administrative units are moving in space (e.g., Poland is here the best known example, Centennia, 1993) or are merging or splitting—causing enormous difficulties for the analysis of statistical data collected for these (changing) units (Openshaw, this volume). The central business districts are changing in size, form and location over time—even without having clear-cut boundaries (Burrough and Frank, 1996).

2.2.4 Approach

The past discussion of time in GIS can be classified in several groups:

a. snapshots and differences between states (Langran and Chrisman, 1988);
b. discussion of work time, data flow and database time (Langran, 1988);
c. formalists' discussion of space/time as a multi-dimensional continuum (Worboys, 1994).

These approaches were comprehensive (or generalising), but did not allow going beyond the obvious because too many dissimilar phenomena were treated at once. What is advocated here is an approach where different aspects of changes are analysed and formally described, starting from the dominant human experience with motion of rigid bodies (but also liquids) and following the cognitive metaphorical transformation of these cognitive categories to describe abstract objects like socio-economic units. It is asked for which real-world situation they are applicable and which application areas use models for *life* or *motion* of objects of this type.

One can differentiate different types of *life and motion* of objects:

a. motion of physical bodies in small scale space, or in geographic space, along a predetermined path or in a field;
b. motion of large physical bodies in geographic space, where the objects follow the rules of rigid body motion or the rules of liquids;
c. motion of collectives in geographic space, where the motion of the collective is the sum total of the motion of the individuals (Cheylan and Lardon, 1993);
d. motion of non-physical objects, e.g., administrative units, which glide shadow-like over the landscape.

It is assumed here that each of these kinds of motion follows its own set of rules—its own ontology (Frank, 1997). Understanding what is similar and dissimilar for these types of changes will be useful, in particular to model the more abstract changes in socio-economic objects. It becomes clear from this analysis that the kinds of *lives* and *motion* objects may have are bound in a set of logical schemata, best described as algebras (or categories in the sense of mathematical category theory, see (Frank, 1996).

Different applications not only differ regarding the models for change they use, but also regarding the scale of time or space they employ. Geological time and space scales are very different from the temporal and spatial scale to describe urban changes. The classification attempted here is then to be combined with these differences in scale.

2.3 STATE OF THE ART

The difficulties to include model data with respect to change in time are multiple, some generic and some specific for spatial databases.

2.3.1 Temporal Databases

Database theory for temporal data exists, for a review see (Snodgrass, 1992), but none of the current commercial DBMS include extensive support for time varying data. Lotus Notes and other systems support replicated databases with time stamps for any change (Dennig, 1994). This can be used to model the flow of information within the organisation (database time) and thus contributes to supported collaborative work; it does not apply to data describing an exterior, changing world (real-world time).

Temporal database theory covers mostly a static or snapshot view of time, and changes cannot be analysed at the level of user semantics. It can be used to reconstruct a database after data loss, but reasoning about motion and change is limited.

2.3.2 Temporal GIS

Geometric models are static. In particular, Euclidean geometry models static situations. Newtonian physics provides a sort of *dynamic geometry*, but it is appropriate only for the table-top object space (Montello, 1993). GIS need support for temporal change in environmental and geographic space. For raster data, the operations of map algebra can be used to compare images from two different points in time (epochs), and to identify cells which have changed. This is not an appropriate model to describe the movement of an object in space in general, but can be used in certain application areas. Chrisman and Langran have investigated land use changes (Langran, 1988; Langran and Chrisman, 1988; Langran, 1989), following a snapshot approach related to the database viewpoint.

A number of workshops discussed support for temporal data in GIS (published reports are available for (Barrera, Frank *et al.*, 1991; Egenhofer and Golledge, 1998)). Generally, these meetings reaffirmed the importance of support for temporal data in a GIS—known at least since (Abler, 1987; NCGIA. 1989). The discussion invariably also shows a wide variety of application areas and possible solutions. Frank has concluded—in analogy to the discussion of the spatial domain—that there are many different types of time, which must be handled differently in a GIS (Frank, 1994). Different situations lead to different experiences of time—walking over a hill is a different experience than observing a city sprawl—which lead to different formalizations.

2.3.3 Change in Philosophy and Cognitive Science

Philosophy looks back over two thousand years of discussion of the notion 'time'. The duality between duration and event, which has occupied the best minds since Zenon

(Fraser, 1981) has been very fruitful (Hofstadter, 1979). It attempted to find a uniform and all encompassing explanation of time—to correspond to the human belief that there is a single notion of time as there is assumed to be a single type of space. Time is crucial to science because time is intricately linked with causation. *Post hoc, ergo propter hoc* is not a permissible deduction rule, but temporal precedence implies so often causality that the inverse conclusion has become a common mistake. For scientific analysis, but also for many administrative applications, an ordered time is sufficient because causation implies precedence, lack of precedence rules out causation.

Experientialism (Lakoff, 1987, 1988) takes human direct experience and perception as fundamental. Other domains of human endeavour that cannot be directly experienced are conceptually organised in analogy to the domains that can be directly experienced (Lakoff and Johnson, 1980). There are at least two different base experiences of time and motion: the observation of motion of objects in figural space (Montello, 1993), for example, moving an apple from the left side to the centre of the table, and bodily experience of walking in open space, e.g., over a hill.

The perception of time is dependent on the situation. Resolution of time is according to the task: only differences that can be of importance are observed and recorded. Human perception limits the resolution of two events to few tenths of a second, but in most situations, much larger concepts of instantaneous are used: start points for meetings are understood as five minutes later or earlier; in business, most duration is measured in days. This is similar to space, where resolution depends on the task at hand.

2.4 RESEARCH PROGRAM: DEFINE DIFFERENT TYPES OF CHANGES RELEVANT TO GIS

2.4.1 Abstract From Scale of Resolution

We start with the assumption that for each description of a process that links causes and effects a specific resolution for space and time is appropriate: a scale on which necessary differentiation can be made and irrelevant details disappear. These scales vary over several orders of magnitude (Morrison, 1982) from the time/space scales for galactic or planetary movement, to those used to describe the daily migration of humans in a city. Fraser assumes that reality is divided in several clusters of interaction, which all have similar time/space scales (Fraser, 1981). Such clusters form the subject matter of scientific disciplines.

Certainly there are a number of combinations of time and space scales relevant to geography (Frank, 1994). The effort here should abstract from these, as they are trivial to characterise; it is sufficient to indicate the resolution in time and space for a process.

2.4.2 Abstract From the Particular Type of Space and Time

From a perceptual perspective, there are different types of time as there are different types of space. Perception of space is different for small-scale space (table-top) and geographic space (Montello, 1993). This is the same for time, where time in administrative processes is discrete and structured as containers, whereas time for other social and natural processes has a continuous aspect and can even be cyclic (Frank, 1994).

For the effort here, we do not focus on these differences as they have been dealt with previously. We assume:

a. geographic space, defined as too large to be perceived at once from a single vantage point;
b. a discrete, linear time.

For both, time and space, we assume a resolution fine enough for the process of interest.

2.4.3 Separate Change in Life and Motion Aspects

A possible hypothesis to structure change in geographic objects is that the aspects of *life* and *motion* can be separated.

By *life* we understand all aspects of the existence of an object in time. It is *created*, and lives on till it is *destroyed*. It may be *killed* and later *reincarnated*. Objects may *fuse*, *aggregate*, etc. Also the non-spatial properties of the object may change in time (Al-Taha and Barrera, 1994).

Motion covers all aspects of the movement of the object or change in the form, i.e., all geometric changes. A change in the form can be seen as a movement of the boundary of the object.

2.4.4 Start With Prototypical Situations

The prototypical situation for motion is the movement of human or animal bodies in space, but also the motion of liquids in small-scale space. The best formalization is for movement of rigid bodies, as taught in every high school physics class. For life, the prototypical situation is the life cycle of a person (or animal), but also the life cycle of a rigid body or of liquids.

We assume that the abstract objects of geographical space—in particular the socio-economic spatial units—are perceived and represented in terms very similar to these prototypical situations. Metaphorical transformation allows for the change from small-scale experience to large-scale space and from the physical experience to the general abstract (Lakoff and Johnson, 1980; Lakoff, 1987).

2.5 PROTOTYPICAL SITUATIONS

2.5.1 Life Styles

The notion of *life* is closely related to persons. A person is identical with himself from birth to death. This requires a notion of identity for an entity, where two things can be tested if they are identical, even if they differ in some descriptive values: a person at age 10 and at age 30 is the same legal person, despite the apparent change and the likely complete change of all the substance (molecules) it consists of. It is proposed to use the operations *create* and *destroy* to start and end an entity with identity (Al-Taha and Barrera, 1994).

Further, one may allow that entities get *killed* and *reincarnated* (Al-Taha and Barrera, 1994)—customary for gods and heroes of the ancient sagas.

Objects can *evolve*, where one object disappears and a new one appears at a specified time.

An object may be *identified* as being the same as another one or two objects may *aggregate* where the identity of the second object is preserved within the first one. Two objects can be *fused* to form a new one.

An object may *spawn* a new one or two previously aggregated objects may become *disaggregated* again. *Fission* breaks an object into parts, which form new objects.

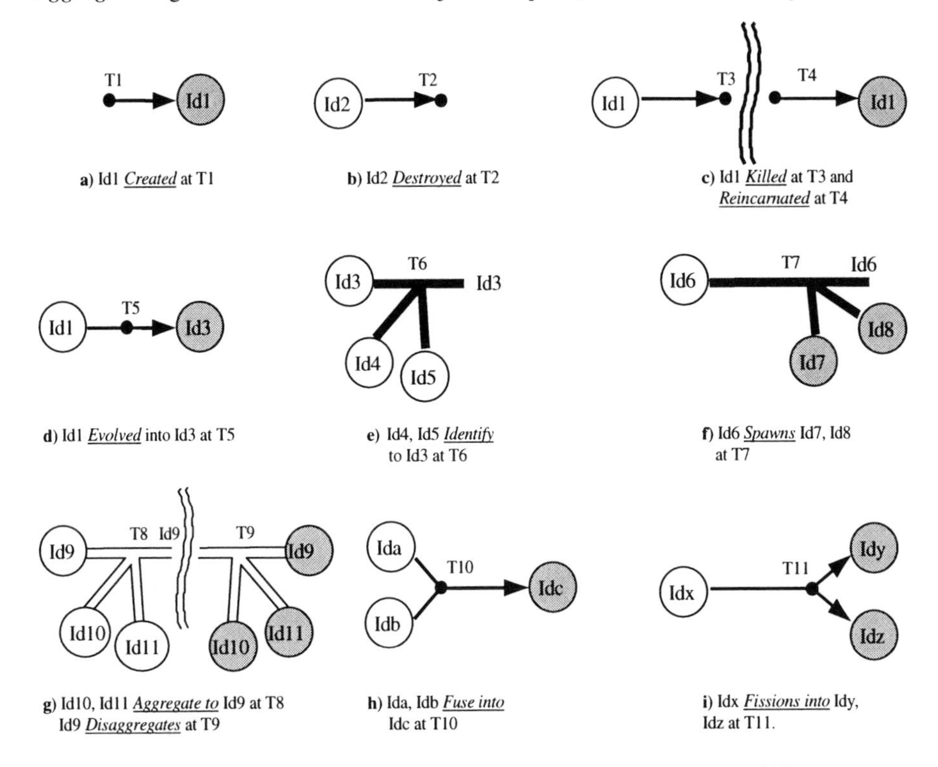

Figure 2.2 Temporal constructs of identities (from Al-Taha and Barrera, 1994)

Not all these changes can occur to all objects, they form logical clusters and depend on each other. Formal methods can be used to analyse these dependencies and to document them. By lifestyle we understand coherent sets of life operations: cars and similarly manufactured objects are created, parts are aggregated and disaggregated from them and they are destroyed. It is very unusual for cars to be killed and reincarnated, to evolve or spawn new objects. Thus the operations *create, aggregate, disaggregate* and *destroy* form the lifestyle 'manufactured goods'. For lifestyles also see (Medak, this volume).

2.5.2 Movement of (Rigid) Bodies in Space

Movement of Small Objects in Small-Scale Space

The prototypical case is the observation of a physical body in space. This is one of, if not the primary experience of change as motion. A physical object is found at position a1 at time t1 and position a2 at t2, and at any time ti at a position ai between a1 and a2. It is implied that the size of the object is relatively small compared to the movement and the temporal resolution of the movement. Galton has formalised an ontology of movement of physical objects, which reconciles the event and the duration position (Galton, 1995; Frank, 1997).

It appears that movement of small objects is typically perceived as instantaneous change of position; the path of the change is usually not described nor noticed, probably because it is not consciously planned, but an appropriate path for the movement is left to the non-conscious motor planning.

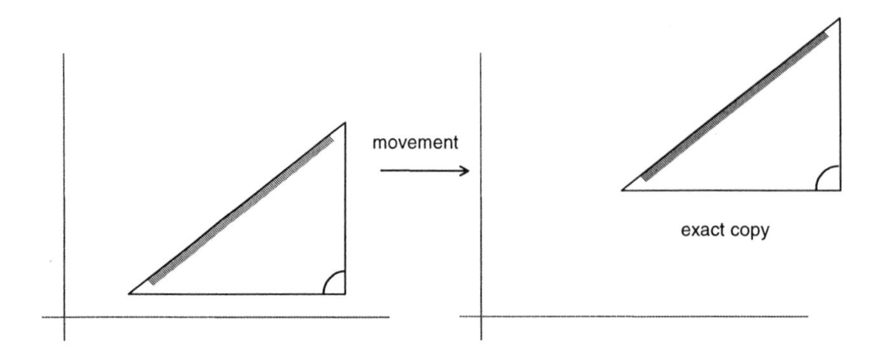

Figure 2.3 Movement of a small object

Movement of a Person

Generally, movement in large-scale space (or space larger than small-scale) of a self-moving body, too large to be moved.

The movement of a person along a path is a fundamental experience. It is experienced by every human as movement along a path with duration.

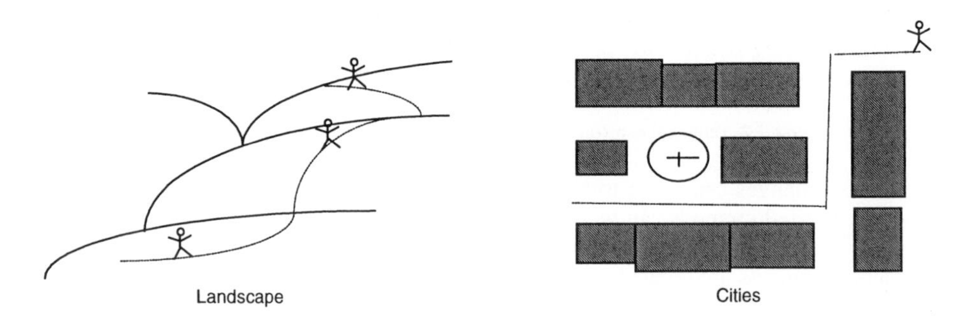

Figure 2.4 Movement of a person

The movement can be across a field in the landscape, visible from a distance. The standard situation is on a surface in a gravity field, which influences the effort necessary for movement along the path; gives a component of resistance/acceleration along the path and one orthogonal to it. Related is the movement of a liquid in space: it follows the vector field. The movement of water in the landscape is of utmost importance to human economy.

Change of Geometric Form: Movement of a Boundary

The change can result in a change in the boundary, which is often seen as a movement of the boundary. One says the forest advances, meaning that the forest boundary changes (the only forest actually moving is the Birnam Wood in Macbeth). It appears as if the boundary was moving—but mind: boundaries are not physical objects, they are limits between physical objects and as such have a complex ontology (Casati and Varzi, 1994).

Change in the geometric form can have multiple origins. It can occur through accretion of the small pieces the object consists of or the loss of components, which leads to erosion. To this the lifestyle of kill/reincarnation fits: a lake (for example, the shallow Neusiedlersee in Austria) can disappear through draught and reappear.

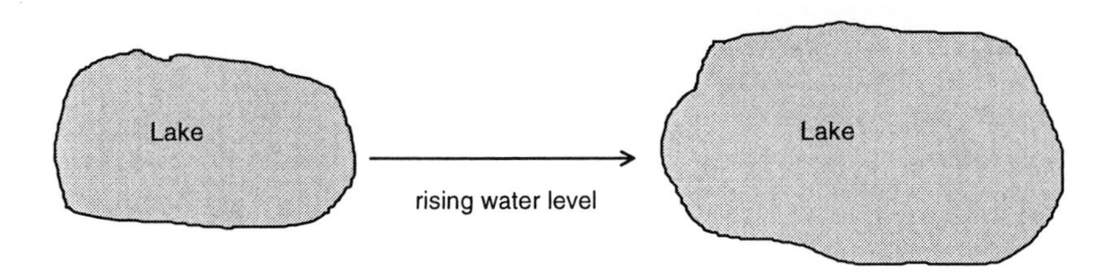

Figure 2.5 Change of geometric form

2.5.3 Changes Which are Understood as Movements

Many changes are understood as movement—e.g., the movement of a forest—even if in reality the forest as an object does not move, and even if no physical objects move at all.

Collective Bodies

By collective bodies we understand objects which are composed of individuals that are separated at the next level of resolution (Talmy, 1983; Cheylan and Lardon, 1993). Compare with collective nouns in language (Langacker, 1991; Langacker, 1991).

The movement of individuals in a collective body can be ordered, i.e., the same individuals remain in the advancing boundary (military formation), or those most back advance to the front (flocks of sheep).

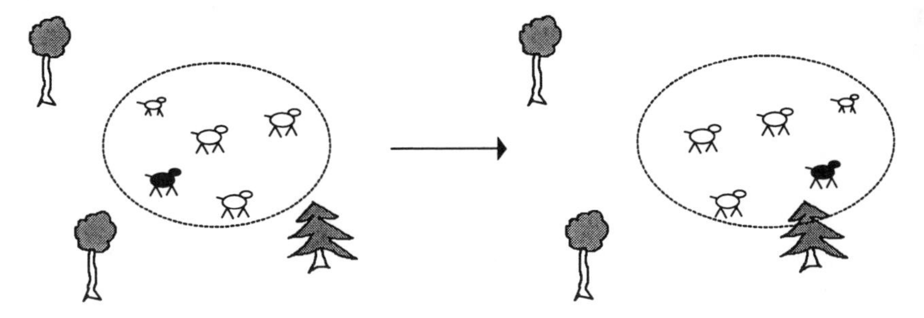

Figure 2.6 Movement of a flock of sheep

Liquid-Like Things

The movement of liquid-like things is similar to an ordered collective movement, but the form of the entity always follows the limits of other objects. In geography, cities seem to flow around 'forbidden areas' (like lakes, forest, if legally protected, etc.).

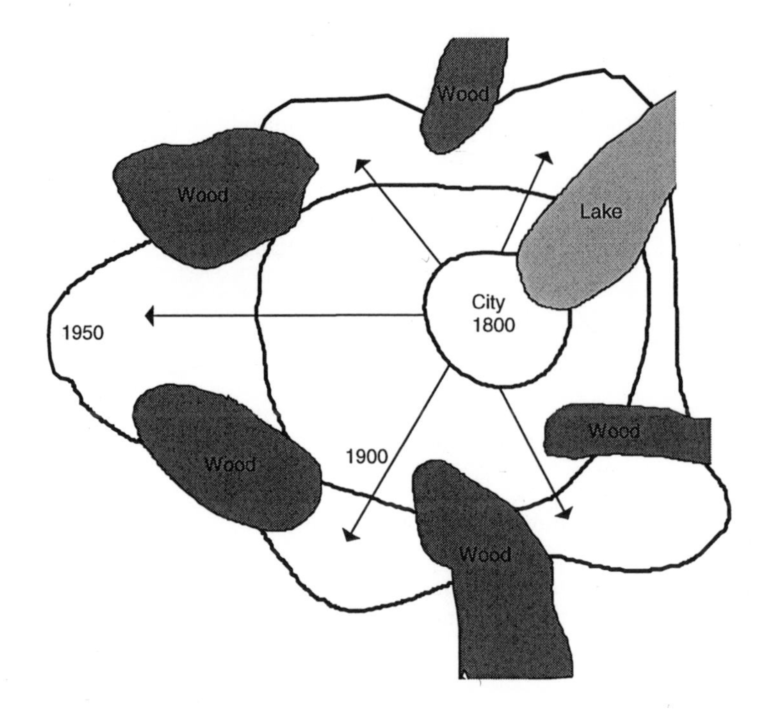

Figure 2.7 Growing of a city (e.g., Zurich)

The movement may include evaporation (e.g., glacier—where movement is mass movement plus the movement of mass through the boundary (df/dt + integral v x P/dl) or accretion from diffuse distributed material, which is collected at a location and thus increases the object.

Apparent Movement Resulting from Individual Changes of Properties

If individual areas of space change their property individually, then a movement of the area with the same property may appear. A tree in a forest does not move, but trees on one boundary of the forest may be cut consistently and be preserved on the other boundary. Thus the forest recedes on one side and advances on the other, without any physical movement. If the change of an individual cell is strongly correlated to the properties of the neighbours, individual change may appear as a movement. All habitats seem to belong to this group.

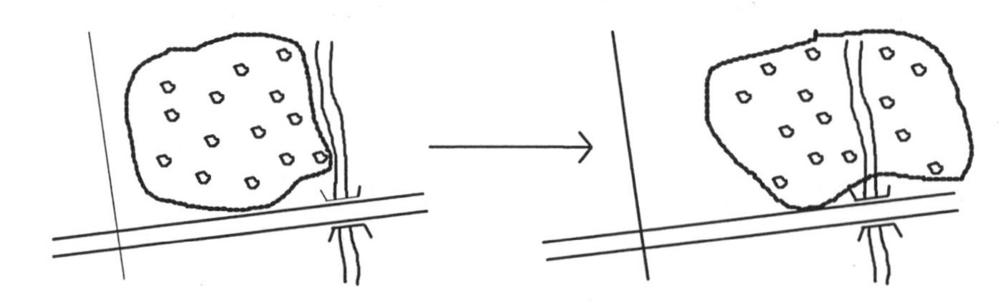

Figure 2.8 Movement of a forest

Movement of Shadows

By shadows we understand areas of uniform, non-essential properties of the objects which make up space (Casati and Varzi, 1994). Shadows are cast on the object and can move without affecting the object. This is ontologically similar to legal assignments of areas to urban zones, to protected areas, etc. Ownership rights are similar cases or the assignment of a country to an area; in this sense, Poland (or any other nation) is a shadow and as such can move (this is to be differentiated from the area populated by people speaking the polish language, who move as collective bodies).

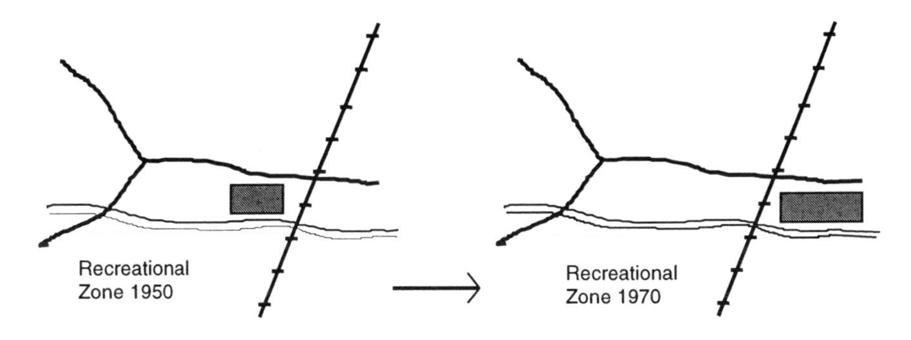

Figure 2.9 Movement of recreational zone

2.6 REPRESENTATION OF CHANGE IN GIS

In a current GIS we cannot describe the changes directly but indirectly: change is the difference between two snapshots, representing states. A direct representation of movement of physical bodies is a vector of their speed (defined as difference in position divided by difference in time). With a starting point one can compute the points this object will be at in future. Other forms of change are very different to describe and no such simple and general method is known (for example, differential equations).

In a GIS we must be able to represent directly the different kinds of change. Each of the lifestyles and kinds of motion leads to its own representation schema. Therefore the classification effort described above is crucial to identify appropriate representations. One may follow the example set by natural sciences: differential equations are a powerful method for a wide variety of kinds of changes in attribute values, which can be described as fields $a = f(x)$ (Goodchild, 1992), and thus are applicable for natural science objects in a GIS. Kuipers (1994) shows how to translate differential equations to the qualitative domain and how to use them in cases where no exact measurements but only qualitative (i.e., more, less, etc.) descriptions are available. These methods could be extended to spatial objects.

Representations are not only necessary for the internal storage in a GIS, but are required over a whole spectrum of methods of communication. Graphical and natural language methods are necessary to communicate effectively with humans, whereas the formal and internal storage methods support the design and implementation of GIS software.

The traditional methods for communication of change used by humans must be studied to understand the cognitive categories humans use to discuss change. Especially natural language expressions for change can be analysed to understand the image schemata (Lakoff, 1987). The cartographic tradition contains a series of examples to represent change on maps.

2.7 TOPICS TO WORK ON

From this 'program' a series of specific topics to work on can be identified. These topics appear relevant for today's GIS industry and ready to work on, and small enough that there is a good chance of success within one year's effort.

The research topics can be seen as a set of theoretical questions:

a. formalising the prototypical cases listed (Section 2.4);
b. description of the prototypical cases for natural language expression or graphical expression of change, life or motion;
c. formal representation methods and effective storage methods, including particular data structures;
d. connection between change, life and motion and general description of data quality.

In each case, a specific subset of the total questions should be tried first and justified by a particular real case. For example, the description of sheep movements on a mountain (Alps) could serve as an interesting test case, many different methods could be applied to. The paper by Cheylan and Lardon (1993) could serve as the base description of the case.

In general, work in the area of change of socio-economic units should be motivated by specific and realistic cases. It is far too easy to construct complex cases which cannot be solved in general, but each specific case of which the complex one is an amalgamate could be solved.

REFERENCES

Abler, R., 1987, The National Science Foundation—National Center for Geographic Information and Analysis. *International Journal of Geographical Information Systems*, **1**, pp. 303–326.

Allen, K.M.S., Green, S.W. and Zubrow, E.B.W., 1990, *Interpreting Space: GIS and Archaeology*, (London: Taylor & Francis).

Al-Taha, K. and Barrera, R., 1994, Identities through time. In *Proceedings of International Workshop on Requirements for Integrated Geographic Information Systems, New Orleans, Louisiana.*

Barrera, R., Frank, A.U. and Al-Taha, K., 1991, Temporal relations in Geographic Information Systems: A Workshop at the University of Maine. *SIGMOD Record*, **20**, pp. 85–91.

Burrough, P.A. and Frank, A.U., 1995, Concepts and paradigms in spatial information: Are current geographic information systems truly generic? *International Journal of Geographical Information Systems*, **9**, pp. 101–116.

Burrough, P.A. and Frank, A.U., Eds, 1996, *Geographic Objects with Indeterminate Boundaries*, GISDATA Series Vol. 2, (London: Taylor & Francis).

Casati, R. and Varzi, A.C., 1994, *Holes and Other Superficialities*, (Cambridge, MA: MIT Press).

Centennia, 1993, (Chicago: Clockwork Software Inc).

Cheylan, J.P. and Lardon, S., 1993, Towards a conceptual data model for the analysis of spatio-temporal processes: The example of the search for optimal grazing strategies. In *Spatial Information Theory—A Theoretical Basis for GIS, European Conference COSIT '93*, Lecture Notes in Computer Science 716, edited by Frank, A.U. and Campari, I. (Berlin: Springer-Verlag), pp. 158–176.

Dennig, J., Ed., 1994, *Lotus Notes—Das Kompendium*, (Markt & Technik).

Egenhofer, M.J. and Golledge, R.G., Eds, 1998, *Spatial and Temporal Reasoning in Geographic Information Systems*, Spatial Information Systems Series, (New York: Oxford University Press).

Frank, A.U., 1990, Spatial concepts, geometric data models and data structures. In *Proceedings of the Conference on GIS Design Models and Functionality*, Leicester, UK, (Midlands Regional Research Laboratory, University of Leicester).

Frank, A.U., 1994, Qualitative temporal reasoning in GIS-ordered time scales. In *Proceedings of the Sixth International Symposium on Spatial Data Handling, SDH '94*, Edinburgh, UK, (IGU Commission on GIS), **1**, pp. 410–430.

Frank, A.U., 1996, An object-oriented, formal approach to the design of cadastral systems. In *Proceedings of the 7th International Symposium on Spatial Data Handling, SDH '96,* Delft, The Netherlands, (IGU Commission on GIS).

Frank, A.U., 1997, Spatial ontology: A geographical information point of view. In *Spatial and Temporal Reasoning*, edited by Stock, O. (Dordrecht: Kluwer Academic Publishers), pp. 135–153.

Fraser, J.T., Ed., 1981, *The Voices of Time*, (Amherst: The University of Massachusetts Press).

Galton, A., 1995, Towards a qualitative theory of movement. In *Spatial Information Theory—A Theoretical Basis for GIS, International Conference COSIT '95*, Lecture Notes in Computer Science 988, edited by Frank, A.U. and Kuhn, W. (Berlin, Springer-Verlag), pp. 377–396.

Goodchild, M.F., 1990, A geographical perspective on spatial data models. In *Proceedings of the Conference on GIS Design Models and Functionality*, Leicester, UK, (Midlands Regional Research Laboratory, University of Leicester).

Goodchild, M.F., 1990, Spatial information science. In *Proceedings of the 4th International Symposium on Spatial Data Handling, SDH '90*, Zurich, Switzerland, (IGU, Commission on GIS).

Goodchild, M.F., 1992, Geographical information science. *International Journal of Geographical Information Systems*, **6**, pp. 31–45.

Hofstadter, D.R., 1979, *Gödel, Escher, Bach: An Eternal Golden Braid*, (New York: Vintage Books).

Kuipers, B., 1994, *Qualitative Reasoning: Modeling and Simulation with Incomplete Knowledge*, (Cambridge, MA: MIT Press).

Lakoff, G., 1987, *Women, Fire, and Dangerous Things: What Categories Reveal About the Mind*, (Chicago: University of Chicago Press).

Lakoff, G., 1988, Cognitive Semantics. In *Meaning and Mental Representations*, edited by Eco, U., Santambrogio, M. and Violi, P. (Bloomington: Indiana University Press), pp. 119–154.

Lakoff, G., and Johnson, M., 1980, *Metaphors We Live By*, (Chicago: University of Chicago Press).

Langacker, R.W., 1991, *Foundations of Cognitive Grammar—Descriptive Applications*, (Stanford, CA: Stanford University Press).

Langacker, R.W., 1991, *Foundations of Cognitive Grammar—Theoretical Prerequisites*, (Stanford, CA: Stanford University Press).

Langran, G., 1988, Temporal GIS design tradeoffs. In *Proceedings of GIS/LIS '88, Third Annual Int. Conference*, San Antonio, Texas, (ACSM, ASPRS, AAG, URISA).

Langran, G., 1989, Accessing spatiotemporal data in a temporal GIS. In *Proceedings of Auto-Carto 9*, (Baltimore, MD: ASPRS & ACSM).

Langran, G. and Chrisman, N., 1988, A framework for temporal geographic information, *Cartographica*, **25**, pp. 1–14.

Montello, D.R., 1993, Scale and multiple psychologies of space. In *Spatial Information Theory—A Theoretical Basis for GIS, European Conference COSIT '93*, Lecture Notes in Computer Science 716, edited by Frank, A.U. and Campari, I. (Berlin: Springer-Verlag), pp. 312–321.

Morrison, P., 1982, *Powers of Ten*, (San Francisco, CA: W.H. Freeman).

Mounsey, H. and Tomlinson, R.F., Eds, 1988, Building databases for global science. In *Proceedings of the IGU Global Database Planning Project*, Tylney Hall, Hampshire, UK, (London: Taylor & Francis).

NCGIA, 1989, The research plan of the National Center for Geographic Information and Analysis. *International Journal of Geographical Information Systems*, **3**, pp. 117–136.

Sinton, D., 1978, The inherent structure of information as a constraint to analysis: mapped thematic data as a case study. In *Harvard Papers on GIS*, Vol. 7, edited by Dutton, G. (Reading, MA: Addison-Wesley).

Smith, L.K., Boyle, A.R., Dangermond, J., *et al.*, 1983, Final Report of a Conference on the Review and Synthesis of Problems and Directions for Large Scale Geographic Information System Development, NASA Report, available from ESRI, Redlands, CA.

Snodgrass, R.T., 1992, Temporal databases. In *Theories and Methods of Spatio-Temporal Reasoning in Geographic Space*, Lecture Notes in Computer Science 639, edited by Frank, A.U., Campari, I. and Formentini, U. (Berlin: Springer-Verlag), pp. 22–64.

Talmy, L., 1983, How language structures space. In *Spatial Orientation: Theory, Research, and Application*, edited by Pick, H. and Acredolo, L. (New York: Plenum Press).

Worboys, M., 1994, Unifying the spatial and temporal components of Geographical Information. In *Proceedings of* the *Sixth International Symposium on Spatial Data Handling, SDH '94*, Edinburgh, UK, (IGU Commission on GIS).

CHAPTER THREE

Time and Spatial Database, a Conceptual Application Framework

Jean-Paul Cheylan

3.1 INTRODUCTION

The first group of interest in 'time and spatial representation' emerges from *genealogical* hypotheses: the knowledge of the active and past processes underlying the observed *reality* directly provides a good interpretation of the space organisation. Interaction between processes, for example, nature and society, or in another way, between social actors and decision-makers managing space, bears important explanatory factors—the way and the kind of urban growth explain one part of the urban structures showing successive rings. In the same way, management measures that aim at producing a required situation must be based on the ability to direct the active processes toward this purpose. There appears a need for information or knowledge about the active structures, their dynamics and inertia conditions.

A second group of interest in the representation of spatio-temporal dynamics came from the need for updated information, mainly in the situation of permanent data services. An updating mechanism should be more than a file transfer: how to manage user's information linked to disappearing features?

The first part of the chapter gives an outline of the difficulties encountered in producing a pertinent *real-world* representation leading to the spatio-temporal database design. How to observe the real world?, With what kind of rules?, How to organise the time?, Is the past (respectively the future) unique or multiple?, With what kind of aim in mind?, Is a unique interpretation of the notion of *time* sufficient, or do we need multiple time representations?, Are we faced with continuous or discrete processes?, Is the disjunction between them grounded in the real world, or is this an observation artefact?

The second part proposes to establish a pragmatic classification of the spatial features dynamic. We first show that a set of four situations—permanent, modifiable, varying and changing features—seem to be sufficient to describe most of the dynamic situations. A more technical interpretation of the classification, in terms of spatial and temporal concepts, is also given as well as a first semantic classification of the queries.

The conclusion enlightens some elementary transformations and shows their convergence with other approaches of the spatio-temporal dynamics modelling.

3.2 CONCEPTS NEDED FOR SPATIO-TEMPORAL REPRESENTATION

3.2.1 Observation Processes, Data Acquisition Rules

Representation of dynamic or changing situations may be seen as the interaction between two relatively independent processes: the specification process and the acquisition process. In other words, the quality of the representation depends largely on the adequacy of the *protocol*—the observation specifications (Gautier, 1995)—to the semantic rules of the acquisition (*Real World* building).

3.2.2 Time Sampling and Reality Representation Protocol

Regular Time Sampling Protocol

The first and maybe most usual situation is that of a regular time sampling protocol with, generally, a loose link to the process, or no link at all. This is the case for census (arbitrarily fixed interval of approximately 10 years), or remote sensing (hourly or daily for meteorological data, monthly for land cover change, but the images having a sufficient quality are *randomly* distributed). Assuming some knowledge about the general shape of the process, useful to define sampling rules, about periodicity or trend, a semantic interval of validity can be reached. We are here in the domain of *signal processing* and can use the Nyquist sampling theorem. The data produced by such protocols look like versions, or snapshots separated by unknown or inferred situations. (Cheylan and Desbordes-Cheylan, 1979; Snodgrass, 1987).

Released Observation

The observation is linked to some kind of explanatory moment or situation. The process may be neither periodical nor regular, showing erratic or highly non-stable (bifurcation, chaos, or unknown rules). This kind of situation is very frequent in social sciences; a good example is the electoral behaviour. In this way, the observation could only be linked to an explanatory knowledge of the observed process. The observation may allow to decide or to start another observation or a change of protocol. The switch between an interpreted observation and a real-time protocol is an example. An observation of an urban growth process shows that the emergence of a secondary centre (satellite city) may lead from an annual aerial photograph protocol to a real-time permanent analysis of building permits.

Organically Linked Observation

All the significant events or dynamics that affect the observed process are known, or at least registered as in an analogical observation protocol: a limnigraph, or quite analogical: the motion picture of the observed time and space behaviour of individuals in the Hägerstrand view of urban space use.

3.2.3 Time(s) Structure(s)

Order

In history as well as in archaeology, most of the time processing aims at the hypothetical reordering of facts, not well or not totally ordered in the observation. The reasoning may be based on partial order, due to the lack of *dates* mapped onto a unique and reliable calendar. The time is then a set of constraints, observed independently, which must be logically consistent. The same partial order is used in linguistics, where the discourse is only producing pre-ordered or partially ordered time stamps (Asher and Sablayrolles, 1995; Bernard *et al.*, 1991). When an observation protocol includes textual or interview sources, we need to process that kind of time structures (Allen, 1983; Bestougeff and Ligozat, 1989; Frank, 1994).

A totally ordered structure allows a simple and efficient time representation and processing through arithmetical analogy.

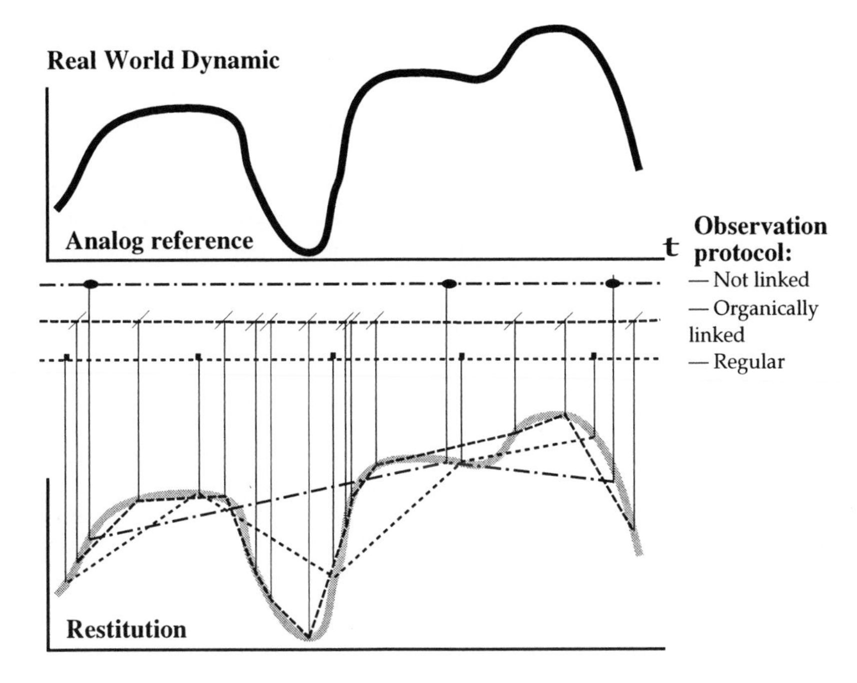

Figure 3.1 Sampling effect on the *real-world* representation

Structure of the Time Line

At the time of processing tasks, many interpretations of the *real world* may coexist. The representation of these alternatives—in the past as contemporary interpretative hypotheses, or in the future as hypothetical scenarios—is frequently needed. One needs a versioning mechanism, used here to express different hypothetical states at the same date (Gançarski, 1994). The management of data consistency and redundancy between those versions is also needed.

Time Reversibility

The idea is to point out some retroactions: a posterior document may be used to re-interpret a dynamic situation (a recently finished building may later give an interpretation about the strategy underlying the work since its beginning). We may use that kind of reversibility to 'enrich' a former situation, without any change of a structure or an object. With destructive updates, the propagation of the knowledge, back in the past, should be very difficult. Hence, we only assume here a new past version.

Time Semantics

In temporal database approach we need to distinguish at least between the *real world* processes time—the *valid time*—and the database processing time—*procedural time*—(Figure 3.2).

Without any time description the database is *static*. With valid time only, the database is called *historical*, allowing to represent the time progress. With procedural time only, the database is called *rollback*, which only allows going back to the former state. If both valid and procedural time are present, the database is called *bitemporal* (Worboys, 1994).

User's Time

Beyond the two main concepts of valid and procedural time, many other user's interpretations of time may be assumed. We mainly point out three of them which appear to be useful:

1. *observation time*: data are only known through observation. We need a time stamping mechanism, distinct from the valid time (observation does not directly define the valid time) and from the procedural time (the time when data is registered is independent of the date of collecting).
2. *interpretation time*: organises and links the different information obtained in an analytical activity. It translates the cognitive activity of the user when validating hypotheses and elaborating interpretations. The interpretation time should be seen as tracing the reasoning process which uses the database.
3. *uncertainty time*: the semantic definitions of the real-world objects may evolve (taxonomy, object classes, quality, confidence levels) and this evolution is given through some meta-information which possesses its own dynamics. The existence of a forest in its maturity or a factory today does not lead to the same confidence in its existence twenty years later. A representation of that kind of process showing the confidence evolution should be useful in a quality assay approach.

In the classical temporal database, time definitions, observation time, interpretation time and uncertainty time are generally considered as *user's time*, more directly linked to the semantics and slightly controlled (Snodgrass and Ahn, 1986; Jensen *et al.*, 1992; Yeh, 1995).

Because of complexity and performance, we generally disconnect valid time and procedural time (Cheylan and Lardon, 1993). In most temporal analysis applications we assume that information about valid time is complete and logically consistent. We work with the *Closed World Assumption*, which is supposed to give complete knowledge of the real world facts.

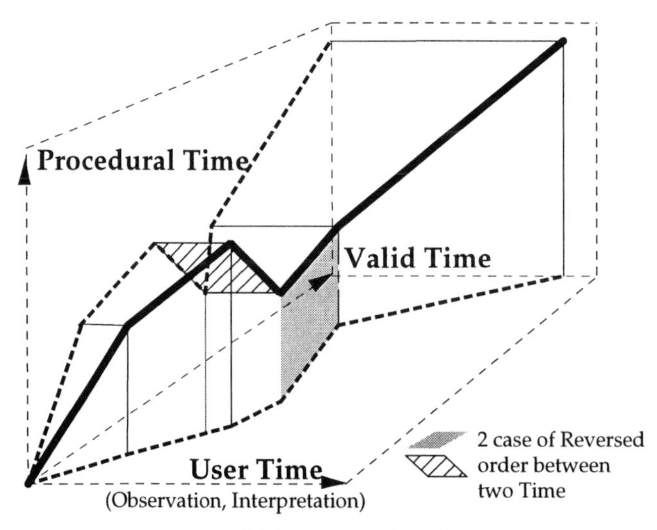

Figure 3.2 Three semantics of time

State, Events

Two time structures are generally shown as two dual forms of representation:

- The first one defines events as instantaneous time points changing the state of the universe. It generally leads to a discrete view of time and an understanding of associated transformations incorporated as mutations (Latarget, 1994).
- The second one defines stable states (or sometimes dynamic states described as quite continuous), bounded by events. This approach leads to a representation with versions (Cellary and Jomier, 1990) or to the definition of dynamic interval with a homogeneous functional definition (Yeh and Viémont, 1992; Galton, 1995).

These two representations are not exclusive. Theoretically, we are able to deduce one from the other (Allen, 1983), and it would be useful to have a mixed representation, putting together states and events. The problems of time consistency and integrity between a state (version) representation in extension (the database represents enumeratively all the situations), and a representation of mutation through a procedural mechanism (transaction journal) are not well defined and remain unsolved.

States Set

Most of the applications assume that the process states sets are *intrinsically* linked to the nature of the underlying processes. Some processes are not *intrinsically* continuous or conversely discrete. The crystal structure rupture can be seen as a propagation process along certain pre-existing default and taking a long time; conversely most of the *continuous* processes such as climatic changes should be seen at the observation stage; as a chronological series of discrete observed states. The structure of the states set appears to be defined only by the observation protocol and the semantics specifications.

Temporal Operators

The temporal queries implementation needs some operators. In an instant logic equality and inequality are the only two. If we assume temporal intervals, relations between the time intervals imply the definition of a precedence operator, associated with the sets operators (union, intersection, inclusion, equality). Allen (1983) has shown that the equality and seven relations (with their symmetrical counterpart) are sufficient:

```
X before Y:     XXX  YYY          X equal to Y:     XXX
                                                    YYY

X meets Y:      XXXYYY            X overlaps Y:     XXX
                                                      YYY

Y during X:     XXXXXX            X starts Y:       XXX
                  YY                                YYYYY

X finishes Y:     XX
               YYYYY
```

3.3 SPATIAL FEATURES DYNAMIC

What is changing? Qualities, space, geometry, spatial relationships and identity. A real-world process may be represented under such an assumption by different data structures and dynamics, depending almost only on the decision of representation. We propose first to attempt classification of the dynamic situation representations. Our classification (Cheylan *et al.*, 1994; Gayte *et al.*, 1997) is mainly grounded on application needs and experience. We tried to gain some more explicit (pre-formal) specifications in terms of temporal database problems arising from the different goals assigned to the representation.

3.3.1 Elementary Spatial Change Situations

Permanent Features

In a first, quite simple case, a spatial set of features is permanent. The geometry and the topology (if present) remain the same, only the attribute values may change through time. The price of the cadastral parcel, the population of the commune, the number of commuters between two towns are classical examples of this situation. The evolution of some links between geographic features (the case of dynamic interaction matrix: Sanders, 1993) fall under the same category and could also be expressed as a functional expression of the time. Most of these applications here work with a set of inter-constrained geographical objects: a set of districts partitioning the space, a set of hydrologic arcs represented as a set of connected and ordered spatial features (Figure 3.3).

Modifiable Spatial Features (Split-and-Merge Processes)

A large number of problems should be viewed as the re-composition of a unique space —in terms of a unique semantic layer affected by some *split-and-merge* processes (Frank, this volume). Elementary space entities, having the same dynamics, should be the result of the observed process, or alternatively, should be seen as the re-composition of predefined elementary units. In any case, we face a dynamic space-partitioning process. If we observe a historical cadastre or the history of crop succession in an agricultural settlement, the whole space is always affected by one of the descriptive terms (owned by somebody, or covered with a crop). At any date the state of the process is a spatial partition. Chronologically varying relationships occur between the different and successive partitioning states of the same space. A parcel or a crop immediately succeeds (precedes) its *parent* (*children*) state; the link may be denoted by a temporalisation mechanism of the identifiers and a pointer to the parent(s).

An important part of the semantic knowledge about occurring dynamics is carried by the time relations between the successive, but not concordant, spatial units partly covering the same space (Al-Taha and Barrera, 1994).

The description problem is here analogous to the genealogical analysis of family dynamics. In terms of database problems, it leads to the management of a temporal graph, linking features between successive situations. The graph is the skeleton of the chain of mutations having a tree or a lattice structure. Queries about the graph structure seem to encompass the power of the set classical relational systems, leading to iterative or recursive processing.

Varying Features (Creeping Features)

Some observations lead to a family of different spatial features: the same semantic objects (the urbanised area, the whole region affected by rabies or AIDS, a tropical storm going through an ocean) have a changing coverage, or perimeter, but do not produce new units. Different spaces affected do not share any specified relationship; spatial features stay the same, within different successive configurations. Users consider spatial units as the same, only changing in location and shape. In a spatio-temporal database approach, spatial features keep the same identifier all along. Changes are produced by geometric transformations, seen as a series of geometric versions, or as a functional space transformation.

We face here a multiple representation problem showing the same spatial feature at different dates; we do not need any management of spatial feature identifier changes. A snapshots or a versions approach could be adopted (Cellary and Jomier, 1990), and needs an interpolation mechanism between the successive known states.

If two spatial features merge together, we shift back to the former situation of *modifiable spatial features*, and have to manage the temporal identity links.

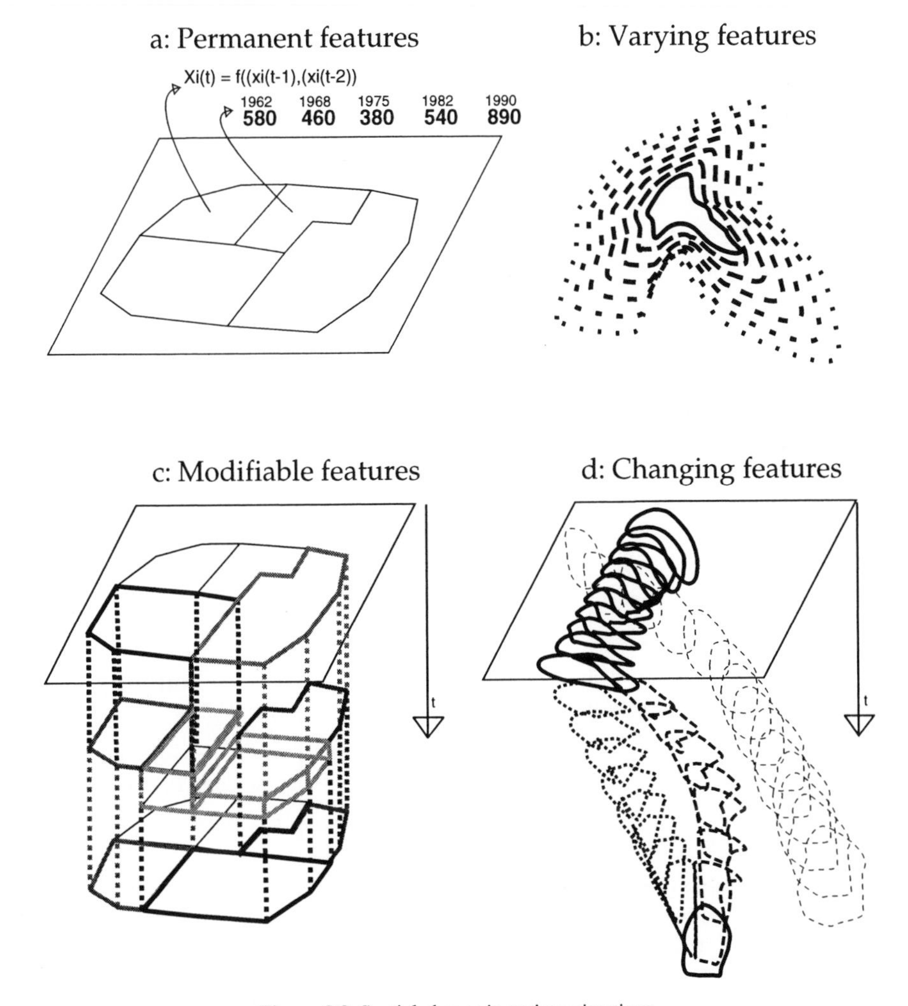

Figure 3.3 Spatial change in various situations

Changing Spatial Features (Creeping + Split-and-Merge Features)

We focus here on spatial features changing in shape, location and able to produce new features or to merge two preceding features into a unique one. The process representation must show all the significant changes (according to the semantic code of the observables) in the processes to be significant.

Here we find the restitution by symbolic objects of weather situations, or the observation of time-space behaviour, as approached by Hägerstrand (1968) or strategies in time and space (Cheylan, 1993) or in the induction of behavioural rules grounded on *expert* behaviour observation (Deffontaines, 1989).

3.3.2 Representation of Changes

At a first glance, the classification of changes proposed is based on a distinction between the features spatial relationships occurring, shown in both columns. A second disjunction describing the identity temporal behaviour occurs in both rows. A second point of view, more conceptual, goes to a *prototypical* approach, pointing out some convergence with other proposals (Frank, this volume; Claramunt *et al.*, 1997).

GIS Domain Application

Table 3.1 Representation of spatial features dynamics

SPACE Spatial features classification **TIME** features succession	**A** Mutually constrained features within space	**B** Unconstrained features in space
1 Permanent feature identifier	**Permanent** 1 spatial topology *Static or 'cartoon' mapping*	**Varying** 1 space/time topology *Graphical versioning (CAD, morphing)*
2 Temporalised identifier temporal links	**Modifiable** n spatial topologies *GIS + genealogy graph*	**Changing** n space/time topologies *GIS + genealogy graph + graphical versioning*

The first column (Table 3.1) is dedicated to features associated by a strong integrity inter-dependence, as in classical topological planar layers under partitioning rule, or as in a river network. Typically, we refer here to situations as administrative partitioning of space, viewed as static (upper left) or dynamic (lower left). In the second case, the partitioning rule leads to a situation analogous to the partitioning of a time–space *volume*.

The second column describes situations with a less constrained space. Some of the features or behaviour traces evolve and move in space without any temporal or spatial constraints except some local integrity rules (a vehicle, or a person always exists and is located somewhere).

The first row describes situations with a less constrained pace. Some of the features, or behaviour traces evolve and move onto space without any temporal or spatial constraints except some local integrity rules (a vehicle, or a person always exists and is located somewhere).

The second row deals with features that can split and merge together in order to produce new situations, and so for new spatio-temporal identifiers.

Change as Life, Genealogy and Motion

The same representation of change could be viewed as the combination of three notions: life, motion and genealogy. Life and motion were proposed by Frank (this volume), as a way to differentiate the vague concept of change. In our view, Frank's concept of life has to be refined as follows:

1. *life* (in Frank: change in the object) affects objects, or features, along their own life span. A tree grows, the colour changes, but the tree remains the same (in the view of the database semantics); we still call that kind of change *life*. The evolution of the population of census units is a simpler example. We could also think of the same population as dynamic, occurring in non-changing spatial units, but modelled with the same functional expression or simulation of the population growth. At a more complex level, the population of the units could also be expressed as a functional dependence of some neighbourhood description, or inter-units interaction (Sanders, 1993). We understand that the lack of spatial in-depth change in these situations does not imply the absence of change.
2. *genealogy* (in Frank: life), a second kind of life occurs when features *give birth* to new features, like parcels splitting. The temporal link between the successive situations is partly described by the kind of transition process occurring. We call this kind of life *genealogy*. The distinction, which seems semantically sound (Cheylan and Desbordes-Cheylan, 1979; Al-Taha and Barrera, 1994), is also directly related to the management of the identifier (for example, by time stamping) and multi-versions queries expression problems in temporal databases (Snodgrass and Ahn, 1986; Cellary and Jomier, 1990).
3. *motion* gives the third concept, as explained by Frank. A more precise classification, which defines the following cases: stability, expansion or contraction, deformation (heteromorphic change in shape, size and orientation), translation and rotation, as minimal and orthogonal case have recently been given (Claramunt *et al.*, 1997).

Life, Genealogy, Motion and Representation in the Database

The three concepts are easy to link to the dynamics situation of the four spatial features (Table 3.1) by the part of the features description affected by the change:

1. the descriptive, non-spatial part of the description linked to the *life* concept and involved everywhere, but sufficient only in the permanent feature case;
2. the spatial (geometrical and topological) part of the description involved in motion and genealogy, and only sufficient for the *varying* case;
3. the chaining of identifier as the description of the features historical span, only needed in the two *genealogical* cases, and sufficient for the *modifiable* case).

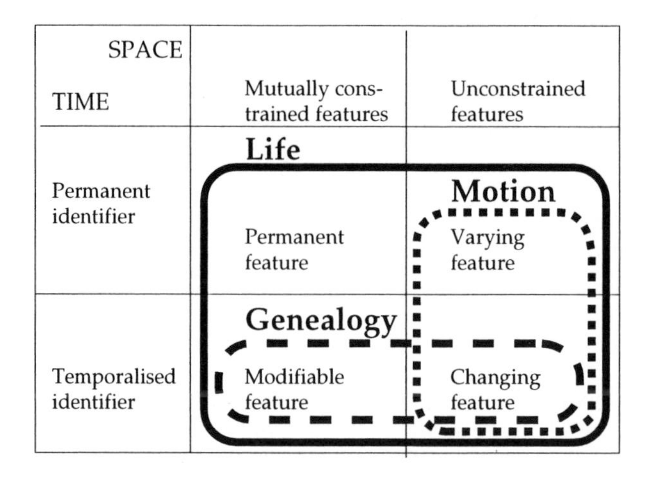

Figure 3.4 Spatial change as life, genealogy and motion (alternative definitions to Frank)

3.3.3 Elementary Spatio-Temporal Processing

It seems (Segev and Shoshani, 1987) that only a small number of query categories is sufficient to allow more complex spatio-temporal processing. The classical—SQL-like—or more specialised query languages (TSQL, TQUEL: Snodgrass, 1987; Postgres: Rowe and Stonebraker, 1987; TSQL2: Snodgrass *et al.*, 1994) are unable, or not powerful enough, to implement the whole set of proposed queries.

We assume that the topological relations (Egenhofer, 1989) as well as the temporal interval relations (Allen, 1983) are both available within the language. The database is a *historical* one, supporting valid time representation, without any interaction between valid time and procedural time (no past updates). Time is assumed to be discrete, linear, and in total order. Queries may be applied to each of the four situations described above.

Queries about a State

What was the situation (the spatial configuration) of *the space* S at date *t*?

A1: What was the district population in 1982?
B1: What were the regions infested by rabies in July 1992?
A2: What was the crop organisation in spring 1995?
B2: Where was the tropical storm TS1 on the 20th of July 1994, at 10 a.m.?

The result is a static map, a spatial time slice, a snapshot. In case of A1 and B1 from Table 3.1, time intervals are associated to attribute values or versions; the answer is a selection of the time validity interval as argument. In case of A2 and B2, the query is not as simple if all the identifiers are time stamped with their validity interval. It is necessary to select all the features born before the date and to restrict the result by deleting all the features dead before this date.

Queries about the History of Space

A1: What are the successive population values for the commune C?
B1: What are the urban/rural intervals of location L?

A2: Who are the successive owners of location L?
B2: What are the successive grazing rates at location L?

The query is as follows: what did happen here? The expected result is the temporal vector of local attribute values, showing the successive states of local entities covering that space, like a cartoon or a movie, depending on the states set structure (respectively discrete or quasi continuous). All the situations in our classification should support that kind of query. The resulting ordered set of states with their time validity intervals, should be consecutive or not (showing gaps), depending on the kind of temporal integrity rules assumed.

Queries about the History of Features

A2: Which are the land parcels owned by owner O and sold to a public owner?
B2: Which are the mixed flocks (sheep and goats) coming from the merging of flocks?

This kind of query is only relevant for the cases of A2 and B2 (from Table 3.1), where successive and geometrically different features occur in the same place and are historically linked by some temporal links (parcels split, a new crop covering two precedent crops). Ancestor concept, a set or a structured identifier list of the preceding feature of each feature is needed here. The processing leads to the explanation (in extension or in intention) of a state and transition graph linking temporalised identifiers. The relative depth of the path along the graph leads to an iterative or recursive processing. The answer to such a kind of query may take two forms: the first one, which is relatively simple, is a set of temporalised identifiers (queries such as: who are the previous owners of a parcel?). The second is much more complex, involving structures or patterns of the desired answer sub-graph such as: which are the parcels coming from the split of public ownership, the last mentioned having been established by the gathering of private parcels. Temporal completeness, or other temporal constraints, should be integrated if semantic integrity rules contain such information (Yeh and Viémont, 1992).

Queries about Time Instants or Intervals

1. When did the population of the town T become greater than...?
2. During which period (time interval) does the ownership of Mr. X regularly increase?

The last group of queries calls for a temporal element as answer: a date or a time interval. Here the space is only present by means of identifiers, so that the kind of queries is the same as the one defined in temporal database research.

These four classes give only a rough classification. It is mandatory to explore the consistency and completeness of the propositions. Further, important problems such as spatio-temporal integrity, multi-versions querying, or long and complex spans of historical versions remain unsolved.

3.4 CONCLUSION

The first part of the chapter put in evidence the large number of design decisions we are faced with to organise a spatio-temporal application. The importance of the observation specification and *real-world* representation processes establishing the representation of the perceived real world. In that context technical, conceptual and *formal building* lead to define:

a. the classes of time structure;
b. the organisation of the time line (branching in the past or the future, or not branching);
c. the needed complementary time semantics;
d. the decision to establish a state set of the process restitution (discrete or pseudo-continue), and also
e. the kind of temporal processing aimed at.

The second part starts from an empirical and applied spatial change classification. Spatial features organised by a strong spatial consistency are distinguished by those showing a less *constrained* behaviour. The distinction between the part of the spatio-temporal features affected by the transformation processes: the semantic-descriptive part only; the spatial part only; the linkage between features along the time line and their genealogy is shown as the main design criteria. The resulting change classification of elementary changes has a complementary GIS-database interpretation and a conceptual one, pointing out the concept of life, motion and genealogy.

A first attempt to classify the spatio-temporal queries on a functional basis is then given, leading to further works in the direction of the specification of a spatio-temporal query language.

ACKNOWLEDGEMENTS

The work reported here has been funded by the French CNRS-IGN common research project PSIG. Despite the collective work of the A1 Axe here reported, and the permanent and highly productive work with Sylvie Lardon and Thérèse Libourel, the author is solely responsible for the content of this publication.

Thanks are extended to Andrew Franck, Nicholas Chrisman, Robert Jeansoulin and Roswitha Markwart for their helpful comments on the preliminary version of this chapter and their help in the task of improving the translation from French into English.

REFERENCES

Allen, J.F., 1983, Maintaining knowledge about temporal intervals. *Communications of the ACM*, **26**.

Al-Taha, K. and Barrera, R., 1994, Identities through time, ISPRS Working Group II/2, Workshop on the Requirement for Integrated Geographic Information Systems, New Orléans.

Asher, N. and Sablayrolles, P., 1995, A typology and discourse semantics for motion verbs and spatial PPs in French. *Journal of Semantics*, **12**, Lexical Semantics Part 2, edited by Boguraev, B. and Pustejovsky, J.

Bernard, D., Borillo, M. and Gaume, B., 1991, From event calculus to the scheduling problem, semantics of action and temporal reasoning in aircraft maintenance. *Journal of Applied Intelligence*, **1**, pp. 195–221.

Bestougeff, H. and Ligozat, G., 1989, *Outils logiques pour le traitement du temps, de la linguistique a l'intelligence artificielle*, (Masson).

Cellary, W. and Jomier, G., 1990, Consistency of versions in object-oriented databases. In *Proceedings of the 16th VLDB Conference*, Brisbane, Australia, pp. 432–441.

Cheylan, J.-P. and Desbordes-Cheylan, F., 1979, From data sources to questioning in temporal data base with cognitive aims. In *Proceedings of IFIP Conference: Data Bases in the Humanities and Social Sciences*, (North-Holland).

Cheylan, J.-P. and Lardon, S., 1993, Towards a conceptual model for the analysis of spatio-temporal processes. In *Spatial Information Theory—A Theoretical Basis for GIS, European Conference COSIT '93*, Lecture Notes in Computer Science 716, edited by Frank, A.U. and Campari, I. (Berlin: Springer-Verlag), pp. 158–176.

Cheylan, J.-P., Lardon, S., Mathian, H. and Sanders, L., 1994, Les problématiques temporelles dans les SIG. *Revue Internationale de Géomatique*, **4** (3–4).

Claramunt, C., Thériault, M. and Parent, C., 1997, A qualitative representation of evolving spatial entities in two-dimensional spaces. In *GIS Research UK, Proceedings of 5th National Conference*, Leeds, UK.

Deffontaines, J.P. and Lardon, S., 1989, Grasslands and agrarian systems, methodological considerations on space in the management of grasslands. In *Grassland Systems Approaches. Some French Research Proposals. Etudes et recherches sur les systèmes agraires et le développement*, (INRA Paris), pp. 209–218.

Egenhofer, M.J., 1989, A formal definition of binary topological relationships. In *Foundation of Data Organization and Algorithms*, edited by Litwin, W. and Schek, H.-J. (Berlin: Springer-Verlag).

Frank, A.U., 1994, Qualitative temporal reasoning in GIS–ordered time scales. In *Proceedings of the 6th International Symposium on Spatial Data Handling*, Edinburgh, UK, Waugh, T.C. and Healey, R.C. (London: Taylor & Francis).

Galton, A., 1995, Towards a qualitative theory of movement. In *Spatial Information Theory—A Theoretical Basis for GIS, International Conference COSIT '95*, Lecture Notes in Computer Science 988, edited by Frank, A.U. and Kuhn, W. (Berlin: Springer-Verlag), pp. 377–396.

Gançarski, S. and Jomier, G., 1994, Managing entity versions within their context: a formal approach. In *Proceedings of DEXA '94*.

Gautier, D., 1995, Analyse des dynamiques spatiales d'un massif de chataigniers. In *Revue Internationale de Géomatique*, **5** (1).

Gayte, O., Libourel, T., Cheylan, J.-P. and Lardon, S., 1997, *Conception des systèmes d'information sur l'environnement*, (Paris: Ed. Hermès).

Hägerstrand, T., 1968, *Innovation diffusion as a spatial process*, (Chicago: University of Chicago Press).

Jensen, C.S., Clifford, J., Gadia, S.K., Segev, A. and Snodgrass, R.T., 1992, A glossary of temporal database concepts. In *ACM-SIGMOD Record*, **21** (3).

Latarget, S., 1994, Un modèle orienté-objet pour la gestion de l'histoire des bases de données géographiques, mémoire de D.E.A. des sciences de l'information géographiques.

Rowe, L.A. and Stonebraker, M., 1987, The Postgres Data Model. In *Proceedings of 13th International Conference on VLDB*, Brighton, UK, pp. 83–96.

Sanders, L., 1996, Dynamic modelling of urban systems. In *Spatial analytical perspectives on GIS*, edited by Fischer, M., Scholten, H.J. and Unwin, D. (London: Taylor & Francis).

Segev, A. and Shoshani, A., 1987, Logical modeling of temporal data. In *Proceedings of ACM SIGMOD Conference* 1987, pp. 454–466.

Snodgrass, R.T. and Ahn, I., 1986, Temporal databases. *IEEE Computer*, **19**, pp. 35–42.

Snodgrass, R.T., 1987, The temporal query language Tquel. In *ACM Transactions on Database Systems*, **12**, pp. 247–298.

Snodgrass, R.T. *et al.*, 1994, TSQL2 language specification. In *ACM SIGMOD Record*, **23** (1).

Worboys, M.F., 1994, Unifying the spatial and temporal components of geographic information. In *Proceedings of the 5th International Symposium on Spatial Data Handling*, Vol. 2, Charleston, SC, (IGU Commission on GIS), pp. 602–611.
Yeh, T.S. and Viémont, Y.H., 1992, Temporal aspects of geographical databases. In *Proceedings of EGIS '92*, Munich, pp. 320–328.
Yeh, T.S., 1995, *Modèlisation de la variabilité des entités dans un système d'information géographique*. Thèse de Doctorat de l'Université Paris.

CHAPTER FOUR

Understanding and Modelling Spatial Change

Marinos Kavouras

4.1 INTRODUCTION

The great majority of GIS developments have dealt with reality and related problems utilising multiple static databases. *Time* has only been used as an attribute to index specific facts. Imprecise, nevertheless very useful and frequently used knowledge of time (including order and intervals) has not been possible yet. Furthermore, formalisation of change between two different states (facts or situations) in time has not been developed. There still seem to be major conceptual problems in dealing with spatial change in GIS. Spatial change is intrinsically associated with time.

This chapter addresses a number of important issues, in an attempt to stimulate scientific research towards the development of a unified spatio-temporal framework. In this framework, different application areas make entirely different demands on the temporal GIS. The differences are the result of: (a) the way objects are formed, and (b) the spatial reasoning required. A taxonomy of spatio-temporal aspects related to spatial change is first presented, based primarily on application but also on logic. A similar application-induced classification has been proposed by Burrough and Frank (1995), where the differences between the capabilities of current GIS software and their requirements are compared. Two important modelling issues follow next—the examination of space, motion, time and change at different levels of detail (scale/generalisation concept), and the motion of objects with non-definite boundaries.

A very important application, that of planning, is in essence a temporal effort. An effort, which goes well beyond understanding (scientific reasoning), towards prediction, which will then enable possible planning of a desirable future. The dependence of urban development and planning on spatio-temporal structures and the difficulties involved are presented in the chapters by Dupagne and Salvemini (this volume). The importance of understanding spatio-temporal change as well as effects and causes in rational planning of the use of space is elaborated by Gautier (this volume).

From the wide variety of applications requiring advanced but very different spatio-temporal modelling, three diverse cases are presented. The first refers to spatial change in one of the most common and established socio-economic applications—the cadastre. The second refers to change of socio-economic units, which is not obvious but created (limited) by definition (e.g., zone classification). This type of application and particularly the methodology and consequences of designing zoning systems for representing socio-economic data is detailed by Openshaw (this volume). Finally, the last example addresses the problem of building temporal GIS to record uncertain information, as it can be found in historical texts and link these to known locations while deciding for a proper spatio-temporal reference framework.

4.2 SPACE, TIME AND PLACE

Human experience of space ('choros'), place ('topos') and time ('chronos') is mostly subconscious. Humans, influenced by landscapes, pictures or maps, learn to organise visual elements into a spatio-temporal structure. According to Tuan (1977), every activity generates a particular spatio-temporal structure. Time and space intertwine and define each other in personal experience. Place is essentially a static concept—an organised world of meaning—a pause in the temporal current (ibid.) (Figure 4.1).

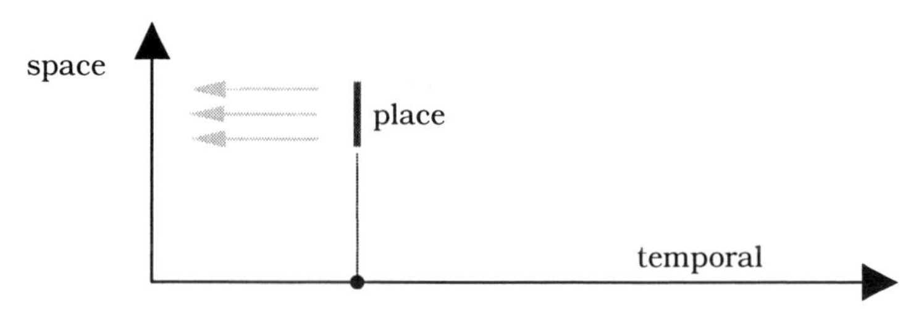

Figure 4.1 Place as a static spatial meaningful organization in time

Time and space are important in planning. This is because in planning we set goals. A goal has both temporal and spatial meanings. Time clearly connotes effort (resources, energy) required to reach a goal. Distance, for example, unlike length, is not a pure geometric concept. It is expressed as cost of movement along a path on a friction surface based on some criteria which usually imply time. Even when the distance is expressed in length units (e.g., 2500 km to drive), human experience finds it more helpful (Tuan, 1977) to quickly convert distance units into time. 'Three days' drive' tells us more directly the money required to purchase resources and energy (e.g., gas, food, hotel). When goals are set, they refer only to the future; thus time is usually portrayed as an arrow pointing to the future. There is, therefore, an implicit reference of time in the ideas of movement, effort, freedom, goal and accessibility. The purpose of this chapter, however, is to relate time explicitly to space.

Spatial planning is concerned with "the avoidance of unintended repercussions while pursuing intended goals over time, space and function" (Harris and Batty, 1993). It has been stated (ibid.) that traditional GIS are too simplistic to deal with the complex planning process, one reason being their crudeness in handling temporal data and also data on interactions and flows. It is further advocated that pragmatic planning is likely to embody many different models, theories and data systems; and "the notion that there will be a plurality of information systems, models, and planning processes relevant to any single problem is the reality we have to live with" (Couclelis, 1991). Although this may seem valid for solving complex spatial planning problems, it is our strong belief that the present inadequacy of current GIS to support such a task is largely due to the lack of a concrete theoretical foundation, which among others, has not found acceptable ways to represent generically temporal data, processes, and data on flows and interactions associated with socio-economic applications. This chapter touches upon some important aspects associated with spatio-temporal change.

4.3 TIME AND CHANGE IN GEOGRAPHY

Main constituents of geography are *space*, *relations in space*, and *changes in space* (Morrill, 1970). Geography questions about *location* and *place*, *structure*, and *processes* in space. It examines regional characteristics, similarities and differences, interrelations, behaviour and evolving, leading to distributions, patterns and structures about organisation in space. It is obvious that mere static sets of descriptive data—what most GIS systems offer today—cannot serve such goals. In the same context, an important requirement of advanced temporal applications is the ability to deal with definite and indefinite, finite and infinite temporal information. In cartography, there have been some seemingly static graphical representations, especially maps, which instruct us to see time 'sweeping' through space. There is currently no database model, however, which offers functionality in a single unified framework (Koubarakis, 1994).

An early diagrammatic conceptualisation about the incorporation of temporal information in geography is that of the *geographic space–time matrix* (Berry, 1964) (Figure 4.2). The columns of the matrix refer to places, whereas the rows refer to characteristics. The intersection between a column and a row defines a cell containing a geographic fact. Spatial changes over time are depicted as cross-sections (cards) from every time instance of interest.

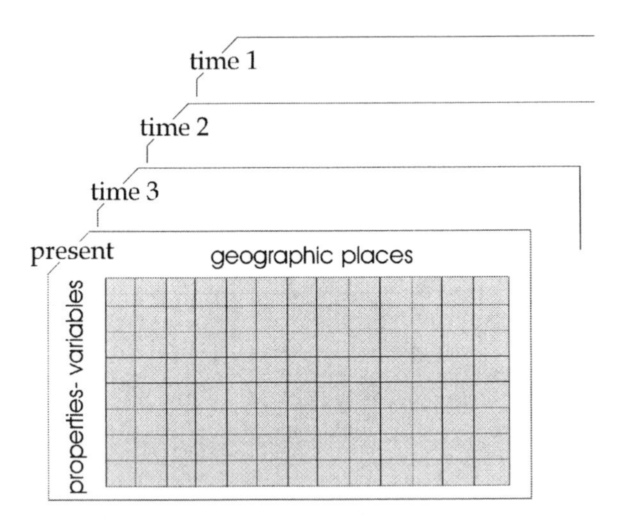

Figure 4.2 Geographic Matrix (from Berry, 1964)

A comparable diagrammatic conceptualisation has also been used by Dangermond (1990), in which time is treated as one of the three main constituents of a GIS. Besides these early conceptualisations, numerous more recent efforts attempt to tackle the issue of time, distinguishing absolute or relative time, point or interval time, valid or transaction time, quantitative or qualitative temporal reasoning, database issues, etc. (Allen, 1983; Al-Taha & Frank, 1993; Frank, 1994; Langran, 1989; Peuquet, 1994; Whigham, 1993; Worboys, 1994).

A successful and unified approach to spatio-temporal modelling faces many conceptual problems to solve prior to implementation issues. In this course, it identifies a number of issues for scientific discussion and research. The emphasis is not put on the list of issues, which does not claim to be exhaustive or complete, but on the argument that space and time should be considered conceptually equivalent, treated as such from the modelling point of view. In such fundamental departure from the static model (Figure 4.3), starting from what is descriptively known (*what–where–when*) (e.g., the triad framework proposed by Peuquet, 1994), the next challenge is in reaching understanding, i.e., knowledge about processes and rules (i.e., answering to *how–why* phenomena occur, relate and evolve). The level of understanding is tested by the ensuing scientific process of prediction (answering to *what–if* questions), that is, tested against the ability to make conditional predictions based on alternative hypothetical scenarios. All the above steps are necessary to move beyond science to planning which is concerned with the design of actions (plans) to accomplish a specified objective (answering the *what–design* question), while avoiding undesirable consequences. Finally, the control of the action itself, known as management process (answering the *what–control* question), is concerned with the standards and instructions necessary for the realisation of the decided plans.

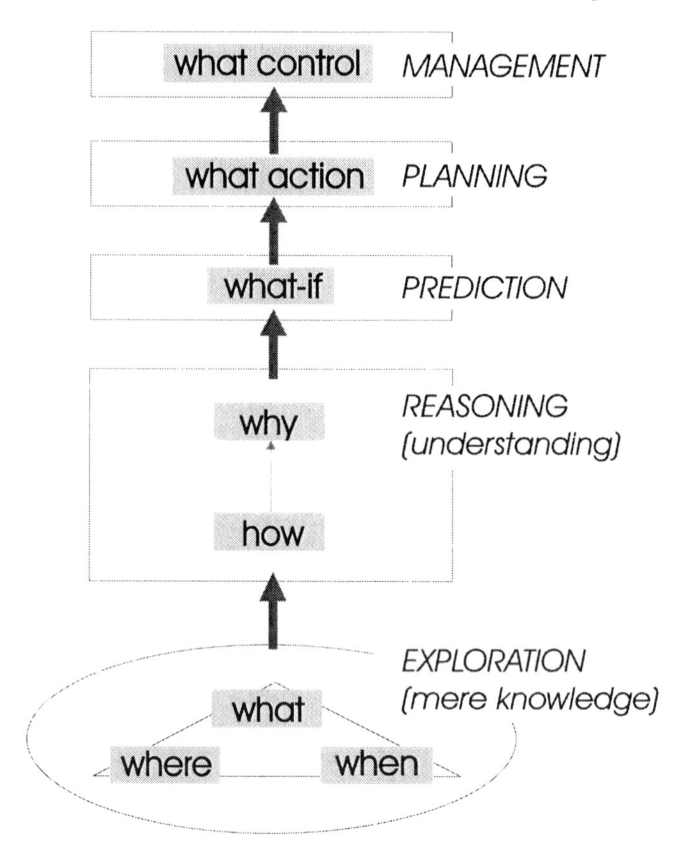

Figure 4.3 From exploration (from Peuquet, 1994) to understanding, prediction and planning

4.4 A CONCEPTUAL TAXONOMY OF SPATIO-TEMPORAL CHARACTERISTICS

Changes affect any phenomenon attributed (directly or indirectly) to space. In the present context, changes refer to a single or more spatial phenomena, their internal and external static or dynamic properties, life characteristics, behaviour and quality. More specifically, on the basis of spatial applications affected by change, spatio-temporal characteristics of a phenomenon can be conceptually classified/structured by five major components:

1. INTERNAL characteristics (not related to other phenomena)
   static properties–characteristics (shape, orientation, descriptive, etc.)
   movement (motion)

2. EXTERNAL characteristics (related to other phenomena)
    location (absolute or relative)
    dynamic topology
    non-geometric relations–dependencies (to other concurrent phenomena[1])
3. TEMPORAL characteristics
    life duration (time intervals)
    temporal order (explicit/inferred)
    created-by & transformed-into (inheritance)
4. BEHAVIOURAL–METHODOLOGICAL characteristics
    resistance to changes
    rules–operations applicable
5. QUALITY characteristics (meta properties)
    degree of detail
    relative importance (ranking, order)
    certainty–reliability

In the above taxonomy, it is first important to realise that all characteristics are subject to *change*. For example, given a phenomenon, there may be change in its descriptive internal characteristics, change in its resistance to change (behaviour), change in its dependence to other phenomena, change in its representation detail, etc. The second important realisation is that *change* itself is expressed by characteristics of the other components. For example, the degree of detail (a meta-property) applies to an internal static property but also to temporal resolution. Similarly, temporal information (e.g., life duration) can refer to the life of a spatial entity, to the duration of a particular internal property of the entity, to the duration of a particular operation valid on this entity, etc.

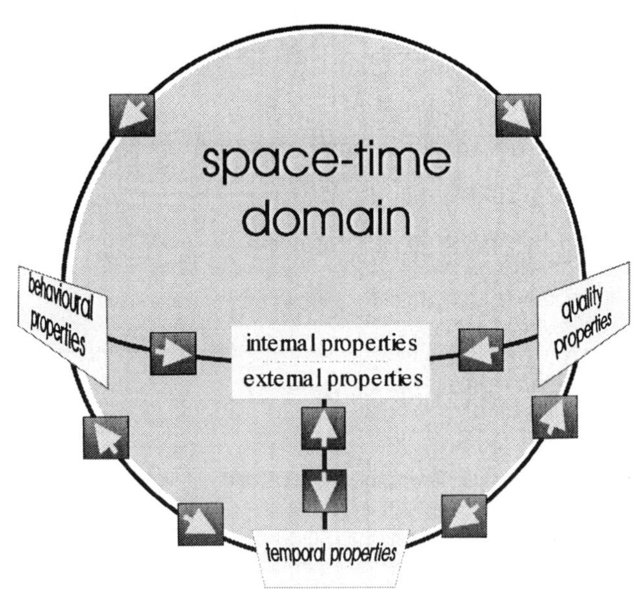

Figure 4.4 A schematic representation of major components related to spatio-temporal change

[1] Although it may sound a bit paradoxical at first, absolute location definitely constitutes an external characteristic; for it requires definition of a reference system. This implicitly relates to all other phenomena.

This interrelation between the above characteristics with respect to spatio-temporal change is schematically shown in Figure 4.4 by the arrow connections between the components. In practice, most of these interrelations are totally absent from current GIS capabilities that are merely descriptive. One such research field of particular interest (and difficulty) is that of having and formalising quality (metadata) describe temporal or spatial aspects. Quality metadata among others include resolution (degree of detail) of both spatial and temporal aspects, degree of reliability (certainty) on the available spatio-temporal knowledge, relative importance of objects/relationships/properties/operations, etc.

4.5 SCALE, RESOLUTION AND GENERALISATION IN SPACE AND TIME

So far, in GIS, the notion of scale, resolution and generalisation has been mostly dealt with as a cartographic (visualisation) problem. A limited effort has been made towards the modelling importance of scale, that is, the generalisation of internal representations of geographic objects, which is necessary for spatial computations and analysis, combination of data sets at different levels of detail (resolution levels), etc. In the present context of temporal GIS, generalisation is presented as a generic modelling concept of essential importance (Figure 4.5). It refers to all spatial concepts, not only to models physical objects, processes, and their visualisation, but furthermore to their internal and external relations, and variable resolution of behavioural, quality and temporal characteristics, including reasoning.

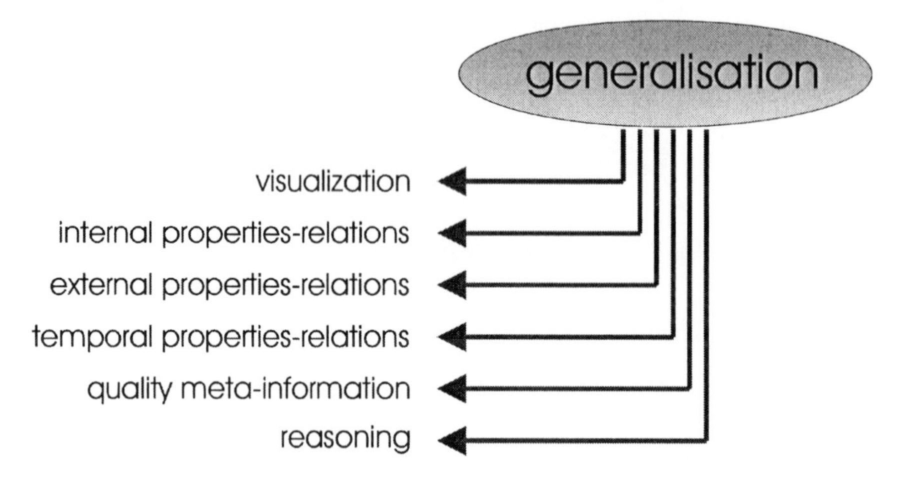

Figure 4.5 The modelling importance of generalisation on spatio-temporal concepts

Temporal generalisation constitutes an open research field for GIS. It refers to working in the spatio-temporal domain at different time scales supporting time-related data or queries such as: "What was the spatial distribution of unemployment in country A in the 1930s?" "What was the population of country B in the beginning of the 18th century?" "What will be the shape of the European coastline (spatial generalisation) in the 2050s (temporal generalisation)?" Currently there are no formal models to deal with complex queries, i.e., queries on information other than those explicitly reposited in the database as snapshots limited by their temporal resolution.

In several application domains, entities are spatially defined not by physical but by descriptive properties such as statistical data representing socio-economic phenomena. In such cases, any (usually time-based) change in the descriptive characteristics results in a new spatial description. While in most practical implementations, every new spatial description is represented as a new object, the transition between the different spatial representations of a phenomenon corresponding to time epochs, can also be viewed (and therefore modelled) as *motion* from one state to another. This is particularly applicable in the case of 'definition limited' (or 'field-based') phenomena, that is, phenomena defined by a classification of a field domain. The advantage of the second approach is clearly the ability to describe not only the states but also the process. This topic is extensively discussed in the rest of the book.

In principle, temporal resolution can be related to spatial resolution in two ways:

1. by using a phenomenon-based description as reference to spatial and temporal attributes (vector approach), or
2. by using a space–time discretisation as reference, with description of states of phenomena (raster approach).

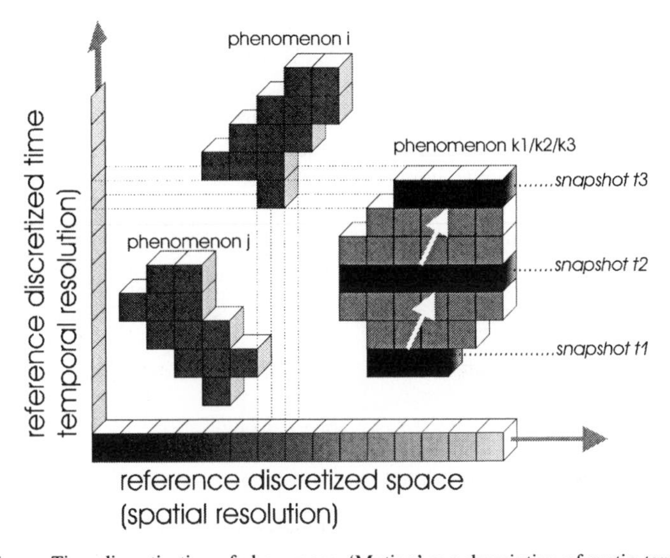

Figure 4.6 Space–Time discretisation of phenomena. 'Motion' as a description of spatio-temporal change

Figure 4.6 depicts schematically the space–time discretisation of spatial phenomena. Information about the status of a phenomenon k is depicted with at least three not continuous time instances t1, t2, t3. The phenomenon can be represented either as three different phenomena (static representations) k1, k2, k3, or as three states of a single phenomenon (dynamic representation) moving from one state to the next according to some behavioural rules supporting or not persistence.

With respect to temporal issues, there is an important differentiation between conventional GIS applications and spatial planning. The first (GIS) focus on data of the present, while in some cases they keep record of information of the past. On the other side, planning starts from the present (current problems) or sometimes from the future (forecast outcomes), but it is mainly concerned with solutions for an ideal future, the events still to come. Temporal scale is important to planning—first in developing and calibrating the forecasting models and secondly in defining the necessary spatio-temporal resolution for the specific planning problem. Planning for the immediate future definitely requires models of higher spatio-temporal resolution than strategic planning addressing longer-term goals. When dealing with socio-economic applications and spatial planning, it is also important to establish:

a. appropriate representations for change (gradual or not) based on the occurring events, and
b. variable spatio-temporal resolution capability.

4.6 APPLICATIONS DEMANDING SPATIO-TEMPORAL MODELLING

Various spatial applications pose different spatio-temporal demands. To appreciate the differences, three indicative diverse applications are presented. The first refers to changes in widely known and well-established bounded SEUs created by cadastre. The second refers to other SEUs whose spatial definition varies greatly. Finally, the third application exhibits the difficulty of formalising uncertain spatio-temporal information such as found in historical texts.

4.6.1 Modelling Cadastral Changes

Cadastral policies have intrinsic spatial, legal economic political and social perspectives. Change in cadastral applications can take various forms—the most fundamental (but not necessarily the most important) being that of boundaries. What is to be temporally monitored is whether and when a boundary change results in a new object (land parcel). A second issue related to land parcel change is that of generation (creation) and deletion (transformation into others) (Figure 4.7).

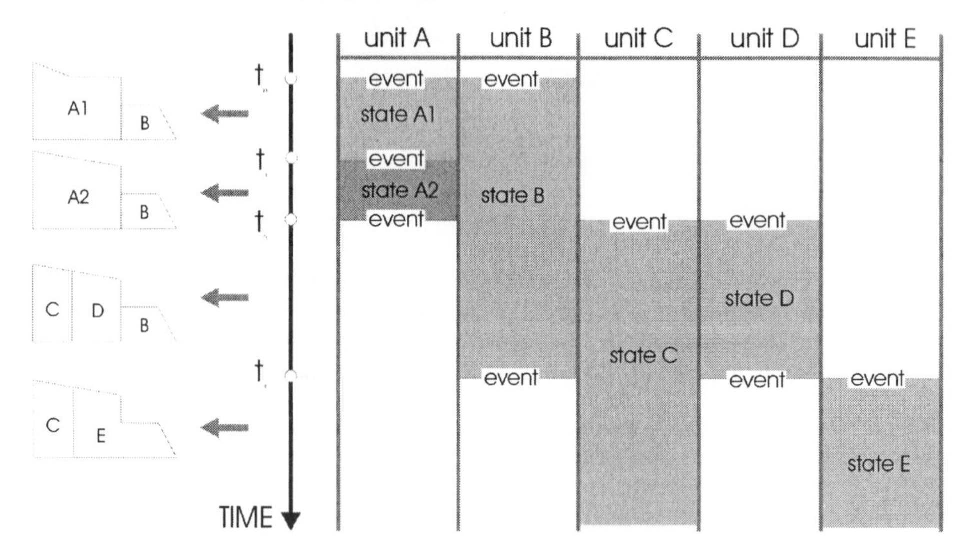

Figure 4.7 An example of spatial changes in cadastral parcel units

Continuous temporal monitoring is essential for users to check the property status at any time in the past. Also for taxation purposes, it is often required to check and justify the source of assets for investment, etc. The important characteristic of these types of applications, exhibiting a strong legal character, is their spatial and temporal discontinuity. Namely, cadastral characteristics are mostly bounded and spatial interpolation rules do not apply. Furthermore, the rule of persistence holds, that is, unless formally legal change (transaction) takes place, the previous state is considered as valid. Various levels of transaction and valid times at the different data-flow gates also find a wide application in cadastral databases.

A thorough consideration of temporal issues in cadastral applications can be found in the Ph.D. thesis of Al-Taha (1992). A formal database approach to object-oriented modelling of changes and events is presented by Worboys (this volume). The necessary database design issues to facilitate temporal GIS are introduced by Stefanakis and Sellis (this volume). Finally, a practical example of the necessary spatio-temporal query language is developed by Yuan (this volume).

4.6.2 Modelling Change in Field-Based Socio-Economic Units

In contrast with cadastral applications, spatial change in other socio-economic units exhibits different characteristics. First of all, phenomena are often continuous distributions (even within boundaries) and interpolation rules normally apply. Persistency rules often do not hold, in the sense that no formal transaction is needed for changes to take place. Raster-based models are often used to represent a discrete view of a phenomenon, and object definition is normally based on classification schemes. Subsequently, when descriptive information changes between two time epochs, the defined object may be shifted from the initial one. This constitutes a movement of the phenomenon from one spatial location to another (Figure 4.8). Generalisation rules on attributes and meta-attributes (quality descriptions) apply widely to this type of phenomena.

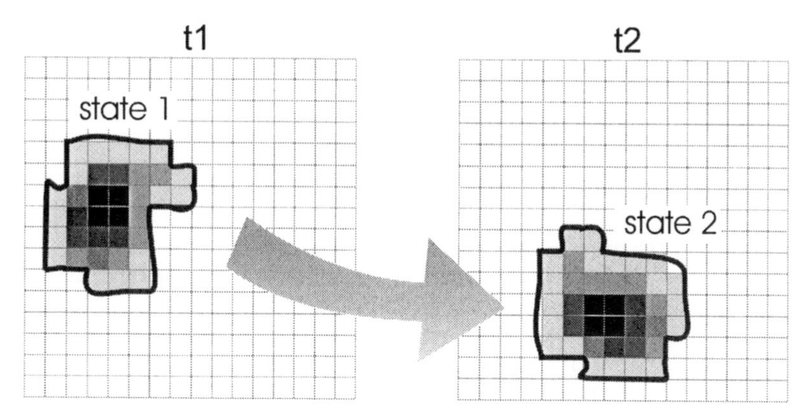

Figure 4.8 Spatial change in a field-based SEU between two time epochs

4.6.3 Spatio-Temporal Structures About History

The selection of efficient spatio-temporal structures is also very important when dealing with the past. The formalisation of historical knowledge provides an excellent base for exercising spatio-temporal change. An interesting application is that of formalising geographic descriptions of historical interest, cited in ancient documents. Pausanias' journey, for example, teems with quantitative and qualitative descriptions of sites, places, events, courses, monuments, etc., with spatial-temporal reference framework of variable completeness, uncertainty and imprecision. In such not only interesting but also scientifically valuable descriptions, knowledge about space and time is mostly relative and rarely depends an absolute positions or dates.

Points of spatial reference and toponyms change often or they refer to different location or area extend. Uncertain spatial descriptions such as 'to the east, 'near, to the left of river Ladon', 'after the crossing', 'down (south)', are very frequent. Less frequent are more quantitative spatial descriptions such as "... a cape on the mainland called Messati, exactly in the middle of the voyage from the port of Erythraia to the island of Chios". Temporal references are even more vague: The writer refers to the period he describes using descriptions such as 'today', 'in my days', 'in the old days', etc. On the other hand, there is significant 'absolute' information (e.g., location of ruins) that lacks descriptive information. Developing a system to incorporate such knowledge, enable correct relation between historical-archaeological facts, test hypotheses and allow reasoning, is a very challenging task.

The study of historical spatio-temporal structures shows also in other applications. For instance, in the History of the Peloponnesian War (431–404 B.C.) written by a great historian Thucydides, Finley (1972) refers to Thucydides' simple yet unique scheme of dating events. One can also recognise the need for spatio-temporal topology and variable resolution, since "in a large-scale war, with many things happening in different places at the same time, dating by years would not give the right kind of picture for Thucydides. All the little connections and sequences, the day-to-day causes and consequences, would be lost". Of course, there were cases not requiring precise time, where descriptions such as 'once upon a time' served well enough. Formalising however, chronologically imprecise historical information from the distant past with descriptions such as 'later' or 'much later' on one hand, while elsewhere there is "a passion for the most minute detail-minor commanders, battle alignments, bits of geography and the like", proves to be extremely interesting.

CONCLUSION

To epitomise, this chapter introduces an application-induced classification of spatio-temporal characteristics affected by change. It also addresses a number of important issues, in an attempt to stimulate scientific research towards the development of a unified spatio-temporal framework. Finally, it demonstrates how different applications make entirely different demands on temporal GIS.

The choice of Thucydides, in the light of what has been said so far, was not accidental. It was done to illustrate that geography and history besides using spatio-temporal structures, also share a more general goal—one hidden in Finley's notes (1972):

> "Greek intellectuals like Thucydides were in dead earnest about the conviction that man is a rational being. As a corollary, they believed that knowledge for its own sake was meaningless, its mere accumulation a waste of time. Knowledge must lead to understanding. In the field of history, that meant trying to grasp general ideas about human behaviour, in war and politics, in revolution and government. Thucydides' problem, in short, was to move from the particular to the universal, from the concrete and unique event to the underlying patterns and generalities, from a single revolution to revolution in essence, from a demagogue like Cleon to the nature of demagogues, from specific instances of power politics to power itself."

REFERENCES

Allen, J.F., 1983, Maintaining knowledge about temporal intervals. *Communications of ACM*, **26**, pp. 832–843.

Al-Taha, K., 1992, *Temporal Reasoning in Cadastral Systems*. Ph.D. thesis, University of Maine, USA.

Al-Taha, K. and Frank, A.U., 1993, What a temporal GIS can do for cadastral systems. In *Proceedings of GISA '93*, Sharjah, UAE, pp. 13.1–13.17.

Berry, B.J.L., 1964, Approaches to regional analysis: A synthesis. *Annals of the Association of American Geographers*, **54**, pp. 147–163.

Burrough, P.A. and Frank, A.U., 1995, Concepts and paradigms in spatial information: Are the current geographic information systems truly generic? *International Journal of Geographical Information Systems*, **9** (2), pp. 101–116.

Couclelis, H., 1991, Geographically informed planning: Requirements for planning-relevant GIS. *Papers in Regional Science*, **70**, pp. 9–20.

Dangermond, J., 1990, A Classification of software components commonly used in Geographic Information Systems. In *Introductory Readings in GIS*, edited by Peuquet, D.J. and Marble, D.F. (New York: Taylor & Francis), Chapter 3, pp. 30–51.

Finley, M.I., 1972, Introduction and Appendices to Thucydides—*History of the Peloponnesian War*, translated by Warner R. Harmondsworth in 1954, (Middlesex, UK: Penguin Books).

Frank, A.U., 1994, Qualitative Temporal Reasoning in GIS–Ordered Time Scales. In *Proceedings of 6th International Symposium on Spatial Data Handling, SDH '94*, Edinburgh, UK, edited by Waugh, T.C. and Healey, R.G., pp. 410–430.

Harris, B. and Batty, M., 1993, Locational models, geographic information and planning support Systems. *Journal of Planning Education and Research*, **12**.

Koubarakis, M., 1994, Database models for infinite and indefinite temporal information. *Information Systems*, **19** (2), pp. 141–173.

Langran, G., 1989, A Review of temporal database research and its use in GIS applications. *International Journal of Geographical Information Systems*, **3** (3), pp. 215–232.

Langran, G., 1992, *Time in Geographic Information Systems*, (London: Taylor & Francis).

Morrill, R.L., 1970, *The Spatial Organization of the Society*, (Belmont, CA: Wadsworth).

Peuquet, D.J., 1994, It's about time: A conceptual framework for the representation of temporal dynamics in Geographic Information Systems. *Annals of the Association of American Geographers*, **84** (3), pp. 441–461.

Tuan, Y.-F., 1977, *Space and Place. The Perspective of Experience*, (Minneapolis: University of Minnesota Press).

Whigham, P.A., 1993, Hierarchies of space and time. In *Spatial Information Theory—A Theoretical Basis for GIS, European Conference COSIT '93*, Lecture Notes in Computer Science 716, edited by Frank, A.U. and Campari, I. (Berlin: Springer-Verlag), pp. 190–201.

Worboys, M.F., 1994, Unifying the spatial and temporal components of Geographical Information. In *Proceedings of 6th International Symposium on Spatial Data Handling*, Edinburgh, UK, edited by Waugh, T.C. and Healey, R.G., pp. 505–517.

PART TWO

Ontological Background

INTRODUCTION

Analysing the 'life and motion' of spatial socio-economic units reveals a baffling set of complex questions regarding their foundation in space and time as fundamental categories. The chapter by Carola Eschenbach opens the discussion with an apparently simple question—how did the population of European capitals change in the last 90 years—and shows how such an innocent query raises many deep philosophical problems. Such questions include: What are the capitals? How to deal with countries where the capital changed in the past 50 years? How to deal with countries splitting and merging?

It is necessary to address these questions from a fundamental position to make good use of the insight that philosophers have gained in exploring the problem of change during the past 2000 years. Firstly, the object and its identity must be separated from the space it occupies—what cartography cannot do—and to identify the referential links implied in properties such as 'capital of', both requirements which will be voiced again by May Yuan (Chapter 15). Eschenbach points out different options with regard to the persistence of a spatial socio-economic unit: Is it its name (or legal identity), which persists? Is it the location, which must not change? Should we consider the space occupied, independently of how many individual spatial socio-economic units are located within at a specific point in time? Each option leads to a different answer to simple queries. What seems to be a hair-splitting philosophical question has practical consequences for query processing in a spatial database. We further understand that there is not a single correct answer, but there are different contexts for the same query, which make one of the alternatives more appropriate than others. As a consequence, a temporal GIS cannot implement just one solution as they currently do (Stefanakis and Sellis, Chapter 12), but has to offer various solutions and find a way to select one for each query, as required for legal applications (Al-Taha, Chapter 17).

The second contribution by Barry Smith reviews the metaphysical foundation of our understanding of 'geographic object' and extends the analysis by Eschenbach into the foundations of ontology. Smith starts with an overview of Aristotelian categories and their properties and defines carefully the classical terminology to discuss questions of material existence (substances) and properties of objects (accidents). He quickly introduces some concepts from mereology (Simons, 1987)—the science of wholes and parts—and extends the difference between fiat boundaries, created by human action and not physically existing, and bona-fide boundaries, which are physically existing, to fiat objects, which are created by human order. Most spatial socio-economic units fall into this category, as the political boundaries are very often fiat boundaries and only rarely coincide with discontinuities in space. Smith then goes on to investigate scattered objects—objects that consist of disconnected parts—and behavioural objects. These are steps to identifying the ontological base for spatial hierarchies, hierarchies of behaviours and ultimately to shed some light on Gibson's concept of an ecological niche.

The following chapter by Roberto Casati explores the particulars of a non-substantial phenomenon like shadows, which are dependent on a substance for its existence—a shadow cannot exist without something it is cast upon. Through the analysis of the properties of shadows the chapter explores if shadow is an appropriate metaphor for the explanation of spatial socio-economic unit. Spatial socio-economic units are not substances; a parcel is not the dirt located at a certain place, but an abstract property associated with the material present—removing the earth is not removing the parcel (and the same for all other spatial socio-economic units). In this sense, spatial socio-economic units are like shadows cast on the land (conceived as substance). Casati develops a *theorita* for shadows and then compares it with our common-sense understanding of spatial socio-economic units to see if it is applicable. The theory developed follows a common-sense structure and it is in itself interesting to ponder legal questions of shading cast by buildings on neighbouring parcels. It is also an example of a method to capture common-sense understanding in a theory. In the final comparison of spatial socio-economic units—for example, a country like Italy—and shadows, Casati points out a surprising number of similarities. These observations are confirmed by the fact that 'shadow' is often used as a metaphor to express concepts related to countries (or political spatial socio-economic units in general).

After addressing the ontology of space and spatial units, the forth chapter of this part investigates the notion of time and its inherent complexity. The concept of time useful for physical science—here called parametric time—is perfectly homogenous and symmetric: it can be travelled in either direction and the processes so described can be reversed. The fundamental laws of thermodynamics predict that processes increase entropy and can therefore not be reversed. This is true—as has been pointed out by many—for all biological processes, but also for most processes in physical geography, and, as Georg Franck points out in this chapter, for economic processes. Time is paid in two forms in economy: as interest and as wage—the two for completely different forms of time. These two forms also appear in historiography and generally in the theory of temporal databases; they can be linked to the ontology and the epistemology of the world. It is a valuable confirmation that the distinction of valid time and database time we are all familiar with is not the result of some accident, but deeply linked to the nature of time and how people perceive time and temporal events.

The issues in this part, especially the question introduced by Smith and Casati, will be reconsidered in the next part: the initial chapter by Worboys explores the different possibilities for the modelling of changes and a typical succession of events occurring to objects.

REFERENCE

Simons, P., 1987, *Parts—A Study in Ontology*, (Oxford: Clarendon Press).

CHAPTER FIVE

On Changes and Diachronic Identity of Spatial Socio-Economic Units

Carola Eschenbach

5.1 INTRODUCTION

The systematic treatment of time and data about changes is a challenge to any current approach to data storage and retrieval. The complexity and importance of this topic is reflected by the growing amount of literature (cf. Tsotras and Kumar, 1996). The related problem this chapter tackles is the representation of individuals that change over time and the retrieval of reports on their development with respect to selected features.

The background of these considerations is given by the idea of representing spatial socio-economic units (like states, cities, and parcels of land) in 'historical geographic information systems'. *Geographic information systems* are information systems that are especially equipped for representing spatial data, and evaluating, accessing and presenting data about entities by means of spatial relations or properties and in accordance with the underlying spatial structure. The aggregation of data on the basis of spatial interrelations as well as the retrieval of entities located in a given region is an example of such specific demands (cf., for example, Frank, 1991; Medeiros and Pires, 1994; Güting, 1994). *Historical information systems* are especially equipped for representing changes in (the represented segment of) the world. They thereby reflect the structure of time and contrast with atemporal information systems that at all times represent a state or snapshot of the world, ignoring what has been the case before (cf. Snodgrass, 1992; Jensen *et al.*, 1994). The different structures of space and time correspond to different requirements on methods of evaluation and access. Still, there are some structural similarities on the level of topological structure, which might lead to similar treatment. Historical geographic information systems need to be equipped with both means of evaluation and access (cf. Langran, 1992).

The basic view on the relation between space and spatial entities such as material bodies or socio-economic units employed here is that space serves as a container for spatial entities (cf. Smith, this volume). Space in this abstract sense can be described by geometry and topology. In this chapter, the term *individual* is used to refer to what are commonly called individual bodies, individual things or individual substances, entities that have a certain unity and completeness and that are generally movable and manipulable. The term *entity*, on the other hand, is used in a wider sense including, for example, parts of space and socio-economic units independently of whether they are individuals. The term *region* refers to segments or parts of space, especially those which can be occupied by material bodies and collections of material bodies. Thus, the term *region* is used purely spatially. It is not restricted to any specific dimensionality, although it may be that only two- or three-dimensional regions are of interest in specific applications. In addition, regions need not be connected and being a region shall not presuppose any idea of being homogenous. In contrast, the term *area* is used here to refer to homogeneous zones, like wooded or rocky stretches of land, and the term *unit of area* refers to connected and maximal (homogeneous) areas, like fields, forests or lakes.

Entities that we ascribe spatial properties and relations to (for example, material entities) have a *location*, which, at any time, is a region. Spatial properties and relations are reducible to the locations of the entities in question. For example, an entity is disconnected if and only if its location is disconnected. It is two meters away from another entity, if and only if their locations are two meters apart. The basic relation between material bodies and space is that of occupation. Material bodies occupy parts of space, their position, place or location and thereby exclude other material bodies from occupying this region. At different times, the same body can occupy different regions, and the same region can, at different times, be occupied by different material bodies. A material individual can be destroyed, while the space it occupies cannot be destroyed. In addition to occupation, there are further ways of ascribing regions to material individuals, especially people, for example, their place of birth, of habitation, or of work. When we look at socio-economic units, we have to consider further ways of being located. We should not expect different ways of being located to exhibit the same behaviour and justify the same inferences. For example, several people may live at the same place, but they cannot occupy the same region at one time.

While the spatial relations between regions themselves are fixed once and forever, spatial relations between other spatial individuals may change when they change their locations. Any change in location I will call a *movement of the individual.* The region where I am now will stay here forever and keep its spatial relations to all the other regions. But if I change my location, my spatial relations to other material individuals change. The use of *movement* introduced here may seem to be too wide (including growth, shrinking, deformations) and too narrow (excluding rotation of perfect discs or spheres) at the same time, but it will do for the purpose of this chapter.

The problem we will mainly investigate in this chapter is the possibility of accessing the history of a spatial socio-economic unit that moves in space by tracing its representation in a historical GIS through time. We will consider problems arising from the intuitive notion of identity when applied to entities like socio-economic units and natural units existing and changing in space. The main goal is to identify aspects of the general problem that tend to obscure the difference between changes of persisting individuals and the replacement of individuals one by the other. Therefore, we will not specify how to re-identify a socio-economic unit after some change occurred, but try to spell out the difference between tracing an individual through time and evaluating expressions or queries at different moments of time.

5.2 CHANGE AND DIACHRONIC IDENTITY OF SPATIAL UNITS

5.2.1 An Example of Tracing Spatial Socio-Economic Units Through Time

(Q1) How did the size of the population of the European capitals develop from 1900 to 1996?

In order to answer this query, it is not only necessary to have information about how many people lived at different times at different places. What is needed in addition, is knowledge about what the capitals are, where these cities are located and whether their location changed over time (for example, when they grow), and a lot more. Furthermore, the assumption is fundamental to this query that there are some things (cities), which persist (or endure), i.e., exist over a period of time, and change, i.e., have different properties at different times, and that these cities can be traced through time or re-identified at different points of time as being the same city. The problem whether an individual at one time is the same as a given individual at another time, I call the problem of *diachronic identity*.

While common sense seems to have good intuitions about what is to be considered the same individual and what is considered another one, a thorough analysis reveals several difficult problems. This is especially true if time and the possibility of change over time is taken into consideration (cf., above all, Locke, 1690; more currently: Gabbay and Moravcsik, 1973). The discussion on identity across different moments of time is mainly based on examples from the domain of material (or physical) individuals and especially persons and artifacts (cf., for example, Sagoff, 1978; Scaltsas, 1981; Lowe, 1983; Thomson, 1983; Chisholm, 1976). This chapter is meant to shed some light on the problem of diachronic identity of spatial units and its relation to language and representation in historical geographic information systems (cf. Egenhofer and Mark, 1995).

As a motivation for this section, let us point out some of the problems related to query (Q1): From 1871 to 1945, Berlin was the capital of the German Reich. In 1920, Berlin was united with several smaller cities (one of those being Spandau) and villages under the name '(Groß)Berlin'. In 1948 Berlin was divided politically into Berlin (East) and Berlin (West), in 1961 the division was established physically by building a wall. From 1949 to 1990, Bonn was the capital of the Federal Republic of Germany (FRG) (founded in 1949), while Berlin (East) (also called 'East-Berlin') was the capital of the German Democratic Republic (GDR) (founded in 1949 as well). In 1990, the Federal Republic of Germany and the German Democratic Republic were united, including the political (and physical) union of Berlin (West) and Berlin (East). Until 1992, Bonn was the capital of the united (Federal Republic of) Germany, and since then Berlin has been the capital of Germany.

With this background, there are different options of tracing the German capital in order to answer the query. All options yield the same result for 1996, but they differ with respect to their response regarding, for example, 1910 and 1970.

(A1) *Tracing a persisting city*: This is based on the assumption that we can trace cities through time. The report generated reflects the development of individual cities. Two different cities are not included in one report, even if they exist at different times at the same or overlapping places. Such a report would include data of Berlin as a whole only, according to an independently determined knowledge of what was the city of Berlin at different times. For Berlin (East) and Bonn separate reports would be generated. For 1910, the population of, for example, Spandau is not included and not reported, since it was not part of Berlin at that time. If Berlin is assumed to be a city in 1970, the report about this date adds the populations of Berlin (East) and Berlin (West).

(A2) *Tracing the location of a city*: This is based on the assumption that we can trace spatial regions through time but needs only snapshot criteria of what is a city. An entity is included in the report of capital C at some time if it is a city at that time and its location overlaps or is part of the region that is the current location of C. Since in 1910 there were more than one city in the region that now is the location of Berlin, data about all these cities are reported but not added.

(A3) *Tracing a region*: This is based on the assumption that we can trace spatial regions through time, independently of there being any cities or states located at them. The report is about the development in the region that is the current location of a city, independent of the temporal development of the city in question. For 1910 the population data of all cities and villages that were later united with Berlin are added, as well as the data for Berlin (East) and Berlin (West) in 1970. (Cf. the solution described by Ryssevik, this volume.)

(A4) *Tracing the capital of a persisting state*: This is based on the assumption that we can trace states through time. Two entities at different times are included in one report, if they are capital of the same state at the respective times. Tracing the Federal Republic of Germany, such a report would include Berlin data and Bonn data, according to when each of them was the capital. Thus, for 1985 and 1991 the report would include Bonn data, while the report on 1993 presents Berlin data.

(A5) *Tracing the location of a capital*: This is again based on the assumption that we can trace spatial regions through time. Two entities at different times are included in one report if they are capital at the respective times and the regions that are their location at those times overlap (or one is part of the other). Such a report would include data of Berlin and of Berlin (East). For 1910, the population of Spandau is not included and for 1970 the data of Berlin (East) are reported.

As we see, all options depend on the possibility of tracing an individual through time. Options (A2), (A3) and (A5) might seem to be more easily performed than (A1) and (A4). But it might be necessary to be able to answer according to (A1) or (A4), depending on the general task to be performed. As we have seen, (A2) and (A5) lead to reports on more than one city at certain times, even though some of them never have been capitals on their own. The strategy of (A3), in contrast, yields constant results, when employed for queries on the size of the location of the city.

All these problems of how to trace individuals through time are not specific to the example of (Q1), which seems mainly to be a query about the development of the size of populations. The problems have in general to be dealt with, when the representation of changing individuals is attempted for the purpose of evaluating their development over time. Tracing the development of an object over time depends on judging diachronic identity, i.e., identity across different moments in time.

The case sketched here is a rather complex one, since the options outlined all produce different results. It demonstrates that we have to distinguish between the sequence of which city was the capital of a state and how these cities and their location developed through time. We could decide to exclude the German case in answering the query. But even to do this, one needs to have a strong background to decide that there was something unusual going on. Thus, to exclude the German case, diverse options of tracing have to be available to detect that different results can be obtained.

5.2.2 An Example of Tracing Natural Units of Homogeneous Features

Figure 5.1 shows a series of snapshots of a region with some stable features (depicted by dots), which gives a spatial reference frame, and some changing feature F, such as being wooded (depicted by a pattern). In 1a, there is a unit of area which has a regular distribution of feature F, i.e., a unit of F. This unit we might call A. In 1b, there are two units of F, while 1c again has one.

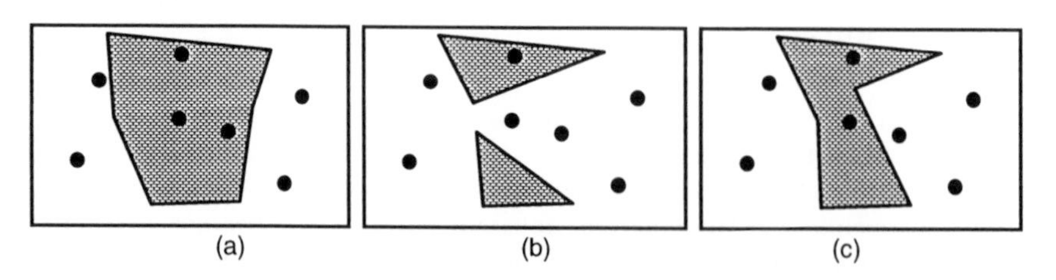

Figure 5.1 Series of snapshots of the same region that changes with respect to some feature

Similar to the example above we might inquire about *A*'s development by query (Q2):

(Q2) How did the number of squirrels in *A* develop?

If enough data about squirrels has been collected, to answer this query seems to be simple. But there are several ways of understanding and representing this query and the data:

(A6) *A* is the maximal area of *F* in the region depicted in the three snapshots. In (a) and (c), *A*'s location is spatially connected; in (b) it is not.

(A8) *A*, being the topmost unit of *F*, exists throughout. In (b) some part has been split off, i.e., an additional unit started to exist, and in (c) it has been incorporated into *A* again. (Alternatively, we could consider *A* the lower unit of *F*.)

(A7) *A* is a spatial region that exists throughout. In the situation depicted by (a) this region is homogeneously *F*. In (b) and (c) it is not homogeneously *F* anymore and therefore its boundaries are not perceivable completely.

(A9) *A* only exists in (a). In (b) it has been split, i.e., replaced by two other individuals, which have been united in (c), i.e., commonly replaced by another individual.

Figure 5.1 is neutral with respect to the question of diachronic identity and the same might be ascribed to the represented part of the reality. Considering *A*, a unit of area in (a) is based on *A*'s being homogeneous *F*, *A*'s being connected and *A*'s being maximal with respect to these properties. But the criteria that allow the unique identification in (a) do not lead to a unique way of re-identifying *A* after changes occurred. Criteria of being a spatial unit that are applicable at one time do not necessarily include a criterion for diachronic identity. This might in many cases lead to rejecting the assumption that a unit of area persists and changes, preferring analysis (A9), which assume the successive replacement of units in time.

5.2.3 The General Problem of Tracing Individuals Through Time

Identity is fundamental to our understanding of the world. Our common-sense mastery of identity over time is mainly based on two sources: The ability to judge identity or difference of material and spatial objects at one time, and our ability to judge identity or distinction of complex organisms like persons at different points of time. The philosophical discussion on diachronic identity shows that the criteria underlying our judgment in these two fields are unclear and in general not easily transferable between different domains.

If an individual (like a person) changes over time (in location, weight, mood or hair colour), then there also is something constant about this individual. Otherwise, one would rather assume one object to be replaced by another (or by several others). This constancy can serve as a basis for the re-identification of something afterwards as the same individual as that of before. This problem is elaborated in the philosophical discussion of identity and persistence in many respects. In this discussion, two points of view can be identified that differ with respect to whether diachronic identity (or being the same individual at different times) is an irreducible, primitive concept, or whether diachronic identity is based on other concepts (cf., for example, Merricks, 1994).

The first viewpoint is bound to the assumption that diachronic identity is objective and independent of our understanding and conceptualization of the world. The second viewpoint allows for the assumption that diachronic identity is subjective, relative to our conceptual system, man-made, and even context dependent. Individuals are assumed to be made of (different) stages, versions or temporal parts, which exist at different times and are bound together by criteria of persistence. Thus, as a possible basis for defining identity for material objects, spatio-temporal continuity is taken into consideration, for artifacts their purposes, and for persons their memory or consciousness. For legal individuals such as cities and states, constant rights and responsibilities seem to be better candidates as a basis to define (diachronic) identity than, for example, constant location or constant population.

The two examples presented above differ with respect to the basic assumptions on what spatial units are. The first emphasizes the idea of an individual (the city of Berlin) that exists, is identifiable on its own and has a location in space. A spatial unit in this sense is the location of an individual. The second example suggests that being a spatial unit is a matter of feature distribution in space. A spatial unit in this sense is an area that homogeneously has a feature, is connected, and enclosed by regions that do not exhibit the feature in question. For shadows, both views are available: A shadow as an individual caused by another individual vs. a shadow as a unit of area with respect to illumination (cf. Casati, this volume).

Under an atemporal or synchronic perspective, i.e., if only snapshots are represented, the difference is not crucial. But taking change and the development over time into consideration leads to important differences. The spatial criteria for being a unit mentioned above only apply to the states of the world at one time and between changes. They do not apply to the question of re-identifying a unit across different times. Diachronic identity of spatial units is not exclusively based on its spatial properties. Rather, if diachronic identity is applicable, as in the case of spatial socio-economic units, this is a consequence of our respective understanding of non-spatial aspects of the socio-economic units. Since this might be a matter of a general conceptual framework and of the specific task to be performed, the possibility of differences with respect to diachronic identity needs to be taken into consideration. Still, restrictions might apply to the possible locations of the individual and its movements such that space can contribute to the re-identification of the individual.

5.3 REPRESENTING SOCIO-ECONOMIC UNITS

The spatial socio-economic units considered in the subsequent discussion have the following characteristics: they are man-made, strongly conventionalized and relevant for social communities. They have a location or territory and people belonging to them, and there may be different ways for people to belong to the same or different socio-economic units. These units are not just aggregations of people (like, for example, the set of all car owners), but the people belonging to them interact and the unit affects them in some way. The locations of socio-economic units can be constrained by the physical environment but are not determined by it. The specific status of a socio-economic unit can lead to preferences about spatial properties of its location: to be connected (to support interaction and communication), and not to overlap the locations of other units of similar status. Modes of belonging of people and location might interact in different ways; consider nations that define belonging to the nation on the basis of the place of birth being part of the actual location of the nation. Spatial socio-economic units can start and cease to exist. They can change over time in who belongs to them, their location and their effects on people (cf. Frank, this volume).

Some examples of spatial socio-economic units are legal units like cities and states (modes of belonging of people: habitation, nationality; mode of location of the unit: matter of control), legal units like parcels of owned land (belonging: ownership, habitation; location: matter of control), higher order legal units aggregating other socio-economic units like state unions (European Community or NATO), cultural aggregations like language communities or herds (belonging: exhibiting a defining property; location:

mediated by the location of the members) and institutions with relatively stable locations (and legal connections to other socio-economic units) like embassies, universities, departments of universities. This collection should make clear that socio-economic units are related to different individuals in several ways and a single and uniform way of representing them should not be expected.

Spatial socio-economic units have a location. Although they do not occupy the location in the way material bodies do, most socio-economic units exclude the co-existence of another unit *of the same status* at the same time at the same place. Cities, states or parcels of owned land do not overlap units of the same status, although units of different status can overlap or be spatially included one in another. This spatial inclusion is very important for the internal hierarchical structure of socio-economic units: (usually) a city belongs to a state if its location is part of the state's location.

There are several reasons for representing socio-economic units as individuals in addition to their location and population. Socio-economic units in contrast to parts of space, which exist once and forever, exist only for some restricted time. They can change their location, which parts of space cannot do. Thus, there has to be a basis to represent temporal restrictions of the existence and the location of socio-economic units. In addition, the time of existence of a socio-economic unit might be different from the time at which it has some location. Thus, it should be possible to represent that a socio-economic unit does at some time exist but has no location (what you might claim happened to Austria between 1938 and 1945). The explicit representation of a socio-economic unit like Berlin is also needed if specific non-spatial information about it shall be represented, for example, that Berlin is a city, who is its mayor, when was it founded, that it is the German capital, that it was divided politically and physically by a wall and that it became united again.

According to the general remarks on changes of individuals, assuming legal spatial units such as cities and states to be moving, i.e., change their location due to change of the physical environment or legal matter, implies the assumption that they are persisting individuals such as persons or portions of matter. Our conception of spatial units determines which changes we accept to be movements and which changes we categorize as the disappearance of one unit (at one place) and as the appearance of another unit (at another place). Such judgments are a basis to specify socio-economic units according to their possible movements. If, for example, a city is destroyed by an earthquake and another city built up afterwards, under which (spatial) conditions do we consider the latter the same as the first or a new city? If not one but two or more cities are built up afterwards, which spatial conditions lead us to re-identify one as the old one? If the decisions are independent of location but solely based on legal and social factors, then location is not considered an essential property for the permanence of socio-economic units of this status.

If for units of a certain status location is assumed essential, then there are still several options available. The strongest restriction is to assume that change of location is impossible. This might be assumed for owned parcels of land. For other kinds of restrictions we can distinguish between global and local restrictions: Global restrictions apply between any times of the existence of a socio-economic unit. If, for example, we assumed that cities could only grow and shrink in the global view, then for any two times of its life span, its location at one time is part of its location at the other time. Local restrictions, in contrast, only need to apply to restricted periods of the life span. If, for example, we assumed that states could only grow and shrink in the local view, then the life span can be subdivided into periods of growth and shrinkage. In this case, for two arbitrary times, the respective locations of a state need not even overlap.

If spatial socio-economic units are represented as individuals that can change in time, then decisions have to be made on what is the criterion of diachronic identity and how diachronic identity is represented. On this basis, the database can be evaluated with respect to the histories of socio-economic units. With respect to databases that are designed to support the representation of space and the location of individuals, attention has to be drawn to the relation between spatial socio-economic units and space, and the changes in this relation.

5.4 UNIQUENESS OF IDENTIFIERS

The development of an individual with respect to some feature can be reported only if the individual can be traced through time. In databases, identity is usually coded by using a unique identifier or key. If the same identifier is used, the same individual is meant; if different identifiers are used, different individuals are meant. Thus, for internal representations of socio-economic units in temporal databases, tracing an individual might not seem to be a problem at all.

Using unique forms of representation is a basic way to encode identity in databases or knowledge bases. Although identifiers in formal languages resemble names in natural languages, identifiers are less flexible in at least two ways. Firstly, several individuals can have the same name; secondly, individuals can change their names. In contrast to this, identifiers in formal languages are employed as unambiguous and permanent representatives of individuals. Correspondingly, I will assume that the representational relation for these terms is independent of the context or expressions they are imbedded in. That is, all contexts considered are extensional. Intensional contexts such as exemplified by *Peter believes that the current capital of Germany is Bonn*, which does not imply that *Peter believes that Berlin is Bonn* is true, are excluded as representational structures from the discussion.

The third property to be mentioned here is that individuals can have several names. If in a formal system an individual is represented by different expressions, additional means have to assure that they are treated as representing the same individual. Therefore, possibilities of expressing identity statements and inferential mechanisms on identity have to be included. For matters of simplicity, one might assume that identifiers in a formal system are unique.

To assume identifiers to be unique seems to be an inexpensive way to handle the problem of diachronic identity. However, it does not solve the problem but only shifts it to the procedure of entering the data. For entering the date we still need the criteria that determine the assignment of database identifiers. Representing a changing world, we have to consider three main aspects:

- Which criteria determine whether to use the same identifier or different identifiers for coding data about the same time point or period of time?
- This aspect concerns the question of synchronic identity, i.e., of delimitation of entities and therefore is connected with asking for criteria of counting: how to distinguish between having one Germany (the German Reich) and having two (or more) German states (FRG and GDR)?
- Which criteria determine whether to use the same identifier or different identifiers for coding data about different time points of periods of time?

This is the aspect of diachronic identity.

- Whether these two sets of criteria match. (And what to do if they do not.) Criteria of synchronic and diachronic identity do not match, if, for example, a at time t_1 is the same as b at t_1, b at t_1 is the same as c at time t_2 and a at t_1 is not the same as c at t_2. If this is possible, we have a relation of temporal identity:

> objects may be the same at some time and not be the same at others. Since nothing can effect *a* differently from *b*, there are good reasons to reject such a possibility.

For regions, the criteria of identity are clear: different regions can be distinguished by spatial relations. There are no different regions that are equally related to all the regions. And since regions do not move, incorporating time yields no additional problems.

In contrast, socio-economic units pose problems, since even the first set of criteria is not clear. Socio-economic units can for some time or always be collocated and still not be the same. In addition, they might move or—for some time—have no location at all, while still existing (this is a possible conception of what happened to Poland between 1940 and 1945). Looking at the example of Germany again, we might decide now that—for many purposes such as legal, cultural and language development—Germany of 1996 is to be assumed the same individual as the German Reich, although changed and developed in several ways. If they are represented by the same identifier, then the question whether the Federal Republic of Germany of 1980 should be represented by this identifier, comes up. Thus, either it is assumed the same individual as the German Reich (i.e., the same identifier is used) or different from Germany of 1996 (i.e., different identifiers used.) Either decision reflects a certain political view (which might be rather unimportant in the context of certain kinds of statistics). In addition, assuming that the information system should be able to include data on actual states of the world, decisions would have to be done without knowledge about the future development. As a consequence, a decision to use the same identifier or different identifiers to represent units in different contexts might turn out unfortunate with respect to how the system of units develops, such that some data need to be corrected afterwards.

An additional problem of encoding identity at the time of entering the data is that there are usually several users and data sources contributing data to a geographic information system. To ensure that decisions upon diachronic identity are based in the same way on the same set of criteria is a difficult task.

Therefore, not to assume different identifiers to represent different entities proves the more flexible basis, at least in the case of socio-economic units and when information about different times is encoded. On such a basis, it is possible to construct the history of socio-economic units based on views or even on the query or its context relative to specific but not universally applicable criteria of diachronic identity. The reconstruction (or construction) of the life and history of a socio-economic unit then needs additional means to recognize those identifiers that represent the same unit.

5.5 TRACING SOCIO-ECONOMIC UNITS VS. TRACING REGIONS

The possibility of evaluating a historical database with respect to the development of an individual over time crucially depends on the applicability of diachronic identity, either when entering data or when accessing them. Criteria of identity (temporal as well as atemporal) belong to the ontology of the domain in question. Thus, the option of diverse criteria of diachronic identity corresponds to the opportunity of choosing among diverse ontologies or modes of being (cf. Smith, this volume). In this and the following section, I will describe how the diverse options of answering queries like (Q1) depend on the ontological background. Accordingly, whether a specific way of evaluation a database can be made available depends on the ontological assumptions encoded in the conceptual model.

Let us again have a look at a query similar to that of the introduction:

(Q3) How did the size of the population of the city of Berlin develop from 1900 to 1996?

This query could in principle be answered by tracing according to the principles (A1), (A2) and (A3) mentioned at the beginning:

(A1) Tracing a persisting city
(A2) Tracing the location of a city
(A3) Tracing a region

The evaluation according to (A1) is based on assuming Berlin to be a persisting individual that changed with respect to its location and it needs some possibility of tracing the city of Berlin through time independent of location. One option is to select the data according to name, i.e., at any time report the population of the city called 'Berlin' at that time. But this method is not reliable either, since city names are also subject to change (cf. cases such as Byzantium–Constantinople–Istanbul, St. Petersburg–Petrograd–Leningrad).

The evaluations (A2) and (A3), in contrast, can be performed independently of any specific criterion of diachronic identity for cities. They rather depend on the possibility of tracing a spatial region. In the context of geographic information systems, we can assume that such criteria are unambiguously specified by the underlying conceptual model of space. If, for example, absolute coordinates of some sort are used in specifying the location of entities or phenomena, then we can assume that their reference is independent of the temporal context of its use. It is much less clear how regions of space can be traced through time in ontologies that assume space not to exist and be individuated independent of the objects occupying it.

(A3) only employs its representation of Berlin to determine which region the evaluation shall be about. (A2) additionally uses knowledge about the location of cities at different times independently of whether the cities are to be assumed the same across different times. Thus, (A2) and (A3) employ the concept of a city only as a criterion of individuation at one moment of time, while (A1) employs the concept of a city also as a criterion of permanence across different moments of time.

The answers following (A1), (A2) or (A3) agree if there is a city that did not change its status or location. Queries as (Q3) can be considered systematically ambiguous with respect to the underlying ontology. The ambiguity can be resolved by the context, the conceptual model underlying the GIS or the user, who is forced to specify the queries including information about how to trace. But the information, which options are available, might not be available to the user. Thus, it might be necessary to supply answers by comments on which ontology was chosen or to first let the user pick the ontology.

5.6 ROLES AND FUNCTIONS

In addition to names like *Berlin*, more complex descriptions like *the capital of Germany* are common means of natural languages to refer to individuals in the world. Corresponding expressions in logical languages are complex terms consisting of function symbols and their arguments.

Names and definite descriptions differ not only in their syntactic complexity. In contrast to names, the reference of complex descriptions is in many (but certainly not all) cases depending on a temporal parameter that needs not be specified explicitly. *The queen of England, the pope, the temperature, the German capital* are descriptions whose reference depends on a time parameter or temporal index. Complex descriptions whose reference is not time-dependent are, for example, *the mother of John Lennon* and *the first man on the moon.* Some time-dependent descriptions of natural language can be used systematically to refer to an entity or to specify the underlying function, as is obvious in the example given by Montague (1973). *The temperature is ninety and rising*, which can be true without ninety rising (cf. Löbner, 1979).

Accordingly, if a sentence such as *the German capital is close to Cologne* was true in 1980 and is false in 1996, then there are, in principle, two analyses possible:

- In addition to Berlin and Bonn there is a socio-economic unit called *the German capital.* It has once been at the same place as Berlin, vanished and reappeared at another site (collocated to Bonn) and then moved back to its old location.
- There is no unit in addition to Bonn and Berlin, but being a capital is a role or function that is fulfilled by cities. Different cities have been capital at different times.

Assuming that the term *capital* specifies a (partial) function (capital-of) mapping states and times to cities, there is no need to take the first perspective. Thus, what we call *the moving of the capital from Bonn to Berlin in 1992* did not involve a change of location of a socio-economic unit, but a change in value of the function capital-of (cf. the distinction between genuine change *de re* and change in a nominal sense by Sharvy, 1968). Thus we have to distinguish between movement of individuals, as defined in the introduction, and more general changes involving space.

The concept of capital is a fundamental or primitive concept that need not be defined in terms of others. But in addition to such expressions, there are others that specify functions definable on the basis of other (types of) concepts. This is especially useful in domains like space, matter or collections. To give an impression of the generality of this phenomenon, the next example involves matter:

I have a cup filled with tea. I empty the cup and refill it with fresh tea.

Corresponding to the example of the capital, we have two options to describe the temporal behaviour of the expression *the tea in my cup.* We can assume that there is one individual referred to all the time, shrinking first, growing then, being fresher in the end than in the beginning. Or we assume that the expression refers to different entities depending on time again. The difference to the example of the capital is that in this case the function we are interested in is definable on the basis of the concepts of tea, in and my-cup, none of them being a function. Still, the function is definable on the basis of an operator that maps a predicate *tea in my cup* to the sum of all the matter the predicate applies to (cf. Sharvy, 1980; Löbner, 1985; Eschenbach, 1993).

Correspondingly, based on properties of individuals, we can uniquely identify the (maximal) collection of individuals the property applies to. That is, based on the predicate *inhabitant of the city of Berlin* we can—at any time—identify the maximal collection of them, i.e., the population of Berlin. Based on properties of regions of space, we are, in a similar way, able to identify the maximal area that exhibits this property. Based on the predicate *covered with trees*, we can identify the maximal region covered with trees, i.e., the sum of all such regions. Since all the properties discussed are time-dependent in that they apply to different objects at different times, the referential term also varies in reference.

The population of Berlin might grow and the region covered with trees might shrink. But to represent these facts we do not need to represent an individual (the population and the forest, respectively) in addition to the persons and the spatial regions. This corresponds to Montague's example of the temperature that rises, without ninety rising.

Thus, natural language descriptions enable us to (unambiguously) refer to entities in the world. But according to changes in the world, at different times they may refer to different entities. Therefore, a statement on the occurrence of change employing such descriptions can reflect a change of the properties of the (constant) referent or a change in what the description refers to. That we are able to refer to an individual at one time and use the same expression with respect to another time does not guarantee the re-identification of the same individual.

Applying this thought to the example given in Section 5.2.2, we find that interpretations (A6) and (A8) of query (Q2) can be derived by employing different terms whose reference varies with time to define what *A* is. Since interpretation (A7) depends on re-identifying a region of space and the failure of re-identification in (A9) needs no explanation, we find that natural units, in contrast to some socio-economic units, need not be assumed to be represented as persisting individuals with specific criteria of diachronic identity.

5.7 TRACING DESCRIPTIONS

On this basis, we shall consider another query, similar to the one we started with.

(Q4) How did the size of the population of the German capital develop from 1900 to 1996?

It is obvious now that there are two different ways of interpreting the expression *the German capital* leading to the diverse possibilities of answering:

(A10) If capitals are assumed to be individuals in addition to the cities, then the expression does not change its reference. It is as constant in reference as *the mother of John Lennon*. But in the German case the capital would be a moving entity, being collocated with Bonn and Berlin at different times.

(A11) If the expression *the German capital* is assumed to refer to cities and change in reference over time, the reference of the expression can be determined for any time index, such that for any time data on the city that was capital at that time are reported.

(A12) Or, based on the same assumption, the expression can be evaluated with respect to only one time index (for example, the time of the query) and the development of the city it refers to at that time be reported, independently of whether this city was capital during all the time.

Since option (A12) depends on tracing a city through time, the same procedures of evaluation as for query (Q3) mentioned in Section 5.5 apply. The result varies with respect to the criterion of diachronic identity applied to cities in general (cf. (A1), (A2) and (A3)).

Option (A11) leads to two other possibilities of evaluation. They correspond to the principles (A4) and (A5) mentioned in Section 5.2.1.

(A4) Tracing the capital of a persisting state (Germany),
(A5) Tracing the location of a capital.

Option (A10) leads to the same result as (A4), as long as capitals are not assumed to persist without the corresponding state persisting and capitals are always (exactly) collocated with a city. But it opens up the possibility for a much more elaborated model of capital and its spatial changes over time, independent of the lives of states or cities in general.

5.8 CONCLUSION

We saw that in order to treat the development of changing individuals we have to consider ontological questions. The collection of options introduced in the beginning can systematically be explained by analyzing the ontological background of:

a. Regions of space,
b. socio-economic units such as cities and states,
c. and, if sensible, entities defined by roles of function such as capitals in addition to cities.

Depending on which of these we consider persisting entities, modes of tracing are to be made available. In addition, relations between entities of different type can be used to transfer criteria of diachronic identity from one domain to the other. This is the case, for example, if we trace the region Berlin is currently located at through time, independently of whether it was the location of Berlin through all this time.

In the discussion of persisting socio-economic units and diachronic identity we managed to identify two groups of phenomena that are connected to diachronic identity but tend to overshadow the basic problems of ontology. Firstly, we considered the relation of spatial socio-economic units to natural spatial units defined on synchronic or snapshot criteria only. We saw that synchronic spatial criteria are not sufficient to characterize diachronic identity if spatial change is not ruled out. Therefore, the re-identification of socio-economic units across different times needs to be based on non-spatial attributes.

Secondly, we saw that the reference of natural language expressions can vary with time. Therefore, a sentence of the form *The A is B* being true at one time and false at another time can depend on a change in what *the A* refers to as well as on a change of the individual *the A* refers to with respect to the property *B*. The truth-value of *The capital of Germany has the name of 'Bonn'* changed between 1991 and 1995, although the city called *Bonn* in 1991 did not change its name. If *B* expresses a spatial property (for example, being close to the city Cologne), then a change in truth-value of the sentence *The A is B* (i.e., the German capital is close to Cologne) may or may not indicate that some individual moved. Thus, movement of individuals is just one type of spatial change and additional types have to be considered—but distinguished—to reflect the variety of space–time interaction.

Legal socio-economic units are man-made and the conception of whether we have the same unit that has changed in some respect or another unit that replaced the first one is purely based on cognitive, social and—in the case of legal units—political factors. It is therefore a conjectural assumption that people agree about diachronic identity of socio-economic units.

In the design of a historical GIS including socio-economic units that shall be used by a variety of users (as well for storing as retrieving data) there are two extreme strategies. The conceptual model can be based on one specific and hard-wired conception of diachronic identity of socio-economic units. This conception should be well documented, since it needs to be understood and employed by the users. The other strategy for the design is to reflect the variability in the conceptual model and allow for the definition of different views or functions that extract histories according to different conceptions on the diachronic identity of socio-economic units.

ACKNOWLEDGEMENTS

Many thanks to Peter Baumann, Andrew U. Frank, Christian Freksa, Christopher Habel, Wolfgang Künne, Michael Oliva Cordoba, Hedda Rahel Schmidtke, Barry Smith and an anonymous referee for comments on earlier drafts of this chapter.

REFERENCES

Chisholm, R.M., 1976, *Person and Object. A Metaphysical Study*, (London).

Egenhofer, M.J. and Mark, D.M., 1995, Naive geography. In *Spatial Information Theory, A Theoretical Basis for GIS, International Conference COSIT '95*, edited by Frank, A.U. and Kuhn, W. (Berlin: Springer-Verlag), pp. 1–15.

Eschenbach, C., 1993, Semantics of number. *Journal of Semantics*, **10**, pp. 1–31.

Frank, A.U., 1991, Properties of geographic data: Requirements for spatial access methods. In *Advances in Spatial Databases, Proceedings of the Second International Symposium on Large Spatial Databases*, Zurich, edited by Günther, O. and Schek, H.J. (Berlin: Springer-Verlag), pp. 225–234.

Gabbay, D. and Moravcsik, J.M.E., 1973, Sameness and individuation. *The Journal of Philosophy*, **70**, pp. 513–526. (Also in: Pelletier, F.J. (ed), 1979, pp. 233–247.)

Güting, R.H., 1994, An introduction to spatial database systems. *VLDB Journal*, **3**.

Jensen, C.S. *et al.*, 1994, A consensus glossary of temporal database concepts. *ACM SIGMOD Record*, **23**, pp. 52–64.

Langran, G., 1992, *Time in Geographical Information Systems*, (London: Taylor & Francis).

Locke, J., 1690, *An Essay Concerning Human Understanding*, (Oxford: Clarendon Press). (First edition 1690, fourth edition of 1700, reprinted 1975.)

Löbner, S., 1979, *Intensionale Verben und Funktionalbegriffe*. Untersuchung zur Syntax und Semantik von 'wechseln' und den vergleichbaren Verben des Deutschen, (Tübingen: Narr).

Löbner, S., 1985, Definites. *Journal of Semantics*, **4**, pp. 279–326.

Lowe, E.J., 1983, On the identity of artifacts. *The Journal of Philosophy*, **80**, pp. 220–232.

Medeiros, C.B. and Pires, F., 1994, Databases for GIS. *ACM SIGMOD Record*, **23**, pp. 107–115.

Merricks, T., 1994, Endurance and indiscernibility. *The Journal of Philosophy*, **91**, pp. 165–184.

Montague, R., 1973, The proper treatment of quantification in ordinary English. In *Approaches to Natural Language*, edited by Hintikka, K.J., Moravcsik, J.M.E. and Suppes, P. (Dordrecht: D. Reidel), pp. 221–242. (Also in: Montague, R. 1974, Formal Philosophy, (New Haven/London: Yale University Press), pp. 247–270.)

Sagoff, M., 1978, On restoring and reproducing art. *The Journal of Philosophy*, **75**, pp. 453–470.

Scaltsas, T., 1981, Identity, origin and spatio-temporal continuity. *Philosophy*, **56**, pp. 395–402.

Sharvy, R., 1968, Why a class can't change its members. In *Nous*, **2**, pp. 303–314.

Sharvy, R., 1980, A more general theory of definite descriptions. *Philosophical Review*, **89**, pp. 607–624.

Snodgrass, R.T., 1992, Temporal databases. In *Theories and Methods of Spatio-Temporal Reasoning in Geographic Space*, Lecture Notes in Computer Science 639, edited by Frank, A.U., Campari, I. and Formentini, U. (Berlin: Springer-Verlag), pp. 22–64.

Thomson, J.J., 1983, Parthood and identity across time. *The Journal of Philosophy*, **80**, pp. 201–220.

Tsotras, V.J. and Kumar, A., 1996, Temporal database bibliography update. *ACM SIGMOD Record*, **25**, pp. 41–51.

CHAPTER SIX

Objects and Their Environments: From Aristotle to Ecological Ontology

Barry Smith

6.1 INTRODUCTION

What follows is a contribution to the theory of space and of spatial objects. It takes as its starting point the philosophical subfield of ontology, which can be defined as the science of what is: of the various types and categories of objects and relations in all realms of being. More specifically, it begins with ideas set forth by Aristotle in his *Categories* and *Metaphysics*, two works which constitute the first great contributions to ontological science. Because Aristotle's ontological ideas were developed prior to the scientific discoveries of the modern era, he approached the objects and relations of everyday reality with the same ontological seriousness with which scientists today approach the objects of physics. We shall seek to show that what Aristotle has to say about these commonsensical objects and relations can, when translated into more formal terms, be of use also to contemporary ontologists. More precisely, we shall argue that his ideas can contribute to the development of a rigorous theory of those social and institutional components of everyday reality—the settings of human behaviour—which are the subject of this volume.

When modern-day philosophers turn their attentions to ontology they begin not with Aristotle but rather, in almost every case, with a set-theoretic ontology of the sort which is employed in standard model-theoretic semantics. Set-theoretic ontology sees the world in atomistic terms: it postulates a lowest level of atoms or *urelements*, from out of which successively higher levels of set-theoretic objects are then constructed. The approach to ontology to be defended here, in contrast, starts not with atoms but with the mesoscopic objects by which we are surrounded in our normal day-to-day activities. It sees the world not as being made up of concrete individual atoms, on the one hand, and abstract (1- and n-place) 'properties' or 'attributes', on the other. Rather, the world is made of you and me, of your headaches and my sneezes, of your battles and my wars.

The essay is divided into four main parts: the first sketches the bare bones of Aristotle's own ontology, an ontology of substances (objects, things, persons), on the one hand, and accidents (events, qualities, actions) on the other. Part 2 is a more technical section devoted to the semi-formal divisions of some of the concepts at the heart of Aristotle's theory. These technical details, which are inspired in turn by Husserl (1970a), are set out more thoroughly in (Smith, 1997), which also deals with one further major component of Aristotle's ontology, his theory of universals or categories. Part 3 extends the ontology to the realm of what is extended in space, and includes in particular a sketch of Aristotle's own theory of places, and the fourth and final part goes beyond Aristotle to give an account of the ontology of the environments which constitute the everyday world of human action.

6.2 THE ARISTOTELIAN ONTOLOGY OF SUBSTANCE AND ACCIDENT

6.2.1 Substances

At the heart of Aristotle's ontology is a theory of substances (things, or bodies) and accidents (qualities, events, processes). As examples of substances Aristotle has in mind primarily organisms, including human beings. But the category of substance as here conceived will embrace also such ordinary (detached, movable) objects as logs of wood, rocks, potatoes, forks. Examples of accidents are: whistles, blushes, speakings, runnings, my knowledge of French, the warmth of this stone.

To repeat, persons, too, are substances. Indeed, one important reason for admitting substances into our general ontology of everyday reality turns on the fact that we ourselves are members of this category. The ontological marks of substances, as Aristotle conceives them, are as follows:

(i) Substances are that which can exist on their own, where accidents (processes, events, qualities, conditions) require a support from substances in order to exist. Substances are the *bearers* or *carriers* of accidents.

(ii) Substances are that which, while remaining numerically one and the same, can admit contrary accidents at different times: I am sometimes hungry, sometimes not; sometimes tanned, sometimes not.

(iii) Substances are 'one by a process of nature'. A substance has the unity of a living thing. Thus a substance enjoys a certain natural completeness or rounded-offness, being neither too small nor too large—in contrast to the undetached parts of substances (my arms, my legs) and to heaps or aggregates, to complexes or collectives of substances such as armies or football teams.

(iv) A substance has a complete, determinate boundary (the latter is a special sort of part of the substance; something like a maximally thin extremal or peripheral slice).

(v) A substance has no proper parts which are themselves substances. A proper part of a substance, for as long as it remains a part, is not itself a substance; it can become a substance only if it is somehow isolated from its circumcluding whole.

(vi) A substance is similarly not the proper part of any larger substance. A substance is distinct in its form or category from any heap or aggregate or collective or complexes of substances. A substance is accordingly never scattered through space; it is always spatially connected (though it may have holes).

(vii) A substance takes up space. It is an 'extended spatial magnitude' which occupies a place and is such as to have spatial parts. It is not merely spatially extended, but also (unlike other spatially extended objects such as places and spatial regions) such as to have divisible bulk, which means that it can in principle be divided into separate spatially extended substances.

(viii) A substance is self-identical from the beginning to the end of its existence. John as child is identical to John as adult, even though he has, of course, changed in many ways in the intervening years. Qualitative change is in this sense compatible with numerical identity. Thus a substance has no temporal parts: the first ten years of my life are a part of my life and not a part of me. It is not substances but accidents that can have temporal parts. The parts of an accident include its successive phases. The parts of a substance, in contrast, are its arms and legs, organs and cells, etc. (These, for as long as they are not detached, are not themselves substances.)

(ix) The existence of a substance is continuous through time: substances never enjoy intermittent existence.

It is (viii) which causes problems for standard modern approaches to ontology based on physics. This is because a substance, such as Elvis Presley, may continue to exist even though there is no physical part which survives identically from the beginning to the end of his existence. The Aristotelian ontology thus forces us to see the physical world in a new way (in terms of demarcations skew to those which are at work when we view reality through the science of physics).

For further details and references to the relevant passages in Aristotle's writings see (Smith, 1997).

6.2.2 Collectives of Substances

A substance is never the proper part of any larger substance, but substances are often joined together into more or less complex collectives, ranging from families and tribes to nations and empires. Collectives are real constituents of the furniture of the world, but they are not additional constituents, over and above the substances which are their parts.

Collectives inherit some, but not all, of the ontological marks of substances. They can admit contrary accidents at different times; they may have a unity which is something like the unity of a living thing. They take up space, and they can in principle be divided into separate spatially extended sub-collectives, as an orchestra, for example, may be divided into constituent chamber groups. While collectives, too, are self-identical from the beginning to the end of their existence, this existence may be intermittent (as a watch, for example, may be disassembled for a time). There are no punctually existing collectives, and collectives have no temporal parts.

Most important for our purposes is the fact that collectives may gain and lose members, and they may undergo other sorts of changes through time. The Polish nobility has existed for many centuries and it will continue to exist for some time in the future.

6.2.3 Accidents

The category of substance is most intimately associated with the category of accidents. Examples of accidents include: individual qualities, actions and passions, Martha Nussbaum's present knowledge of Greek, a bruise, a handshake, an electric or magnetic charge. Accidents comprehend what, in modern parlance, are sometimes referred to as 'events', 'processes' and 'states'. Accidents are said to 'inhere' in their substances, a notion which will be defined more precisely in what follows in terms of the concept of *specific dependence.*

In contrast to Aristotle (and to the majority of scholastic philosophers up to and including Leibniz) we shall here embrace a view according to which accidents may be either relational or non-relational. Non-relational accidents are attached, as it were, to a single carrier, as a thought is attached to a thinker. Accidents are relational if they depend upon a plurality of substances and thereby join the latter together into complex wholes of greater or lesser duration. Examples of relational accidents include a kiss, a hit, a dance, a conversation, a contract, a battle, a war.

Relational accidents are to be distinguished from comparatives (is longer than, is to the east of, is more famous in South Africa than) and from what are sometimes called 'Cambridge relations' (is father of, is third cousin to) (Mulligan and Smith, 1986). Briefly, relational accidents are entities in their own right, with qualities and changes of their own. Comparatives and Cambridge relations, in contrast, if they can be said to exist at all, exist not as something extra, but only in reflection of certain special sorts of demarcation which are imposed upon the underlying bearers or upon their non-relational accidents.

A somewhat different sort of case is illustrated by those accidents of collective wholes which may survive replacement of their bearers. We shall call objects of this sort 'instituted accidents'. Languages, religions, legal systems do not depend for their existence upon *specific* individuals or groups; rather, they depend generically on the existence of individuals or groups fulfilling certain necessary roles.

6.2.4 Transcategorial Wholes

Accidents, too, may form collectives, both via simultaneous compounding, as for example in the case of a musical chord or a pattern of colour, and via sequencing in time, as in the case of a melody or film performance. Some collectives of accidents (for example, a solo whistling of Brahms' 3rd Symphony) are accidents of substances; some (for example, a stage performance of a Wagner opera) are accidents of collectives of substances. The performance of an opera is an immensely complex sequence of complex relational accidents inhering *inter alia* in the singers and members of the orchestra as well as in the stage and its props. Many of the most impressive achievements of human creativity consist in finding ways in which simple accidents can become compounded together to form complex accidents which are then more than (or different from) the sums of their simple parts. Complex accidents such as opera performances enjoy a complexity which embraces constituents drawn from widely diverse material domains. Already an act of promising manifests a complexity of this sort, embracing constituents of a linguistic, psychological and quasi-ethical sort, as well as physical constituents of different types (including vibrations in the air and ear and associated electrical and chemical events in the brain).

The Aristotelian distinguishes substances and accidents as two distinct orders of being. The former *endure* self-identically through time; the latter *occur:* they unfold themselves through time, and are never present in full at any given instant during which they exist. What, however, is to be said about a complex whole such as Poland, or the First World War, or the institution of the British monarchy? Each of these objects seems to involve both substances and accidents as parts. Wholes of this sort, which embrace within themselves objects from distinct ontological categories, we shall call 'transcategorial'. They are such as to cross the boundary between the two orders of (substantial and accidental) being distinguished by Aristotle, and perhaps for this reason they have been neglected in the tradition of ontology.

Note that many transcategorial wholes in the social and institutional realm, including towns, cities, universities and corporate bodies of other types, are like instituted accidents in manifesting the ability to sustain themselves even though they are subject to a turnover of their constituent substances over time: thus they can continue to exist, even though some of their participant substances are removed and others take their place.

6.3 A THEORY OF OBJECTS

6.3.1 Specific Dependence

The term 'object' in what follows will be used with absolute generality to embrace all substances, accidents, and all wholes and parts thereof, including boundaries. Our basic ontological categories will be defined in terms of the two primitive notions: *is part of* and *is necessarily such that.* 'x *is part of* y' is to be understood as including the limit case where x and y are identical. The mereological notions of discreteness, overlapping and proper parthood are defined in the usual way. Thus to say that x *is discrete from* y is to say that x and y have no parts in common.

As we have seen, substances and accidents may be compounded together mereologically to form larger wholes of different sorts. But substances and accidents are not themselves related mereologically: a substance is not a whole made up of accidents as parts. Rather, the two are linked together via the formal tie of specific dependence, which is defined as follows:

> x is specifically dependent on y = df. (1) x is discrete from y, and (2) x is necessarily such that it cannot exist unless y exists.

My headache, for example, is specifically dependent on me. An accident stands to a substance in the formal tie of one-sided specific dependence only. (Thus it is clear that I am not specifically dependent on my headache.) There are also, however, cases where objects are bound together via ties of *mutual* specific dependence; consider for example the relation between the north and south poles of a magnet. Equally, there are cases where an object stands in a relation of specific dependence to more than one object. Thus there are relational accidents—kisses and hits—which are dependent simultaneously on a plurality of substances.

A further formal tie, in some respects the converse of that of specific dependence, is the relation of separability. We define:

> x and y are mutually separable parts of z = df. (1) x and y are parts of z, (2) x is not necessarily such that any part of y exists and (3) y is not necessarily such that any part of x exists, and (4) x is discrete from y.

z is, for example, a pair of stones, and x and y the stones themselves. Separability, too, may be one-sided:

> x is a one-sidedly separable part of y = df. (1) x is a proper part of y, and (2) some part of y discrete from x is specifically dependent on x, and (3) x is not specifically dependent on any part of y discrete from x.

x is, for example, a human being and y is the sum of x together with some one of x's thoughts.

6.3.2 On Being Substantial

How, now, can we define the notion of substance? Substances are tight unities. They do not fall apart into so many separate pieces. Accordingly we set:

> x and y form a partition of z = df. (1) x and y are parts of z, (2) x and y are discrete from each other, and (3) no part of z is discrete from both x and y (with analogous definitions for n-fold partitions for each $n > 2$).

We then set:

> x is unitary = df. (1) x has no one-sidedly separable parts, and (2) there is no partition of x into mutually separable parts.

All substances will turn out to be unitary in our sense. If we know that x is unitary, all we can infer is that all proper parts of x stand in mutual dependence relations to other proper parts of x. As a further step towards a definition of the notion of substance we note that what is unitary need not, according to this definition, be independent (or in Aristotle's terms: 'able to exist on its own'). Indeed we can define:

> x is an accident = df. (1) x is unitary, and (2) x is specifically dependent on at least one substance.

Relational accidents, such as kisses and hits, contracts and promises, are then specifically dependent on a plurality of substances.

We define further:

> x is substantial = df. (1) x is unitary and (2) x is not specifically dependent on any other object.

But not everything which is substantial is thereby also a substance. For our definition of 'substantial' is satisfied also by quantitative parts of substances such as Darius's (undetached) arm. (That the latter is unitary follows from the fact that any pair of mutually separable parts would have to share a common boundary, which is in our terms a part, and thus would not be discrete in the sense required by the definition of mutual separability.) The latter is not a substance, though it would become a substance on becoming detached, for it then acquires its own complete and exclusive boundary. In order to arrive at a definition of substance, then, it is the notion of boundary which we shall need to take as our guiding clue (something that has not been done in standard treatments of substance in the literature of analytic metaphysics—since boundaries and the mereotopological structures that go together therewith are hard to discern when the world is viewed through set-theoretic spectacles).

6.3.3 Boundary-Dependence

To this end we introduce a new sort of dependence (first discussed by Brentano and Chisholm):

> x is boundary-dependent on y = df. (1) x is a proper part of y, and (2) x is necessarily such that either y exists or there exists some part of y properly including x, and (3) each part of x satisfies (2).

x is, for example, the surface of an apple and y the apple itself. Clause (2) is designed to capture the topological notion of neighbourhood. Roughly, a boundary of given dimension can never exist alone but exists always only as part of some extended neighbourhood of higher dimension (see Varzi, 1994 and Smith, 1996 for details of a formal theory of mereotopology on this basis). There are no points, lines or surfaces in the universe which are not the boundaries of three-dimensional material things. Boundaries are then just those objects which are boundary-dependent on some other object, and ultimately on some substance.

The relation of boundary-dependence holds both between a boundary and the substance which it bounds and also among boundaries themselves. Thus zero-dimensional spatial boundaries (points) are boundary-dependent both on one- and two-dimensional boundaries (lines and surfaces) and also on the three-dimensional substances which are their ultimate hosts. Note that the relation of boundary-dependence does not hold between an accident and its substantial carrier. Certainly my current thought satisfies the condition that it cannot exist unless I or some suitably large part of me exists. And certainly each part of my current thought satisfies this condition also. But my current thought is also specifically dependent upon me, and thus, by the definition of specific dependence it is not a part of me.

6.3.4 Substance Defined

The boundary (outer surface) of a metal sphere is a part of and is boundary-dependent on the sphere itself, but not on anything exterior to the sphere. It is this which makes the sphere a substance. We can now define:

> x is a substance = df. (1) x is substantial, (2) x has a boundary, (3) there is no y that is boundary-dependent on x and on some object that has parts discrete from x.

Darius's undetached arm does not satisfy this definition, since the boundary between his arm and his torso is boundary-dependent on the arm and on an object (the torso) that has parts discrete from the arm. To prove that no substance has a proper part which is itself a substance—and correlatively that no aggregate of substances is itself a substance—consider the following *reductio ad absurdum* argument. Suppose that x is a proper part of y, and that both x and y are substances. Consider, now, the boundary of the included substance x. This boundary must, for at least some portion of its extent, lie within the interior of the including substance y. This portion of the boundary of x then however fails to satisfy clause (3) of our definition of substance.

6.4 JOINTS OF REALITY

6.4.1 Categories Deep and Superficial

What is substantial is always part (though not necessarily a proper part) of some substance. From this it follows that the recognition of the category of substantials in a sense adds nothing new to the totality of what exists. Rather, it reflects cuts in reality skew to those which pick out substances and accidents—and it is the latter, we suggest, which reflect the deep structures in reality. We might indeed choose to ignore what is substantial as such, and see reality as being divided into substances and accidents alone. An adequate account of substances and accidents can however be provided only on the basis of a treatment also of what we might think of as the superficial category of substantials and of the corresponding internal boundaries. For it is part and parcel of what it is to be an extended substance that each substance is marked by the possibility of partition along an indefinite number of interior lines of division. This applies, too, to that extended substance which is the planet earth.

6.4.2 The Packaging of Reality

There are, then, different sorts of parsings or articulations of reality. The first and most important type of parsing results when we follow the outer boundaries of substances, the primary joints of reality. These are boundaries *in the things themselves*, boundaries of a sort which would be present even in the absence of all articulating activity on our part. We have seen the need to recognize also *internal* boundaries of substances which yield partitions into substantials.

Unlike outer boundaries, inner boundaries need not correspond to any genuine heterogeneity (natural articulations) on the side of the bounded objects themselves. They may be purely arbitrary. Thus imagine, again, a homogeneous metal sphere. We can speak of articulations here (for example, of the sphere into hemispheres) even in the absence of any corresponding genuine inner boundaries determined either by some interior spatial discontinuity or by some qualitative heterogeneity (of material constitution, colour, texture, etc.) among the relevant object-parts. Hence we might say that there are not only genuine joints in reality, but also pseudo-joints, of the sort which divide, say, the upper and lower femur as these are depicted in atlases of surgical anatomy.

Let us call inner boundaries of the first sort, for example, the boundaries around my heart and lungs, *bona fide* inner boundaries, inner boundaries of the second sort *fiat* inner boundaries—a terminology that is designed to draw attention to the sense in which the latter owe their existence to acts of human decision or fiat (see Smith, 1995; and also Smith and Varzi, 1997). The distinction between genuine and fiat boundaries applies not solely to inner boundaries but also to objects which play some of the roles of outer boundaries, too. National borders, as well as county- and property-lines, provide examples of fiat outer boundaries in this sense, at least in those cases where, as in the case of Colorado, Wyoming or Utah, they lie skew to any qualitative differentiations or spatio-temporal discontinuities on the side of the underlying reality.

6.4.3 Fiat Objects

But now it is clear that, when once fiat outer *boundaries* have been recognized, then the genuine–fiat opposition can be drawn in relation to *objects* also (thus in relation to both substances and accidents). Examples of genuine objects are: you and me, the planet earth, the surface of the planet earth. Examples of fiat objects are: the northern and southern hemispheres, and all geographical objects demarcated in ways which do not respect qualitative differentiations or spatio-temporal discontinuities in the underlying territory—and not the least important reason for admitting fiat objects into our general ontology turns on the fact that *most of us live in one (or in what turns out to be a nested hierarchy of such objects).*

Dade County, Florida, the United States, the Northern Hemisphere, etc., are fiat objects of the geographical sort. Clearly geographical fiat objects will in general have boundaries which involve a combination of *bona fide* and fiat elements—thus the shores of the North Sea are *bona fide* boundaries, though it seems reasonable to conceive the object demarcated by the totality of such boundaries as a fiat object in spite of this. Fiat objects will in general owe their existence not merely to human fiat but also to associated real properties of the relevant factual material.

The recognition of fiat objects can help us to do justice also to the fact that not all objects with which we have to deal, especially in the geopolitical and legal-administrative realms, need be connected in space. Fiat articulation can create not merely fiat object parts within genuine wholes, but also fiat object wholes out of genuine parts. And then, while genuine objects are in general connected, the fiat boundaries which circumclude constituent *bona fide* objects in this way are often boundaries of scattered wholes. Polynesia is a geographical example of this sort; other examples might be: the constellation Orion, the solar system, the Bahamas.

6.4.4 The Aristotelian Ontology of Places

When we attend more closely to what it means to be *in* Miami, or the Lincoln Bedroom, or the Tonawanda Municipal Swimming Pool, then we discover that there is an ingredient missing from the doctrine of fiat objects as, simply, two- and three-dimensional spatial regions. What is it for a substance to be (to fit snugly) *in* a location or context? To arrive at an answer to this question—an answer which is not entirely satisfying, but which will yet point us in the right direction in our subsequent inquiries—it will be useful to examine Aristotle's technical notion of *place.*

Each substance has its place, Aristotle tells us in the *Physics*. The place of a substance is 'neither a part nor a state of it, but is separable from it. For place is supposed to be something like a vessel'. (209b 26f.) Place cannot be a type of body, however, for if it were, then two bodies would be in the same place, and this Aristotle holds to be impossible. Place has size, therefore, but not matter. It has shape or form, but it lacks divisible bulk. But what, then, is place?

> We say that a thing is in the world, in the sense of in place, because it is in the air, and the air is in the world; and when we say it is in the air, we do not mean it is in every part of the air, but that it is in the air because of the surface of the air which surrounds it; for if all the air were its place, the place of a thing would not be equal to the thing—which it is supposed to be, and which the primary place in which a thing is actually is. (211a 24–28)

Place *contains* body: The thing relates to its place in something like the way the liquid in an urn relates to the urn, or the hand relates to the glove, or a Russian doll relates to the immediately circumjacent Russian doll. A place exactly surrounds the thing, but the place does not depend specifically upon the thing, since the latter can be substituted for another, which is then said to be in the same place. A place exactly surrounds the thing, but not in the sense in which the white of an egg exactly surrounds the yoke, for the two are here such as to form a single continuous whole, and the one is in the other (the yoke is in the egg) not as a thing in its place but as a part in its whole. A place exactly surrounds the thing rather where the thing is separate from but yet in perfect contact with its surrounding body, the latter being therefore marked by a certain sort of interior perforation or hole. The external boundary of the thing then exactly coincides with the internal boundary of that which surrounds it. Thus, when a thing is in a surrounding body of air or water then 'it is primarily in the inner surface of the surrounding body'. The boundaries of the two—the outer boundary of the thing and the inner boundary of its surrounding body—exactly coincide. (211a 30–33) This, then, is place, in Aristotle's view: the place of a substance is the inner boundary of the immediate surrounding or containing body.

6.4.5 Generic Dependence

The place is separable from the substance which occupies it in the sense that it can be occupied at different times by different substances. But the place is also dependent upon the substance, in the sense that places are essentially the sorts of objects into which substances can fit. To capture the sense, in which a place is dependent upon a substance, we need to introduce the notion of generic dependence, which can be defined as follows:

> x is generically dependent on objects of sort S = df. x is necessarily such that it cannot exist unless some object of sort S exists.

To say that a place is generically dependent on a substance of a given shape and size is to say that the place cannot exist unless some appropriate occupying substance of that shape and size exists. A father is in this sense dependent upon a child; a dog owner is dependent upon a dog; a king is dependent upon subjects—but not on any specific child, dog, or subjects. For Aristotle a place exists only if there is some body which it surrounds; but the same place can surround different bodies on different occasions, as the water in the river is different from minute to minute and from day to day. A language, religion or legal system is in the same sense generically dependent on the individuals and groups who serve in their actions to instantiate the corresponding rules, beliefs and customs.

6.5 TOWARDS A THEORY OF ENVIRONMENTS

6.5.1 From Substances to Their Settings

There are a number of strange consequences of Aristotle's theory of place. For example, it follows from the theory that proper substantial parts of bodies (for example: your leg) are not, in fact, in place—but only potentially so: they will actually be in place only if they are transformed into substances in their own right. Moreover, Aristotle associates his general ontology of place with the doctrine of natural places, according to which bodies fall down to the floor when dropped because their 'earthy' nature makes them seek the ground as resting place. The importance of his general ontology of place for our purposes here turns on the fact that it points the way to a theory of environments, of the settings or niches in which substances, and especially human and non-human animals, characteristically and for the most part, exist.

Recall, first of all, that fiat boundaries, the non-physical boundaries added to the world via human behaviour and cognition, may constitute unities out of, radically heterogeneous parts. Think of a performance of a Wagner opera, a British general election, a garage sale. The wholes hereby constituted are spatio-temporal unities even though they are marked off from the rest of reality by no autonomous physical discontinuities. Wholes of these types—which the ecological psychologist Roger Barker has called *physical-behavioural units*—are of inestimable importance for an understanding of human cognition and action, since *almost all human behaviour occurs within one (or in what turns out to be a nested hierarchy of such wholes).* Indeed, we are determined through and through as human beings by such physical-behavioural units, exactly as non-human-animals are determined through and through by the ecological niches into which they have evolved.

6.5.2 The Theory of Physical-Behavioural Units

Ontologists interested in theories of categories have not, it must be noted, invested great efforts in the working out of an ontological theory of the objective environments in which we live and move. Thus we shall be heavily dependent in what follows on the work of Barker and also on the independently developed ecological psychology of Gibson. As Gibson points out:

> The world can be analyzed at many levels, from atomic through terrestrial to cosmic. There is physical structure on the scale of millimicrons at one extreme and on the scale of light years at another. But surely the appropriate scale for animals is the intermediate one of millimeters to kilometers, and it is appropriate because the world and the animal are then comparable. (Gibson, 1966, p. 22)

We shall be concerned, then, still more precisely with the ontology of behavioural environments at the mesoscopic level that is determined by the everyday perceptions and actions of human beings.

The mesoscopic environment, the setting of our actions and perceptions, that into which, from moment to moment, we precisely fit, is not homogeneous. It is divided on a plurality of levels into natural units which may repeat themselves (may exist in many copies), from place to place and some of which may be such as to possess definite, salient boundaries of different sorts. Thus the mesoscopic environment is marked by the presence of substances of different sorts (buildings, rooms, walls, bricks, tables) standing in stable relations to each other. But this environment is marked also by the no less overwhelming presence of what Barker calls *physical-behavioural units*, recurrent types of settings, which serve as the environments for the everyday activities of persons and groups of persons. Examples are: my Friday afternoon class, Jim's meeting with the Dean, your Thursday lunch, the plumber's visit, Gloria's drive to work. Each of these is marked by certain stable arrays of physical objects and physical infrastructure, and by certain stable patterns of behaviour on the part of the persons involved. Physical-behavioural units are in this sense two-sided. They are built out of both physical and behavioural parts. As Barker puts it:

> Physical-behavioral units—conversations, speeches, hunt meets, weddings—are common phenomenal entities, and they are natural units in no way imposed by an investigator. To laymen they are as objective as rivers and forests—they are parts of the objective environment that are experienced directly as rain and sandy beaches are experienced. (Barker, 1968, p. 11)

This importance of physical-behavioural units, recognized by Barker in the early 1960s, can hardly be overestimated. Even our daily ablutions, our journeys from site to site, and our loungings in daydream mode between quests, are recognizable as physical-behavioural units in Barker's terms. Even our more or less unsuccessful *attempts* to engage in given activities can be understand for what they are only in terms of a physical-behavioural unit of the corresponding, full-fledged type in relation to which attempt is determined as attempt and success is distinguished from failure. Only in moments of disorientation do we seem to be set free of all behaviour settings, but this is just to imply that it is in relation to settings that we are in normal cases oriented.

Yet leaving aside the work of a small number of psychologists—Barker and Gibson above all, Kurt Lewin, Egon Brunswik and Fritz Heider (1959) and the Berlin Gestaltists to some degree—investigations of physical-behavioural units whether by scientists or by philosophers are almost unknown. Variants of the Gestaltist concept have, it is true, been rediscovered in recent years, under different guises, by linguists such as Fillmore (1975) with his work on scenes, scripts, frames and slots. The idea is detectable also in Talmy's work (forthcoming) on the windowing of attention in language and in Langacker's discussion (1987) of 'settings' and 'regions' (see also Schank and Abelson, 1977). Schoggen (1989, pp. 302ff.) surveys a series of other near relatives of Barker's concept of physical-behavioural unit or setting, including the concept of 'domain' developed by the geographer Torsten Hägerstrand (1978). As Talmy puts it:

> Linguistic forms can direct the distribution of one's attention over a referent scene in a certain type of pattern, the placement of one or more windows of greatest attention over the scene, in a process that can be termed the windowing of attention. (Talmy, 1996, p. 236)

Common to all such processes is the determination of a boundary, which might be a sharp line or a gradient zone, and whose particular scope and contour—hence, the particular quantity and portions of material that it encloses—can be seen to vary from context to context. The characteristics of such boundaries include:

> First… the material enclosed within the boundary is felt to constitute a unitary coherent conceptual entity distinct from the material outside the boundary. Second, there seems to be some sense of *connectivity* throughout the material enclosed within the boundary and, contrariwise, some sense of *discontinuity* or *disjuncture* across the boundary between the enclosed and external material. … Third, the various portions of the material within the boundary are felt to be *corelevant* to each other, whereas the material outside the boundary is not relevant to that within. (Talmy, 1996, p. 240)

None of these thinkers, however, has demonstrated Barker's ontological sophistication in providing an account of scenes, frames or regions as *parts of reality* comprehending not only human behavioural aspects but also substances and accidents belonging to the physical realm. Husserl's theory of the 'surrounding world' or 'life world' (1970b, pp. 103ff., 272) is a first, informal approximation to an ontological theory of the requisite sort. Its central idea that an environment is that into which an organism exactly fits, is present in germ in Aristotle's doctrine of place. Physical-behavioural units (settings, scenes) have, however, been otherwise almost entirely neglected by philosophers. This is due, first of all, to a tendency amongst philosophers to embrace one or other form of ontological monism, a tendency which has made them glide unknowingly over objects of a transcategorial character, or seek to reduce them—for example, in guise of the Wittgensteinian doctrine of 'language games', to objects of a suitably monistic flavour. It is due, secondly, to the fact that behavioural settings manifest a mereotopological character, which is, again, alien to the world view especially of contemporary philosophers, who have been inspired above all by ideas of logical construction (or more generally by ideas based on predicate logic and set theory as instruments of ontology). The formal ontology of settings, niches, or physical behavioural units is thus far completely undeveloped, in spite of the degree to which recent work in analytic metaphysics has been marked by an increasing readiness to admit into its categorial systems objects—such as artifacts, actual and possible worlds, moments, tropes and individualized properties—of non-traditional sorts.

6.5.3 Ontological Properties of Physical-Behavioural Units

Each physical-behavioural unit has two sorts of components: human beings behaving in certain ways (lecturing, sitting, listening, eating), and non-psychological objects with which behaviour is transacted (chairs, walls, paper, forks, electricity, etc.). Each physical-behavioural unit has a salient boundary which separates an organized internal (foreground) pattern from a differing external (background) pattern. This boundary, too, though it is far from simple, is an objective part of nature, though it may change according to the participants involved or according to the nature of the activity. Each unit is circumjacent to its components: the former exactly surrounds (encloses, encompasses) the latter without a break: the pupils and equipment are *in* the class; the shop opens at 8 a.m. and closes at 6 p.m.

A physical-behavioural unit is a unit: its parts are *unified together*, not through any similarity, but rather through interdependence, including that sort of interdependence which flows from the exercise of controlling power. Events within the unit have a greater effect upon each other than do equivalent events beyond its boundary.

The units have an internal structure, and individuals and categories of individuals occupy the various parts or roles within these structures to different degrees. The relation between participant and setting is to different degrees one of reciprocal co-determination. Each participant has two positions within the unit: first, he is a component which thus goes to form the unit; second, he is an individual whose behaviour, and whose very nature, is itself partly formed by the unit of which he is at any given moment a part.

On the one hand is the objective environment in which we live and move. On the other hand is the interior of the organism. At the interface of these is the person, with his behaviour as a person, behaviour that is connected in complex ways with both his inside parts (neurons, muscles, hormones) and with the outside environment. Environment, person and bodily interior are thus combined together topologically in a nesting arrangement, so that the person, too, is in some sense a boundary or separation phenomenon within a nesting arrangement. The person, with its states and behaviour (perceptions, actions, desires, beliefs, judgments, skills) stands between phenomena on its outside and on its inside, and the latter belong to different orders of reality from both the person itself and from each other. As boundary phenomenon the person is then dependent upon these two environments, the inner and outer. In particular, the person is coloured and shaped, is determined through and through, by the behavioural context of the moment. And because this context is subject to constant change, then it follows, as Schoggen puts it:

> A person has many strengths, many intelligences, many social maturities, many speeds, many degrees of liberality and conservativeness, and many moralities, depending in large part on the particular contexts of the person's behaviour. For example, the same person who displays marked obtusiveness when confronted with a mechanical problem may show impressive skill and adroitness in dealing with social situations. (Schoggen, 1989, p. 7; cf. Barker, 1968, pp. 6, 161)

6.5.4 Spatial Shadows of Physical-Behavioural Units

Each physical-behavioural unit occupies a determinate, bounded *locale* having observable geographical, physical and temporal attributes and having boundaries which are coincident with the boundaries of the behaviour which takes place within it. Certain physical-behavioural units may in addition cast spatial shadows which endure through time. The resulting spatial projections of physical-behavioural units are then on the one hand the product of the behaviour taking place inside them; on the other hand, however, this behaviour may itself be determined and constrained by the given spatial locale—as a football game is determined and constrained by the field on which it is played. Standing patterns of behaviour are here projected in enduring fashion onto two-dimensional space (onto the surface of the earth). Enduring locales may in other cases, however, enjoy some of the properties of integrated physical objects, as in the case of schools and monasteries, observatories and shipyards.

In addition to such immediate spatial projections of human activity on a local scale, however, there are also varieties of human activity which cast spatial shadows in indirect fashion out into regions beyond the immediate locus of the behaviour in question. This occurs, for example, in the case of postal districts, land-parcels within property subdivisions, city blocks, protected wilderness zones which are the products of more or less artificial demarcations imposed on the land from without. In certain circumstances such spatial locales may take on a life of their own—they may acquire proper names, serve as the objects of emotional attachments, grow and shrink or merge and split or indeed (as in the case of Poland and Israel) survive periods of annihilation and transpositions in spatial location over time. Such objects are not behaviour-settings in Barker's sense, though behaviour-settings are nested within them, and these may share common characteristics in virtue of this nesting. Consider Husserl's discussion of the marks of 'Europe' or of 'Western civilization' in (Husserl, 1970b, pp. 269ff.).

6.5.5 Laws of Physical-Behavioural Units

Units have their own behaviour, and their own laws which govern this behaviour—laws which are different from those that govern the behaviour of the person or persons involved (this, too, is a consequence of transcategoriality, and has done much to make physical-behavioural units resistant to scientific treatment). For Barker the laws governing such units may best be understood in mechanical or at least artefactual terms:

> The model of an engine seems to be more appropriate to represent what occurs than is the model of an organism or person. For example, this entity can be 'turned off' and disassembled at the will of the operator, the chairman. He can adjourn the meeting (for a coffee break) and call it to order again. While it is disassembled, some of the parts can be adjusted (a discussant replaced). Individuals have no psychological properties like these. (Barker, 1978, pp. 34f.)

The temporal histories of at least many of the physical-behavioural units by which our lives are structured thus have shapes distinct from the temporal histories of individual persons and their individual experiences. Many physical-behavioural units have sharp beginnings and endings (consider the beginning and ending of a race, or of a contractual agreement). Our pains, illnesses, regrets, in contrast, characteristically grow and fade in intensity. Physical-behavioural units are also sometimes marked by spatial borders which are more crisp and more often rectilinear than are the spatial borders of naturally occurring phenomena such as epidemics or storms. The borders of physical-behavioural units need not be crisp in other respects, however. Consider, for example, the question whether the groom's sneezing is or is not part of that physical-behavioural unit which is his wedding.

On the other hand, physical-behavioural units manifest a capacity for self-sustenance which is much more like what we find in the biological realm. They are characteristically self-regulating, and are such as to guide their components to characteristic states and to maintain those states within limited ranges of values in the face of disturbances (Barker, 1968, pp. 154f.). Slight modifications within given dimensions of the unit can be sustained without detriment to its continued existence as a unit of this type. The total behaviour of the unit—for example, a lecture—cannot be greatly changed, however, without its being destroyed. The lecture must contain an introduction; there must be a speech, there must be listening and discussion. Within the meeting, there are the subparts: chairman, speaker, discussant, audience (as within the sentence there are the subparts: subject, verb, noun, rising inflection, etc.).

6.5.6 The Marks of Environments/Niches/Settings

The ontological marks of environments/niches/settings, as Aristotle might conceive them, are as follows:

(i) Environmental settings are complexes of substances and accidents which require a support from certain special sorts of 'participant' substances in order to exist.

(ii) Environmental settings are that which, while remaining numerically one and the same, can be sustained by distinct participant substances at different times: Clinton is sometimes President, sometimes not.

(iii) Environmental settings are 'one by a process of nature'. An environmental setting has the unity of a living thing. Thus it enjoys a certain natural completeness or rounded-offness, being neither too small nor too large—in contrast to the arbitrary undetached parts of environmental settings and to arbitrary heaps or aggregates of environmental settings.

(iv) An environmental setting has a complete boundary which is determinate at least in the sense that there are objects which fall clearly within it, and other objects which fall clearly outside it.

(v) An environmental setting has actual parts which are also environmental settings: the geometry lesson is a part of the total physical-behavioural setting of the school.

(vi) An environmental setting may similarly be the proper part of larger, circumcluding environmental settings. An environmental setting is, however, distinct in its form or category from any arbitrary heap of environmental settings. An environmental setting may be scattered through space; it need not be spatially connected in the topological sense.

(vii) An environmental setting takes up space; it occupies a physical-temporal locale, and is such as to have spatial parts. It is spatially extended, but it is (unlike physical substances, land-masses, parcels of real estate) such as to have divisible bulk.

(viii) An environmental setting has a beginning and an end in time, but the environmental setting is not self-identical from the beginning to the end of its existence. John as child is identical to John as adult, but the first half of John's geometry lesson is not identical to the second half. Environmental settings, unlike substances, have temporal parts, both natural temporal parts (such as the first set of the match) and arbitrary temporal parts (such as the first 10 seconds of the geometry lesson). Environmental settings are in this respect comparable, ontologically, to accidents rather than to substances.

(ix) The existence of an environmental setting through time need not be continuous: some environmental settings (for example, a chess game, a football match) enjoy intermittent existence.

(x) There are no punctually existing environments, as there are punctual events (for example, beginnings, endings, and instantaneous changes of other sorts).

6.5.7 Hierarchical Nesting

Many ecological units occur in assemblies, as a chick embryo, for example, is constructed as a nested hierarchy of organs, cells, nuclei, molecules, atoms, and subatomic particles.

> A unit in the middle range of a nesting structure is simultaneously both circumjacent and interjacent, both whole and part, both entity and environment. An organ—the liver, for example—is whole in relation to its own component pattern of cells, and is a part in relation to the circumjacent organism that it, with other organs, composes; it forms the environment of its cells, and is, itself, environed by the organism. (Barker, 1968, p. 154)

Physical-behavioural units, too, may be nested together in hierarchies in this way. There are typically many units of each lower-level kind within a given locality, which are typically embedded within larger units: the game is embedded within the match, the honeymoon within the marriage, the lecture meeting is embedded within the university. The drawing of the triangle on the blackboard is embedded within the geometry lesson, which is embedded within the school, which is embedded within the city. Each of these is a physical-behavioural unit in Barker's sense. On the other hand, a randomly delineated square mile in the centre of a city is not a physical-behavioural unit, nor is the mereological sum of its Republican voters; the former has no self-generated unity; the latter has no continuously bounded space-time locus (cf. Barker, 1968, pp. 11f., 16; 1978, p. 34).

6.5.8 The Systematic Mutual Fittingness of Behaviour and Ecological Setting

The behaviour and the physical objects that together constitute the totality of a given physical-behavioural unit are intertwined in such a way as to form a pattern that is by no means random: there is a relation of harmonious fit between the standard patterns of behaviour occurring within the unit and the pattern of its physical components. (The seats in the lecture hall are such as to face the speaker. The speaker addresses his remarks out towards the audience. The boundary of the football field is, apart from certain well-defined exceptions, the boundary of the game. The beginning and end of the school music period mark the limits of the pattern of music behaviour.) This mutual fittingness of behaviour and physical environment extends to the fine, interior structure of a behaviour in a way which will imply a radical nontransposability of standing patterns of behaviour from one environment to another. The conditions obtaining in particular settings are as essential for some kinds of behaviour as are persons with the requisite motives and skills (Barker, 1968, pp. 32).

There are various forces or influences which help to bring about and to sustain this mutual fittingness (cf. Barker, 1968, pp. 30f.). Schoggen (1989, p. 4) describes physical-behavioural settings as consisting of "highly structured, improbable arrangements of objects and events that coerce behaviour in accordance with their own dynamic patterning." Forces which flow in the direction from setting to behaviour include physical constraints exercised by hedges, walls or corridors or by persons with sticks; they include social forces manifested in the authority of the teacher, in threats, promises, warnings; they include the physiological effects of climate, the need for food and water; and they include the effects of perceived physiognomic features of the environment (open spaces seduce children, a businesslike atmosphere encourages businesslike behaviour). Mutual fittingness can be reinforced by learning, and also by a process of selection of the persons involved, whether this be one of self-selection (of children who remain in Sunday school class in light of their ability to conform to the corresponding standing patterns of behaviour), or a matter of externally imposed entrance tests. Forces which flow in the contrary direction, which is to say from behaviour to setting, include all those ways in which a succession of separate and uncoordinated actions can have unintended consequences in the form of new types of actions and new, modified types of settings in the future (as the passage of many feet causes pathways to form in the hillside).

6.5.9 Ontological Transcategoriality of Ecological Settings

A physical-behavioural unit such as a religious meeting, a tennis championship or a sea battle is an intricate complex of times, places, actions, and things. Its constituents can include both manmade elements (buildings, streets, cricket pitches, books, pianos, libraries, the bridges and engine-rooms of battleships) and also natural features (hills, lakes, waves, particular climatic features, patterns of light and sound). These features and elements may be further restricted to a highly specific combination of, say, a particular room in a particular building at a particular time with particular persons and objects distributed in a particular pattern. But the total unit may comprehend also a variety of non-physical components in addition to the behaviour of the persons involved. Thus the unit may comprehend, for example, different types of linguistic, legal and institutional elements, all combined together in space and time in highly specific ways. The phenomena involved are in addition diverse not only as concerns their material constitution but also as concerns their ontological form (thus they comprehend substances, events, actions, states and manifold relations between all of these). As Barker puts it:

> The conceptual incommensurability of phenomena which is such an obstacle to the unification of the sciences does not appear to trouble nature's units.—Within the

larger units, things and events from conceptually more and more alien sciences are incorporated and regulated. (Barker, 1968, p. 155)

As far as our behaviour is concerned, therefore, even the most radical diversity of kinds and categories need not prevent integration.

Hand in hand with the ontological transcategoriality of behaviour settings goes also the ontological transcategoriality of the associated boundaries. Physical-behavioural units are set apart from each other, and from the surrounding environment, not only via boundaries of a simple spatial and temporal sort. There are boundaries also of granularity: some events, for example events on a molecular scale, are standardly of too small a scale to function as events (to be salient features of) physical-behavioural units of the types we are here considering. Events within the interior of my body, too, are standardly such as to fall outside the boundary of the physical-behavioural units in which we standardly engage (not, however, outside the boundary of neurologists or gastroenterologists). Only some sorts of noises will be such as to fall within the boundary of a symphony concert. Only some sorts of deaths will be such as to fall within the boundary of a battle or war. From the standpoint of formal ontology, too, the boundaries of physical-behavioural settings are transcategorial. Thus the boundary of a horse race comprehends within its scope certain substances (horses, riders, spectators), a certain region of space and associated fixtures (land, fences, starting-gate, pavilion), but also certain events, processes, relations, and properties of materially determinate types.

6.5.10 The Ecological Niche

For Gibson, too, reality is a complex hierarchy of inter-nested levels: molecules are nested within cells, cells are nested within leaves, leaves are nested within trees, trees are nested within forests, and so on (Gibson, 1986, p. 101). Each type of organism is *tuned* in its perception and actions to objects on a specific level within this complex hierarchy, to objects which together form what Gibson calls an 'ecological niche'. (Gibson's own account of this relationship of tuning—in terms of information pick-up—need not detain us here.) A niche is that into which an animal *fits*; it is that in relation to which the animal is habituated in its behaviour (Gibson, 1986, p. 129). A niche embraces not only objects of different sorts, but also shapes, colours, textures, tendencies, boundaries (surfaces, edges), all of which are organized in such a way as to enjoy affordance-character for the animal in question: they are relevant to its survival. The given features motivate the organism; they are such as to intrude upon its life, to stimulate the organism in a range of different ways.

The perceptions and actions of human beings are likewise *tuned* to the characteristic shapes and qualities and patterns of behaviour of our own respective natural environments. This mutual embranglement is, however, in our case extended further via artefacts, and via cultural phenomena such as language and its associated institutions. To learn a language is in part also to extend the range of objects in relation to which we are able spontaneously to adjust our behaviour and thus to extend radically the types of niche or setting into which we can spontaneously fit.

6.5.11 In Defence of Realism

A science of human environments will look very different from any science of the more standard sort. This has led some philosophers and cognitive scientists to suppose that environments, settings, physical-behavioural units are phenomena only—that they are subjective constructs, properly to be treated within the methodologically solipsistic framework of psychology. The challenge, then, as Gibson saw, is to demonstrate how a science of environmental settings can be '*consistent* with physics, mechanics, optics, acoustics, and chemistry', being only a matter of 'facts of higher order that have never been made explicit by these sciences and have gone unrecognized'. (Gibson, 1979, p. 17) Hence it should be possible to develop a realist theory of the physical-behavioural units and of other types of fiat objects relevant to everyday human cognition in a manner which does not involve the rejection of standard quantitative physics. Gibson uses the term 'ecology' precisely in order to designate the discipline that should encompass these intermediate-level facts; it is presented as 'a blend of physics, geology, biology, archeology, and anthropology, but with an attempt at unification' on the basis of the question: what can stimulate the organism? (Gibson, 1966, p. 21) The science of ecology will differ from these disciplines in that it will focus on cuts through reality of a different sort.

REFERENCES

Barker, R.G., 1968, *Ecological Psychology. Concepts and Methods for Studying the Environment of Human Behavior*, (Stanford: Stanford University Press).

Barker, R.G. and Associates, 1978, *Habitats, Environments, and Human Behavior. Studies in Ecological Psychology and Eco-Behavioral Science from the Midwest Psychological Field Station*, 1947–72, (San Francisco: Jossey-Bass Publishers).

Brentano, F., 1976, *Philosophical Investigations on Space, Time and the Continuum*, edited by Chisholm, R.M. and Körner, S. (Hamburg: Meiner), Eng. translation by Smith, B., 1988, (London: Croom Helm).

Chisholm, R.M., 1984, Boundaries as Dependent Particulars, *Grazer Philosophische Studien*, **10**, pp. 87–95.

Fillmore, C., 1975, An alternative to checklist theories of meaning. In *Proceedings of the Berkeley Linguistics Society*, edited by Cogen, C. *et al.* (Berkeley: Berkeley Linguistics Society), pp. 123–131.

Gibson, J.J., 1966, *The Senses Considered as Perceptual Systems*, (London: George Allen and Unwin).

Gibson, J.J., 1979, *The Ecological Approach to Visual Perception*, (Boston: Houghton-Mifflin), reprint 1986, (Hillsdale, NJ: Lawrence Erlbaum).

Hägerstrand, T., 1978, Survival and arena: On the life history of individuals in relation to their geographical environments. In *Timing Space and Spacing Time*, edited by Carlstein, T., Vol. 2, (London: Edward Arnold), pp. 121–145.

Heider, F., 1959, *On Perception and Event Structure, and the Psychological Environment, Selected Papers, Psychological Issues*, **1** (3), (New York: International Universities Press).

Husserl, E., 1970a, *Logical Investigations*, 2 vols., translated by Findlay, J.N. (London: Routledge and Kegan Paul).

Husserl, E., 1970b, *The Crisis of European Sciences and Transcendental Phenomenology*, translated by Carr, D. (Evanston: Northwestern University Press).

Langacker, R.W., 1987, *Foundations of Cognitive Grammar*, Vol. 1, Theoretical Prerequisites, (Stanford: Stanford University Press).

Mulligan, K., and Smith, B., 1986, A Relational Theory of the Act. *Topoi*, 5/2, pp. 115–130.

Schank, R.C. and Abelson, R.P., 1977, *Scripts, Plans, Goals and Understanding: An Inquiry into Human Knowledge Structures*, (Hillsdale, NJ: Erlbaum).

Schoggen, P., 1989, *Behavior Settings. A Revision and Extension of Roger G. Barker's Ecological Psychology*, (Stanford: Stanford University Press).

Smith, B., 1995, On drawing lines on a map. In *Spatial Information Theory—A Theoretical Basis for GIS, International Conference COSIT '95*, Lecture Notes in Computer Science 988, edited by Frank, A.U. and Kuhn, W. (Berlin: Springer-Verlag), pp. 475–484.

Smith, B., 1996, Mereotopology: A theory of parts and boundaries. *Data and Knowledge Engineering*, **20**, pp. 287–303.

Smith, B., 1997, On substances, accidents and universals, in defence of a constituent ontology. *Philosophical Papers*, **26**, pp. 105–127.

Smith, B. and Varzi, A.C., 1997, Fiat and bona fide boundaries: Towards on ontology of spatially extended objects. In *Spatial Information Theory—A Theoretical Basis for GIS, International Conference COSIT '97*, Lecture Notes in Computer Science 1329, edited by Hirtle, S.C. and Frank, A.U. (Berlin: Springer-Verlag), pp. 103–119.

Talmy, L., 1996, The windowing of attention in language. In *Grammatical Constructions: Their Form and Meaning*, edited by Shibatani, M. and Thompson, S. (Oxford: Oxford University Press), pp. 235–287.

Varzi, A.C., 1994, On the boundary between mereology and topology. In *Philosophy and the Cognitive Sciences*, edited by Casati R., Smith, B. and White, G. (Vienna: Hölder-Pichler-Tempsky), pp. 423–442.

CHAPTER SEVEN

The Structure of Shadows

Roberto Casati

7.1 INTRODUCTION

This chapter deals with two main problems. It starts out from a theory of spatial representation developed with Achille Varzi (Casati and Varzi, 1994, 1996). Can the descriptive tools of this theory be applied to shadows, here conceived of as *holes in light*? Some adjustments of the theory prove necessary, because shadows are unlike holes in being dependent not only upon material objects but also upon processes. Second, it discusses the viability of the shadow metaphor as a means for modelling the dynamics of socio-economic units.

7.2 MODELS OF COMMON SENSE

Modelling common sense, or finding good theories about it, is a piecemeal project, and it has long been recognised as such (Hayes, 1990, p. 187). We must proceed by isolating conceptual clusters. When a conceptual cluster is well-circumscribed and shows clear-cut, interesting and robust interconnections with neighbouring clusters, we should be able to represent its structure by a small theory—a *theorita* (borrowing here a term from the philosopher H.N. Castañeda). Theoritas maximise focus (modulo their presuppositions, they tend to stay within the limits of their subject-matter) and they thus admit of massive revision when they get integrated into *theories* proper. They constitute mid-term examinations of an object matter; on the other hand, however, they still claim to be exhaustive. In the cluster of concepts relating to intuitive space, whose overall theory is far from being complete, analysis of various sub-clusters and their associated theoritas abound. They cover core concepts, such as those of regions of space, parts of objects, objects in space, space–time objects, and more peripheral concepts, such as those of events and of holes. An interesting though little investigated notion is that of *shadows*; its interest derives from the fact that shadows are eminently spatial but also linked to some generating process, thereby bordering on the *causality* cluster (though in an odd way, for they seem to be *acausal*). In what follows I shall discuss some of the issues raised by shadows, and show their place in an integrated theory of naive space and causality.

7.3 A *THEORITA* OF SHADOWS, AND ITS PROBLEMS

Constructing structured databases is our target, and an unstructured database such as a current dictionary of English (The Oxford Dictionary of Current English, 1992) might very well be our starting point, incorporating as it does a fragment of lexical (encyclopaedic) competence. The OED considers a shadow as a 'patch of shade', a 'dark shape projected by a body intercepting rays of light'. Here the patchy nature of shadows is evoked, and so is their relation to light. A *caveat* on light: for the sake of simplification, a 'classical' conception of light is adopted throughout this chapter (thus phenomena such as penumbra, diffraction through little holes or at sharp edges are not taken into account).

Unless otherwise specified, the light source is point-like and located at a finite distance. Besides, space is assumed to be Euclidean. There is obviously more structure around than suggested by the OED. Todes and Daniels (1975) and van Fraassen, in his *Laws and Symmetry* (1989, p. 217), discuss a very simple theorita of shadows (henceforth the TDV theorita) that summarises in a more articulate form our common-sense ontological intuitions with regard to these objects. Let X be any physical (opaque) object. Then:

I. If X casts any shadow, then some light is falling directly on X. (This is pretty intuitive: light is necessary for casting shadows).
II. IX cannot cast a shadow through an opaque object. (To realise this, consider the following case:)

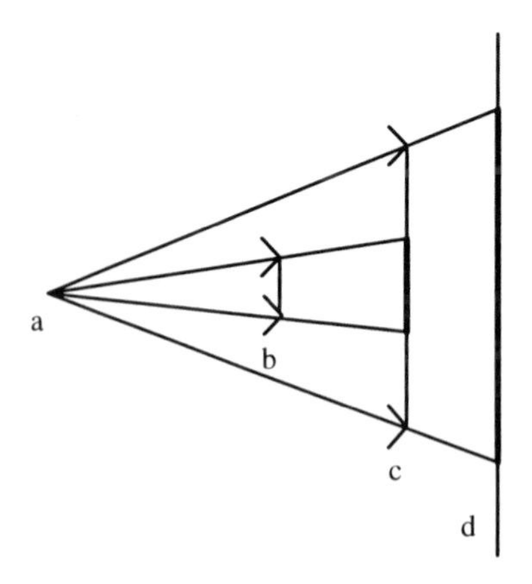

Figure 7.1

Receiving light from a source a, b casts a shadow on a body c, and c casts a shadow on d; but b does not seem to cast a shadow through c onto d.

Finally,
III. Every shadow is shadow of something.

These three principles are not logically incompatible, but the theorita they jointly constitute runs against some empirical facts and is thus inadequate. For take the following case:

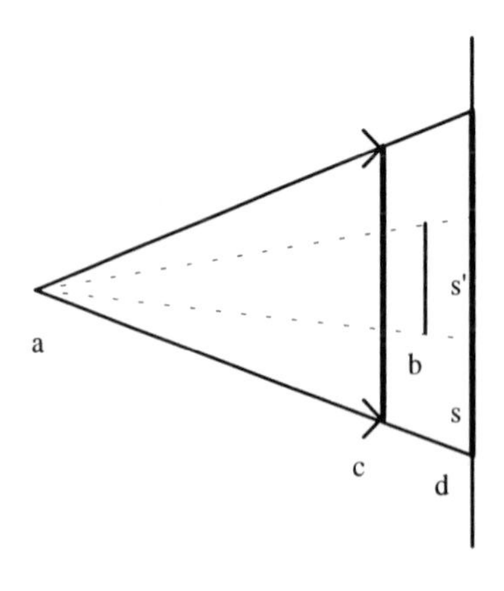

Figure 7.2

Object b has now been moved in the shade projected by c onto d. Consider the imaginary lines linking the profile of b to the light source, and project them onto d, so as to delineate shadow s', which is a portion of the shadow s (of c onto d). What is s' a shadow of? Not of b (by principle I). Not of c (by II). But then, since b and c are the only candidates around, III is false. This is the simplest case. Todes and Daniels (1975, pp. 87–88) consider the more general case of a shadow cast by more than one thing and add the principle (IV): "If a shadow is cast by two things, A and B, it follows that A casts some of it and B casts some of it". Principle IV excludes that the shadow of our problematic case is cast by objects b *and* c.

This case is less academic than it might appear at first sight. There are many examples of *causal pre-emption* of this sort that are a source of conflict in the legal arena. Suppose that your land parcel *a* is in the shade of a skyscraper built on parcel *b*. Suppose now that on parcel *c*, situated between *a* and *b*, another house is being built, which *would* cast an illegal shadow on *a* were the skyscraper not there. On what grounds could you stop the building? For the house on *c* does *not* cast a shadow on your parcel.

Should we give up any of I–III? We will consider this matter more closely in a moment. For the time being, the puzzle shows that we ought to handle shadows with care. As with many other common-sense concepts, if we push analysis too far, we discover hidden inadequacies in them. Common-sense principles can be very *local*, for they are developed for the handling of specific cases, and resist integration in grander conceptual schemes. Should we then get rid of the notion of a shadow altogether?

7.4 AN ELIMINATIVE STRATEGY FOR SHADOWS AND ITS INADEQUACY

One should first show that 'eliminating' shadows is not an easy business—for they are not idle wheels in our common-sense ontology. The basic idea behind elimination is economy: if we can describe a certain portion of the world that *prima facie* is inhabited by entities of a certain kind without even mentioning such entities and without losing descriptive finesse, then we make some savings (savings that can be useful if our theory is to be implemented in some energy-consuming system). The eliminative strategy is thus an interesting move to perform if one wants, say, to reduce the number of primitives in one's database. Eliminativist strategies abound in the social sciences; one example is methodological individualism, the doctrine that social facts are to be explained not by reference to abstract entities such as *the interest of nations* or *class conflicts*, but by reference to the (idealised) psychology of the individual person, usually modelled as a system of preferences. On the other hand, the basic objection to elimination is that, after all, descriptive fine-grainedness might get lost in the process (cf. Casati and Varzi, 1994, for a similar point about holes). The trade-off between economy and descriptive adequacy governs the choice of the types of entities a theory allows for.

How does the *eliminative strategy* work in the case of shadows? One can imagine *paraphrasing* sentences (purportedly) about shadows by means of sentences about *shaded objects*. For instance,

(1) `There is a shadow on the wall.`

will be replaced by

(2) `The wall is shaded.`

How the whole thing develops should be clear: in order to preserve descriptive adequacy, paraphrases are bound to become increasingly clumsy (thus the economy principle is no longer satisfied). This is apparent from some geometrically sophisticated descriptions that are quite straightforward in sentences allowed to refer to shadows and are pretty difficult to translate into sentences that are not:

```
(3) The shadow cast by the table on the wall overlaps
    the shadow cast by the chair in three different
    zones.
```

This suggests that we set aside the eliminative strategy, at least until we obtain a better understanding of the things we want to eliminate.

7.5 IMPROVING THE *THEORITA*: SELF-SHADOWS AND CAST SHADOWS

The TDV theorita could be considered inadequate on another account: it leaves out some interesting facts about shadows that will now be our concern (this is not meant to be a strong criticism, for that very theorita was not produced with the claim to cover completely the phenomenon of shadows in general).

Consider first a neglected side of shadows. It is not only objects which cast shadows: objects also have shadowy parts. In some cases one is unable to make perceptually a clear distinction; still, in most cases the distinction is clear. Consider, for instance, the north face of a building and the shadow it casts on the street and onto the south face of another building. The question now is: "Is the first building shading itself? i.e., Is a self-shadow after all a case of a cast shadow?" Common sense seems to answer this question in the negative. Though proper and cast shadows may be phenomenologically indistinguishable, they are experienced as two types of phenomena, asymmetrically linked to the processual structure of shadows. Thus let us assume this distinction for the time being, and discuss the role of processual asymmetries later.

7.6 IMPROVING THE *THEORITA*: MEREOTOPOLOGICAL FACTS AND PROBLEMS

Where are shadows after all? Let us distinguish three senses of location.

(1) Some things are in other things because they share parts (thus my arm is in me because all of its parts are parts of myself).

(2) Some objects are located relatively to other objects because they are topologically connected to these (without having any part/whole relation to these; thus my arm is on the table because—among other things—it is in contact with it, even though it does not share any of its parts with the table).

(3) Finally, some objects are located 'at' other objects without having any clear part/whole or topological relation to the latter (this is the case of spoons in cups, of fillers in holes, of objects in their places; see Casati and Varzi, 1996; Vandeloise, 1994).

To which locative category do shadows belong?

If shadows are just *darkened parts* of bodies, then they are parts of bodies, and are located in the first sense. What if shadows are outside bodies? There are cases in which shadows are clearly located in the third sense (see below, the shadow that fills a hole: they do not share parts, and are not connected to one another, although their regions coincide—they are simply co-localised). But can we say that shadows are in contact with the bodies they are cast over? It does not seem that common sense has a definite answer to these questions. But I assume that we can conveniently represent common sense as putting shadows in contact with the bodies they are cast over.

7.7 A MORPHOLOGY OF SHADOWS

We shall now abstract from the physical, and even from some of the metaphysical facts about shadows, and concentrate on their geometry, but in a very peculiar sense. We grant that shadows are produced by some projection method and we study their morphology. A morphology is here a sort of very qualitative, very abstract characterisation of some facts concerning shadows—insofar as these facts have a spatial component. Now, geometrical theories of shadow abound, and it is not our concern here to rehearse them—I am not going to dwell on how to draw a shadow, given some facts about a source of light, a surface over which the shadow is cast, and an object situated somewhat between the two. What interests us is a close study of the main characters here.

The basic idea is that the visible shadow is not the whole of the shadow, indeed it is just a boundary of it, and that we have to take into account a *shadow-body* (the technical term 'shadow-body' translates here approximately the English term 'shade' retaining of the latter only the geometrical and morphological connotations). A shadow-body has at least two bound, visible faces (the casting object face and the object face), and at least one unbound, invisible face. In a first approximation—but we shall see some important exceptions—the free, invisible face of a shadow-body is defined by the last array of the photon shower from the light source.

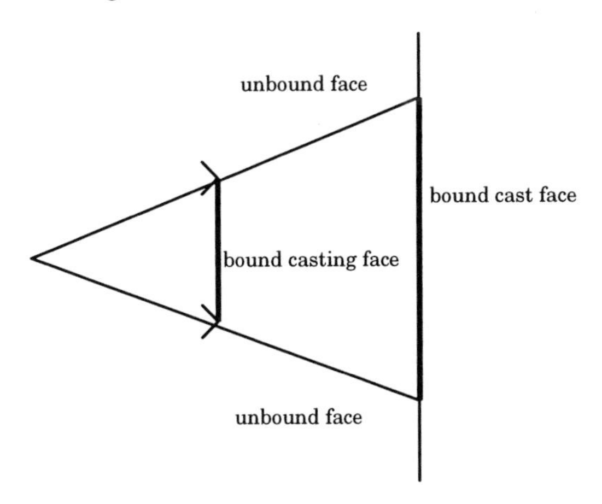

Figure 7.3

7.8 THE TRUE METAPHYSICS OF SHADOWS

Up to now we dealt with common sense—albeit in a refined, mildly regimented fashion. The time has come to present the true metaphysics of shadows, their real nature, and to assess the adequacy of the common sense picture.

Common sense represents shadows as *quasi-objectual entities*. If one were to assign types of objects a position on a continuum from the thingy (chairs and tables) to the unthingy (wind, sounds), shadows would probably appear relatively close to thingy entities. One very thingy feature is that shadows seem to exist continuously over time, in such a way that one can at times judge that the shadow that is now over there is the same that was in another place before. Judgements of identity over time are usually grounded in some conception of an underlying metaphysical structure of the objects at stake. Consider the case of standard spatio-temporal individuals (tables and chairs). Here it seems that three basic metaphysical factors co-operate in ensuring the identity of the object that is now over there with the object that was in another place before. The first two factors are *spatial and temporal contiguity*. The object a that is now over there may not be spatio-temporally contiguous with the object b that was in another place before, but if one is able to track the history of these objects one will see that a = b only if the spatio-temporal histories of a and b are just the same. The third factor is *causal*. In the literature one finds a distinction between *internal* causality and *external* causality, which can be displayed in the following way. External causality summarises the influences that the environment exerts on the object; internal causality summarises the dependencies of the present state of an object from past states (the distinction should not be conflated with the one between self-propelled and hetero-propelled objects). As an example, consider what is the case when you have finished hammering a nail into a piece of wood: the state of affairs holding now (the nail is almost completely inside the wood) depends partly upon the chain of external causality (collisions between the hammer and the nail) and partly upon the internal causality: is the case at the region of space now occupied by the nail depends heavily on what was the case at adjacent spatial and adjacent and previous temporal regions (at least, it depends more heavily on *that* than on *what* happened at non-adjacent such regions).

7.9 SHADOWS AND INTERNAL CAUSALITY

Now, the nature of shadows seems to forbid that portions of a shadow-body (or of shadow-patches, for that matter) are linked to one another across time by immanent causality. That there is shadow now at region r does not depend on there having been shadow anywhere at a previous time.

> Even if a shadow remains constant through a period or undergoes only regular variation, its condition at times through that interval does not causally depend on its condition at earlier times. (...) Rather, the condition of the shadow at any time depends on the way things are with the light source, occluders and surfaces. It does not depend on how things were with the shadow earlier. So unlike physical objects, shadows are not internally causally connected. (Campbell, 1994, 28 ff.)

Thus, given the fact that we do consider shadows as provided with an identity that they (often) retain over time and change, the question is whether our judgements are grounded in anything more than a mere cognitive illusion and have some hope to be true. (The cognitive illusion I have in mind is the so-called ϕ-phenomenon, which is a consequence of gestalt grouping principles in dynamic situations. As a typical example, consider the movement of the cursor on the screen of a computer: there is no moving object there, just a series of pixels that are switched on and off according to a regular pattern. If the pattern is gestaltically sound, the impression of unity across movement arises. The point extends to shadows in a qualified way; for the perceptual unity of a shadow over time is a ϕ-phenomenon even if the shadow does not move. That is, metaphysically speaking there are *instantaneous shadows only*.)

Therefore, what we would describe as the continuous movement of a shadow over a wall hides in fact two metaphysical weaknesses: there is no thing as a single (one and the same) shadow moving from one region to another, and there is no thing as a single (one and the same) movement involving the shadow. The temporally extended shadow is a pseudo-object, and the movement of the shadow is a pseudo-process.

The truth of our identity judgements about (pseudo)shadows and the (pseudo)-processes they might undergo is grounded on something external to shadows themselves. It is the identity of the source and of the casting object that appear to be the most promising candidates—if anything at all is responsible for the unity of the shadow (space forbids here to consider some controversial cases in which the shadow seems to be the same, even though the source of light and the obtruder do not retain their identities: consider a shadow produced by a mirrored source of light and imagine smoothly replacing the light in the mirror with a real light).

7.10 A FURTHER EXTENDED *THEORITA*

If the true metaphysics of shadows construes them as selective gaps in processes, as just 'holes in light' (The definition is to be found in the *Traité des sensations et des passions en général, et des sens en particulier* (1767) by Claude-Nicolas Lecat, as reported by Baxandall (1995: 156). Curiously enough, Lecat thinks that shadows are not 'visible things' because of their being a hole in light (thus endorsing a causal theory of perception that is already found problematic by Locke. See Casati and Varzi, 1994, ch. 11, for an assessment of the relevance of the shadow- and hole objections to the causal theory of perception), we should revise our common-sense intuitions and stretch the theorita a bit further. The resulting theorita is not an extension of the TDV theorita. Shadow-bodies are infinitely extended away from the light source, and have only one face, which does not correspond to the casting face (the self-shadow) but to the hidden face of the enlightened surface of the obtruder. The shadow-body, that is, starts immediately under the enlightened surface of the obtruder (exactly where it physically starts could be a conventional matter at this stage; one can, for instance, assume that this face of the shadow-body is topologically *open*) and extends indefinitely. It 'permeates' all the bodies that are on its way, generating (in correspondence to these encountered bodies) surfaces that are internal to the shadow-body and are oriented. In the extended theorita the target body loses its importance and the obtruder reduces to its enlightened surface. This theory is metaphysically purified (but it is not clear that it serves well the purpose of representing common-sense shadows) and it differs from the common-sense theorita discussed at the beginning on its rejection of principle II. A shadow *can* be cast through an opaque object. This is indirectly proven by considering opaque objects as limit cases of less-than-opaque objects. Obtruders can be non-opaque, or semi-transparent. Putting a non-opaque obtruder in the weak shade of another non-opaque obtruder has the effect of 'strengthening' the shade (i.e., of weakening the light).

Common sense, on the other hand, endorses principle II (possibly) because it mistakenly attributes intrinsic causality to shadows.

The original TDV theorita had a place for the notion of shadow-patches that could allow for a distinction between object- and cast-shadow-patch. In the new theorita the distinction is modal. Shadow-patches are superficial parts of a body that are internal to a shadow-body. *Cast* shadow patches are shadow patches that would not be in the dark if the obtruder were removed.

7.11 THE STRUCTURE OF SHADOWS

To sum up the best theory up to now: shadow patches are internal boundaries of shadow-bodies; shadow-bodies are identified by a source of light and an opaque obtruder. The obtruder is not transparent to light but is transparent to shadow and permeated by it. This is because shading is not a process; it is the absence of one. As a consequence, real shadows are instantaneous, and temporally extended shadows are only pseudo-objects (and processes involving shadows are only pseudo-processes). However, our judgements about pseudo-shadows and their pseudo-processes are not ungrounded, and criteria are available that make their truth depend on the truth of judgements about light sources and obtruders (to a certain extent).

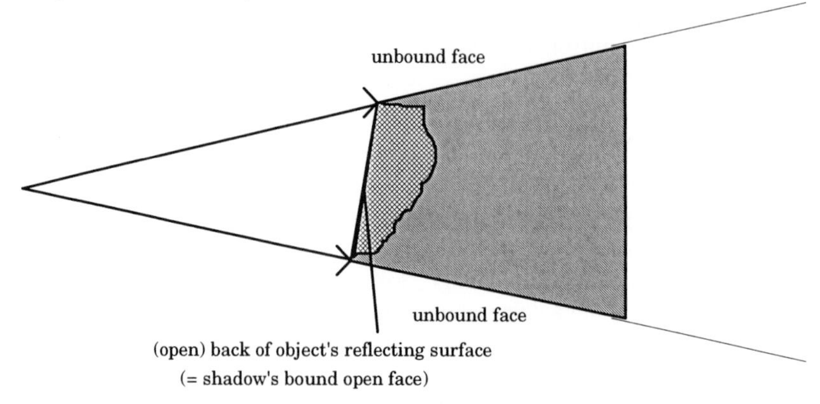

Figure 7.4

7.12 FORMAL PRINCIPLES

This section articulates the theorita. If we ban mirrors, we can identify a shadow-body through the source of light and the obtruder it depends upon, and state that:

1. A shadow-body is externally connected with its obtruder.
2. A shadow-body is sure never to be connected with its light source.
3. An obtruder is wholly located within its shadow-body (unless it is located in some other obtruder's shadow-body relative to the same light source).
4. Successive obtruder's contributions to a shadow-body do not diminish previous obtruder's contributions, unless the following Minimalist Theory is true:

   The Minimalist Theory: given an obtruder and a source, there exists only one shadow-body relative to them (which is the possibly scattered sum of all holes generated by the possibly scattered obtruders in the light of the source).

5. In particular, a shadow-body does not share parts with any material body—not even with those that cast it or onto which it is cast.
6. Only the lightened parts of an object can cast a shadow. It follows that:
   a. Only superficial parts of an object can cast a shadow;
   b. Not all parts of an object that casts a shadow cast shadows;
   c. An object can cast a shadow only in a derivative sense, i.e., only if some superficial parts of it cast a shadow;
7. No two shadows (cast from the same source) ever overlap.

7.13 GENERALISED SHADOWS

I take the question of dependency as representative of a class of more general questions about the existence and the identity of 'incomplete' or 'abstract' objects that are nevertheless clearly localised in time and space and have a *curriculum vitae* of sorts (objects such as nations and geopolitical entities at large). As we have seen, obtruders can be non-opaque, or semi-transparent. This gives us a first generalisation.

It is tempting to consider ordinary shadows as just one among other cases of shadowing phenomena. Ordinary shadows are *light*-shadows, but light is one among other examples of causal flows. A generalised shadow-body is thus whatever area is shielded from a given causal flow. Particularised again, the notion gives us objects such as shadow-bodies in rain (the dry area under your umbrella), but also in crowds, avalanches or glaciers. What really matters is the presence of a flow of sorts, and the possibility of finding shelter.

7.14 SHADOWS AND SOCIO-ECONOMIC UNITS

Although shadows are in themselves an interesting geographic object (consider, for instance, the kind of problem that was addressed by town planners when skyscrapers began to rise), shadows are of some interest to the scholar of the geographical location of socio-economic units, because they provide a powerful and intuitive toy analogy. Both shadows and SEUs are less-than-concrete objects (they are rather different from mountains or lakes); but at the same time they both are *somewhat* localised in space. Moreover, both shadows and SEUs are *dependent* objects of sort—they are what they are because something else (a source of light and an obtruder in the case of shadows; people, their properties, their beliefs, etc. in the case of SEUs) is what it is. Moreover, the way shadows and SEUs are *where* they are, is related to this dependence they both have to the things they depend upon. The present analysis of shadows suggests some of the guidelines for construing the location of less-than-concrete entities.

As an example, consider Italy, taken here as a geopolitical entity (not as just a region of space).

- Italy is a space occupier. It takes some space, in an exclusive way (for instance, it does not interpenetrate with Switzerland). But it does so not in virtue of some 'hard' physical facts (bar ideological nonsense about a sacred physical 'Italian *boden*') but because of a general consensus of a large number of individuals—both Italian citizens and citizens from other countries. Italy is not a metaphysically independent entity. It is causally sustained by the individual wills of a number of people. Italy is *projected.* (So is any shadow.)
- Italy is thus a metaphysically dependent entity. It is a 'lesser' entity. Which does not mean that we ought to endorse an eliminative strategy for Italy. We can accept that it exists, as its citizens and the mountains in it exist, but not as an independent object. (Shadows, although dependent, still exist.)
- The size and shape of Italy depend on how much space can be *controlled* by the citizens of the country, and thereby *sheltered* from other influences or claims. (A shadow's shape and size depend upon those of the obtruder.)
- Italy's shape *partly* depends also upon the physical geography of the space it is projected on. (A shadow's geometry partly depends upon the geometry of the target object.)
- For the purpose of the definition of Italy as a space-occupier, the boundary alone is what matters. Whatever is inside the boundary need not receive any representation on a map. (A shadow is adequately described by its profile. The interior does not represent any feature of the obtruder.)
- The boundary of Italy might be subject to negotiation in some cases—e.g., when individual property rights conflict with international politics. The blurring of the boundary that one might experience here resembles the grey area around a shadow that is projected by a non-pointlike source, and which can be resolved into the sum of many different projections from individual sources.

- Italy does not enjoy any form of intrinsic causality. Where it is now depends much less upon where it was before, than upon the will of the individuals (both of Italian and of non-Italian citizenship) that project it. Barring again mythical nonsense about the magic powers of the *boden*, Italy has no causal inertia. Nor does any other shadow.

7.15 SOME WARNINGS

It is not claimed, of course, that all geopolitical entities are shadow-like. And it is not claimed that *all* aspects of shadows are relevant here either (see the next section for an enrichment). Finally, it is not claimed that the shadow metaphor is good or bad. A more modest claim is put forward here: that the shadow metaphor is a rather intuitive model for our understanding of SEU's way of being in space.

7.16 FURTHER ASPECTS OF THE SHADOW METAPHOR

In this last section I shall list some of the more common metaphorical usages of shadows. Although in this chapter I concentrated mostly on the spatial and causal clusters, it might be argued that other features of the concept of a shadow might be fruitfully employed in modelling SEU's dynamics. An open question is: "Which of these clusters overlap with each other?" (For instance, facts about control and causality-grounded facts about representation.)

- The shadow-as-shelter metaphor. This is a causal/spatial metaphor, with both good and bad connotations. One can be in somebody else's shadow, meaning that one lost one's independence.
- The shadow-as-proxy metaphor. Shadows stand for something projecting them. A causal metaphor, but shadows are bestowed some independence.
- The shadow-as-representation metaphor. Partly overlaps with the shadow-as-proxy metaphor: representations are proxies. Shadows represent only by outline. Their interior is essentially hidden.
- The shadow-as-immaterial metaphor. Probably subordinate to a more general (but not clearly individuated) metaphor, the shadow-as-lesser-entity metaphor.
- The shadow-as-dependent-entity metaphor. (John follows Mary as a shadow.)
- The shadow-as-obscurity, danger metaphor.
- The shadow as silhouette metaphor. (Overlaps with the shadow-as-representation metaphor.) One can abstract from what is inside the shadow, the only relevant thing is the profile.
- The shadow-as-controlled-entity metaphor. Obviously overlaps with the shadow-as-dependent-entity metaphor; it is nevertheless different insofar an entity can be controlled without thereby being dependent.
- The shadow-as-attached thing metaphor ("I have got a shadow"). It is not very clear how this interacts with the dependent feature of shadows. The fact that shadows stick to one suggests that they are relatively independent (they are a nuisance, something one would like to get rid of and cannot.)

ACKNOWLEDGEMENTS

I benefited from discussions with Richard Bradley, John Collins, Carola Eschenbach, Jonathan Raper, Achille Varzi, the participants to the Nafplion meeting, and the students at the Technical University Vienna.

REFERENCES

Baxandall, M., 1995, *Shadows and Enlightenment*, (New Haven and London: Yale University Press).

Campbell, J., 1994, *Past, Space and Self*, (Cambridge, MA: MIT Press).

Casati, R. and Varzi, A., 1994, *Holes and Other Superficialities*, (Cambridge, MA and London: MIT Press).

Casati, R. and Varzi, A., 1996, The structure of spatial localization. In *Philosophical Studies*, **82**, pp. 205–239.

Fraassen, B. van, 1989, *Laws and Symmetry*, (Oxford: Clarendon Press).

Hayes, P., 1990, The Naïve Physics Manifesto. In *The Philosophy of Artificial Intelligence*, edited by Boden, M. (Oxford: Oxford University Press), pp. 171–205.

Todes S. and Daniels, C., 1975, Beyond the doubt of a shadow: A phenomenological and linguistic analysis of shadows. In *Selected Studies in Phenomenology and Existential Philosophy*, edited by Ihde, D. and Zaner, R.M. (The Hague: Martinus Nijhoff), pp. 203–216.

Vandeloise, C., 1994, Methodology and analyses of the preposition. In *Cognitive Linguistics*, **5**, pp. 157–184.

The Oxford Dictionary of Current English, 1992, 2nd edition, edited by Thompson, D. (Oxford: Oxford University Press).

CHAPTER EIGHT

Time, Actuality, Novelty and History

Georg Franck

8.1 INTRODUCTION

Time is a fundamental dimension of experience. The experience we have of time itself is a notorious source of confusion, however. It proves extraordinarily hard to distinguish between the objective and the subjective part of this experience. Evidence of this difficulty is the fact that, in the sciences, there are as many concepts of time as there are distinct bodies of knowledge. It depends on the phenomena a theory is accounting for which concept of time it works with. The concept of time appropriate to account for motion is not rich enough to account for life; the concept of time rich enough to account for biological life is too poor to account for conscious life. A classification of the sciences according to the definition of time implied renders a hierarchy of increasing experiential concreteness and decreasing formal rigour. The richer a theory's account of what we experience as time, the looser become its definitions. The higher the standards of precision in a field of theorising, the narrower becomes its notion of time. This trade-off throws a new light on reductionism. When looked at from the perspective of time perception, reduction to physical reality proves to abstract from an entire dimension of existence. The dimension disregarded by reductive methods is that of *actual* time as opposed to the dimension of *real* time.

8.2 TIME

'Time' is one of the most frequently used nouns in English. Since the meaning of a word is what its use amounts to, it should be an easy exercise to say what time means. Nobody, however, has mastered this exercise satisfactorily until now. Even though we all know what we are talking about when using the word 'time', we run into serious difficulties when asked to say exactly what we mean. The words of St. Augustine, "What is time? If nobody asks me, I know, but if I want to explain it to some one, then I do not know" (*Confessiones*, bk XI, ch. XV, xvii), are as correct today as they were in his own days. Of course, there are clear-cut definitions of 'time'. Time is a well-defined term in the physical sciences, after all. The problem with terminological definitions of time is that they do not express what we mean when using the word in everyday speech. The more rigorous the definition from an operational or logical point of view, the more extraneous becomes its reference to what we experience as the flow of time.

The passage of time is one of the mysteries almost untouched by the progress of science. One can even say that its mysteriousness has grown with the advancement of scientific knowledge. The more secure its march, the more sceptical science grew concerning the objectivity of time's flowing. If this flow is something objective, how can it be measured? Flow implies motion. What is moving when time flows? With the flow of time we mean the movement of the present moment relative to the chronological order of moments. Motion implies speed. What is the speed with which the now travels along the chronological axis? In order to measure the travelling speed of the now, some reference point lying outside the 'moment-just-being' would be required. Moments lying outside the present moment are operationally inaccessible, however. Anything actual *is* present. Hence, there is no way of measuring the speed with which time passes. But that is not all. The speed with which time passes even seems to defy consistent definition.

The dimension of speed normally is meters per second (or whatever units measuring space and time are used). The points to which the flowing of time is relative are not points in space, however, but points in time. The dimension of the speed with which time passes is seconds per second: it takes one second for the now to shift one second along the chronological axis. When divided, these terms cancel each other. By dividing one second by one second we obtain the dimensionless value of 1. Thus, whatever we are going to measure and whatever the subjective impression of the speed with which time passes may be, this passage seems to be bound to have the dimensionless speed of 1. A dimensionless something of value 1 cannot be called a speed at all, though.

The definition of time as something that passes and is centred in the now is contained implicitly in the use of the tenses and temporal adverbs in natural languages. The tenses of past, present, and future, as well as the semantic meaning of temporal adverbs like 'now', 'recently', 'soon', 'today, 'yesterday', 'tomorrow', etc., express the fact that the now is separating two different regions of time. The region lying in front of it contains the totality of events still to come; the region lying behind contains the totality of events that are already gone. The grammar of the *tempora* past, present and future gives expression, moreover, to the fact that the now is ceaselessly travelling. In order to distinguish the concept of time implicitly defined by the use of this grammar from clock time, let us call it *temporal* time.

Clocks do not measure the passage of time but translate temporal intervals into spatial structures. On the basis of the fact that the passage of time is incapable of being measured by clocks (or whatever physical device), the concept of temporal time was criticised by Ernst Mach (1883, pp. 236–241). Mach argued that the concept of a magnitude incapable of being operationalised by measurement is not a physical but a metaphysical one. He pointed out, moreover, that the question whether time goes by is insubstantial for the reality physical theories deal with. Neither classic mechanics nor thermodynamics are changed in any way by how the question is answered. Mach's contention seems to hold even for present-day mechanics. Neither relativity theory nor quantum mechanics contain—let alone depend on—some notion of the *flow* of time.

Shortly after Mach's criticism, Bertrand Russell submitted the grammar of tense to a logical examination (Russell, 1903, §. 442). He pointed out that sentences containing tensed expressions have unstable truth-values. The sentence "Yesterday was Sunday" is true today, Monday, and was false yesterday. The sentence "Now it is night" is true tonight and will be false tomorrow morning. The truth-value of the predicates 'is present', 'is past', 'is future' changes with time. Now-ness is an indexical expression as is here-ness and as are the personal pronouns. The truth-value of sentences containing 'here', 'there', 'I', 'you' may change with time, too. This change, however, is manageable by modal logic, since the truth-value of such sentences does not change if the speaker does not move from the original place of utterance or if nobody else but the original speaker utters the sentence. In contrast, the truth-value of sentences containing 'now' or 'today' changes without further ado. It changes spontaneously and irresistibly by the simple fact that time goes by. As a remedy, Russell proposed to substitute the tenses by the relations 'earlier than' and 'later than'. In a criticism that surpassed Russell's, McTaggart showed that this remedy is inept (McTaggart, 1908; McTaggart, 1927, ch. 33). For an intensive discussion of McTaggart's attack on the concept of (temporal) time as such, see (Franck, 1994).

Sentences with spontaneously changing truth-values are without prospects of scientific approval. Thus, even before Einstein called for a final blow against it, the concept of temporal time was under heavy attack in science. The Newtonian concept of "[a]bsolute, true, and mathematical time" that "flows equitably and without relation to anything external" (Newton, 1687, p. 6) was going to split into three separate concepts. Since time's flow as well as the direction of time is inessential for the formulation of the laws of mechanics and the theory of mechanical clocks, "absolute and mathematical time" turned into simple *parameter* time t. Parameter time is the one-dimensional continuum of datable points. In parameter time, differences in time are reduced to differences in date. Parameter time is homogenous, i.e., without preferred direction, and thus reversible. The direction of time comes in with thermodynamics and the notion of entropy. The entropy of a closed system grows unidirectionally in the sense that an ordering of its states according to entropy renders an order according to date in the direction from past to future. Since the increase of entropy is a statistical law that cannot—at least until now—be reduced to basic mechanics, parameter time and *directed* time are not only different, but may turn out to be incompatible notions of time. It has even been put into question whether the direction of entropy growth is a property of time as such (see Price, 1996). The direction in which entropy grows has to be clearly distinguished, moreover, from what we experience as the passage of time. The growth of entropy does not imply that there is something like the now which travels in the direction of this growth. Or, to put it differently, the notion of a *direction* leading from past to future does not imply the notion of past and future as *regions* of time. Talking of past and future as regions of time presupposes the existence of a moment that, by being actual, separates two domains of non-actual time. Since the distinction between actuality and non-actuality hinges on the notion of fully-fledged temporality, isolating the perplexities surrounding the flow of time requires us to clearly distinguish directed time from temporal time.

Thus, even before the advent of relativity theory, Newtonian time had fallen apart into three concepts, none of which is reducible to another. However, these three concepts fit into a hierarchical scheme whose principles of order are symmetry and symmetry breaking. Parameter time is the most simple and most symmetrical concept of time conceivable. It is, like the continuum of points in a one-dimensional line, perfectly homogeneous, i.e., free from any preference of direction and heterogeneity concerning actuality. *Cause* and *effect* are perfectly equivalent in parameter time. Processes running in parameter time are thus reversible. In order to reverse them, nothing more is needed than the replacement of t by $-t$. As a dimension, parameter time is like another dimension of space. The way of representing a process in parameter time is its so-called trajectory: the sum total of the states the process runs through. Since no difference whatsoever is implied concerning actuality, this totality of states is represented as if the states existed side by side.

In parameter time, instants are ordered like the points in a straight line. There is no relation such as 'smaller than' or 'greater than' between them. Accordingly, talking of an instant being 'earlier than' or 'later than another' is purely conventional. Since there is neither a preferred direction of time nor any difference concerning actuality, the relations of 'earlier than' and 'later than' are perfectly symmetrical. Things change, however, when entropy comes into play. Entropy growth prefers the direction from past to future. With the transition from parameter time to directed time, the order of instants turns into an order like that of the real numbers. In contrast to the points in a line, the real numbers are ordered according to an inherent magnitude. Analogously, the instants of directed time are ordered according to a measure inherent in them. Thus, the transition consists in breaking the symmetry between the two directions discernible in the one-dimensional continuum of instants.

A characteristic example of an irreversible process is black body radiation. Black bodies absorb electromagnetic radiation of any complexity; the radiation they emit depends only on their temperature, however. The radiation absorbed and the radiation emitted are equivalent energetically, but divergent concerning structure. The reflection is oblivious of the colour of the incoming radiation. Colour means structure or, for that matter, order. In black body radiation, order is lost. Loss of order means growth of entropy. Processes enhancing entropy are oblivious of the causes that give rise to the loss of order. Processes of this kind are irreversible because the equivalence of cause and effect no longer holds.

The entropy law or, as it is called, the Second Law of Thermodynamics is a universal law. Entropy growth is a fundamental characteristic of processes running in a world that is far from thermodynamic equilibrium. In such a world, reversible processes can exist only as limiting cases. In such a world, entropy decline or growth of order is not excluded, however. The entropy law is a statistical law allowing for islands of growing order in the ocean of increasing disorder. Each organism represents such a 'dissipative' structure, as the islands are called. Through enforced dissipation of energy—which means through enforced entropy growth in the environment—the self-organisation of structures exhibiting novel orders of complexity becomes possible. Self-organised processes are irreversible not only because causes are forgotten but because instabilities occur that lead to 'bifurcations' in behaviour. Bifurcations make behaviour inherently unpredictable. The symmetry break between the directions of time is fundamental, thus, for the emergence of phenomena such as life, novelty and information (Nicolis and Prigogine, 1989).

The equivalence that is lost when parameter time turns into directed time is the equivalence of cause and effect. On the level of entropy and negentropy growth, cause and effect are no longer exchangeable. The question thus arises what kind of equivalence breaks down with the transition from directed time to fully-fledged temporal time. The answer is not straightforward since no well-developed theory of temporal time is available. The kind of equivalence that no longer holds at the level of temporality became indirectly clear, however, with the advent of relativity theory.

Relativity theory is a principled theory deducted from the absoluteness of the speed of light. With the absoluteness of the speed of light, simultaneity becomes relative to the location of the observer or frame of reference. If simultaneity is relative to that location, the now is relative to the frame of reference, too. Locations that are spatially distant or distinct with regard to relative motion will accordingly differ in time. If the now is the moment of actuality surrounded by regions of what is no more and not yet actual, locations that differ in time cannot belong to one and the same *actual* world. The world as actualised in the now is actual only in the realm of one and the same now. Unsynchronised 'nows' unequivocally belong to different worlds. Thus, temporality splits the universe into as many worlds as there are locations possibly occupied by observers. The rigorous deduction of this result is due to Kurt Gödel (1949).

8.3 ACTUALITY

The symmetrical relation that turns into asymmetry with the transition from directed time to temporal time is the equivalence of *reality* and *actuality*. In relativistic space–time, time assumes its space-like character since relativity theory meticulously avoids any notion of actuality as a mode of existence. In time as the fourth dimension, the states of the universe differing in date are arranged as if they co-existed in time. Accordingly, space–time has been addressed as a 'block' universe, encompassing the totality of its states irrespective of any difference between past, present and future.

In the context of relativity, the now and its travelling turn into a subjective impression to which nothing corresponds except itself. As it is well known, Einstein even called it a subjective illusion—if, however, a stubborn one (Einstein and Besso, 1972). What is less known is that Einstein was seriously concerned with the problem of nowness. The peculiarity of the now within the scientific world view was the topic of a discussion between Einstein and Carnap. "Einstein said that the problem of the Now worried him seriously. He explained that the experience of the Now means something special for man, something essentially different from the past and future, but that this important difference does not and cannot occur within physics. That this experience cannot be grasped by science seemed to him a matter of painful but inevitable resignation." (Carnap, 1963, p. 37)

What Einstein resigned himself to accept as a matter of fact amounts to an existential threat to the social sciences. The social sciences, by being sciences of man, cannot acquiesce in the scientific non-graspability of the now. They cannot help, that is, accounting for the difference between actuality and non-actuality. Without distinguishing the time being from the time past or yet to come, it is impossible to talk of things such as learning, planning, decision and choice. Without temporal time there is no such thing as *understandable* behaviour. It is not before the equivalence of reality and actuality is broken that the concept of *rationality* assumes definite meaning.

The social sciences are condemned to dealing with the perplexities of temporal time. Moreover, they are bound to assume that the individual worlds of the members of society have a common intersection in actual time. People agree on the time being. However subjective the nature of actuality may be, it is of accountable objectivity at the intersubjective level. People agree upon living in one and the same actual world. Splitting this world up into as many separate worlds as there are sentient beings occupying distinct locations in space–time must somehow be prevented from becoming experientially effective. The now must somehow be extended; that is, it must make room for minute differences in simultaneity. Indeed, the now, as we experience it, is not a razor's edge. It spans a kind of interval. This interval has a lower and an upper limit. The lower limit is due to the limited temporal resolution of sensory awareness. The smallest unit of time perception is about 30 milliseconds. Below 30 msec perception of the sequence of stimuli, below a somewhat smaller interval (varying with modality), perception of differences as such comes to an end. The upper limit of the interval experienced as the moment-in-being depends on the ability of focussing attention. The unit of duration that is experienced as a whole may last up to a few seconds (see Pöppel, 1989). The now we are living in is a 'specious' present, as William James phrased it (James, 1890).

The extendedness of the present further enhances the perplexities surrounding the flow of time. It seems to imply that there exists some kind of 'time window' covering a stretch in clock time instead of being point-like. Such a window cannot exist in real time, however. If there were an open time window in real time, it would present a whole package of temporally different instants as being now. There would be an exact, instantaneous present and a 'sloppy', extended present. Within the sloppy present instants could be distinguished as being earlier and later. Moreover, the sloppy present would encompass instants that are already past or yet to come relative to the instantaneous now. A now that encompasses instants that are past or future is a contradiction in terms—as long, at least, as time is supposed to have one and only one dimension.

Depending on the notion of a specious present travelling at an indefinable speed, it seems small wonder that the logical foundations of the social sciences are shaky. Above all, it seems pointless to criticise their models as being restrictive. Only by fairly restricting its scope, can temporality be accounted for in theory. Since the problems of temporality have resisted resolution until today, it has only been by circumvention that they were prevented from becoming destructive. Circumvention has its price, of course. This price rises the higher the greater the safety gained by leaving the dangers aside. Thus, a trade-off can be expected to be operative between the level of formal sophistication of modelling in the social sciences and the scope of temporality accounted for by the models.

Such a trade-off can be observed indeed. The opposing extremes of formal sophistication in the social sciences are mathematical economics and historiography. Characteristically, mathematical economics and historiography are the opposing extremes concerning the scope of temporality accounted for, too. Mathematical economics restricts its account of temporality to the minimum possible. Typically, it abstracts from ignorance and uncertainty, change and discovery, hope and fear, novelty and news (cf. Shackle, 1958, p. 93). The grandest and most sweeping example of this heroic abstraction is the *Walras–Pareto* type of static general market equilibrium. The paradigm of this paradigmatic model is classic mechanics. Accordingly, the decisions leading to market equilibrium are conceived of as being reversible in this model. There is no false trading nor are there contractual liabilities resisting revision. The Walrasian process of ‘t‰ tonnement’ is a process of recontracting that does not end before the equilibrium conditions are met throughout.

Of course, mathematical economics is not restricted to general equilibrium theory. There is also capital and growth theory; both concerned with investment and expectations. However, even capital and growth theory assume that the dynamics of the processes concerned are smooth and differentiable throughout. There is no room for surprise or discovery in these theories, either. In order to account for risk and uncertainty, lacking knowledge about the future is turned into known probability distributions. Genuine ignorance, as well as true novelty is concealed behind a veil of probabilism. (For discussion and criticism of this concealment, see Georgescu-Roegen, 1971, p. 121ff; Shackle, 1990, part I and II; Loasby, 1976; Vickers, 1994.) Mathematically, time in economics is parameter time t. Innovation in the economic process is a conception appropriate to the economic historian rather than to the economic theoretician.

In contrast to time in economics, time in historiography is fully temporalised. The concept of time that historiography works with is the one defined implicitly by the use of tenses and temporal adverbs in natural language. It treats things as incessantly and irreversibly changing. The allowance for change is not even restricted to the present tense in historiography. Historiography accounts for the fact that the constitution of past and future consists in breaking the equivalence of reality and actuality. It accounts, that is, for the fact that the past and future exist only in the present representation. For beings living in the present, the past and future are only accessible through actual recollection or anticipation, respectively. We have no immediate access to whatever may be the reality of world states that have passed or are yet to come. Each piece of information about history, even documents, fossils, and the like, must be in the present in order to be available. Accordingly, all we know, or think to know, about the future is constructed from information available at present. Since this availability is subject to temporal change, the past and future change with the time being as well.

The events whose course historiography is reconstructing carry two dates: the date of their real occurrence and the date of their actual reconstruction. Events that carry two dates cannot be ordered unambiguously in one-dimensional time. A good deal of historiography consists in reviewing, criticising and correcting former historiography. Since historiography has no immediate access to the process it describes, the course of known history is epistemologically encapsulated in the evolution of historiography. This encapsulation means that known history is a process embedded into another process. The processes reconstructed and the process of reconstruction run in different times. An evolution consisting of different processes running in different times is inconceivable on a one-dimensional continuum of instants. Historiography is working with a concept of time that is implicitly defined as having more than one dimension. Since historiography is mainly narrative, relying on the grammar of tense rather than on formalisation, this heavy epistemological implication has rarely been accounted for (see, however, Salamander, 1982). Characteristically, however, the implicature comes to the fore when we try to incorporate history into computer-based geographical information systems (GIS) (Snodgrass, 1992).

In geography, the trade-off between formal sophistication and the scope of temporality accounted for becomes almost visible. There is a characteristic difference between historical geography and time geography. Historical geography is purely narrative. Time geography introduces time in a graphical manner. The concept of time geography is due to Torsten Hägerstrand and the school of Lund (see Hägerstrand, 1970). Typically, time geography visualises sections of space–time by projecting them into 2D + t-'space'. This graphical representation becomes possible by reducing the differences between past, present and future to differences in date. Time in geography is parameter time t, and nothing more than parameter time t. Time geography depicts the states a landscape and its inhabitants are passing through by arranging them explicitly as if in another spatial dimension.

Since time geography reduces temporality to strictly one-dimensional time, there is a sharp line separating time geography from economics. In economics, time may be treated as reversible, but it cannot be defined as strictly one-dimensional. *Homo oeconomicus* is a rational agent. As a rational agent, he or she is concerned not only with the present but also with the future. Since for humans the future exists only in the present imagination, the events which *homo oeconomicus* is concerned with carry two dates, too. The return on investment considered carries the date of the expected accrual as well as the date of the consideration performed. Moreover, the interval between the two dates has economic value. The further the date of the accrual lies in the future, the lower is the present value of the investment (other things being equal). This depreciation even has a market price: its name is the rate of interest.

Interest rates measure the market value that things lose by not being available at present. The rate of interest is the rate at which temporaPl distance is discounted. Discounting for temporal distance means to take goods and evils the less seriously the further they lie in the future. It means, to be specific, lessening the present value of a future good or evil *exponentially* with growing distance. Discounting at a rate of 10% means that the present value of a gain or loss expected after 20 years is only one-seventh, when expected after 50 years no more than $1/148^{th}$. Thus, discounting for distance in time has an enormous impact on what seems to be rational economically. It may even be questioned whether discounting for time is rational at all. It is discounting for time that makes it seem economically rational to deplete natural resources irreversibly. Only by lowering the effective discount rate massively will it be possible to turn our present-day economies into sustainable ones. See (Franck, 1992).

Interestingly, though not surprisingly, the theory of interest has been the most controversial part of theoretical economics. It is the part that suffers most from the restricted concept of time. Nevertheless, it shows something remarkable about the meaning that time has in the social context. It shows that time has at least two distinct meanings in social life. There is the time-preference valued by interest. Interest, however, is not the only market price that time has. There is another price; its name is wage. Wages measure the value of time used for labour. The value of the time valued by wages depends on its actual use. In contrast to this, the time valued by interest cannot be used actually since it is future and thus imaginary.

Thus, time has two different prices. For the capital and distribution theoretic background of this statement see (Franck, 1992, ch. 3). Wages measure the value of time worked, interest measures the value of time deferred. The longer the labourer works the more he or she is paid. The longer the time to be discounted, the lower is the present value of the good or bad in question. The value of time measured by wages is positive. The value of time measured by interest is negative. Time, as it is valued by wages, is a scarce resource, whereas time valued by interest is neither scarce nor can it be called a resource.

Something having two prices means something different in social life. Prices inform about people's preparedness to pay. In expressing their needs and wants through preparedness to pay, people tend to be truthful. It makes no sense to lie in this language. Hence, having different prices means that time practically differs in meaning. Being prepared to pay interest means being prepared to pay for not having to wait. Being prepared to pay wages means to be prepared to pay for services that consume other people's time. The time bridged by credits and loans is future time; the time consumed by the production of useful services is present time. Interest measures the market value of time on the imaginary axis; wages measure the market value of time on the real axis. The fact that time has both prices means that both the imaginary and the real axis of time play an actual role in social life.

8.4 NOVELTY

One of the most striking traits of present-day economic development is the dematerialisation and disembodiment that the economic process is experiencing. Information processing is going to outdo the processing of materials. Information processing differs from the processing of materials in its being nearly immaterial and in its being bound to strict irreversibility. Economically, information means novelty. The economic point of information processing is the *surprise value* it yields.

In the world of classic mechanics there is no such thing as novelty, let alone surprise. Anything to be learned follows from the initial conditions of the processes considered. In order to allow for something novel to happen, the dynamics has to become complex and non-linear. The non-linear dynamics of complex systems is a recent and rapidly growing field of research in physics. It is the dynamics of anisotropic, directed time. It allows accounting for instabilities and bifurcations and, thus, for things not yet contained in the initial conditions of the processes modelled. As promising as non-linear dynamics is in physics and biology, as hard it proves to be incorporated into economic theory, however. For a pioneering venture into the field see (Lorenz, 1993). 'Information economy' is a subject still awaiting treatment by mathematical economics.

The non-linearity of the processes involved is not the only difficulty the hypothetical theory of the information economy is facing. Economically, information is linked to actuality. In physics and biology, the novelty of things can be reduced to novel structures and functions. In economics, the novelty of things has to be turned into surprise value in order to count. Surprise is a phenomenon not reducible to structure and function. It does not make its appearance as long as the distinction between actuality and non-actuality is disregarded. If physics is right in contending that this distinction hinges on subjectivity, surprise is a phenomenon presupposing subjectivity accordingly. If surprise is a subjective phenomenon, the surprise value is subjective, too. Hence, it may well be that information, as it counts in economics, cannot be accounted for without appropriately accounting for subjectivity.

When looked at from this angle, it is no wonder that we are so far from a concept of information capable of being generally consented to. We agree on living in the age of information, but we disagree on how to define information. There are definitions, to be sure. The more precise they are, the more restrictive their scope is, however. The hierarchy of syntactic, semantic and pragmatic information is one of growing scope and diminishing precision in definition, again. We obtain a simular hierarchy when we classify the various definitions—or quasi-definitions, for that matter—according to the concept of time implied. At the level of parameter time, information means structural complexity. At the level of directed time, the definitions of information range from negentropy to what cognition is about. Finally, at the level of temporality, information becomes conceivable as surprise value. This hierarchy, too, is one of growing scope and diminishing precision. Structural complexity is precisely definable and, hence, operationable as algorithmic complexity. Negentropy has the same statistical measure as entropy. In order to define what cognition is about, second order statistics are needed already. Surprise value cannot be defined straightforwardly nor can it be measured directly. Surprise value is what satisfies curiosity; its measure is the subjective preparedness to pay, be it attention or money. It seems that information theory is itself subject to the trade-off between attainable formal sophistication and accountable temporality.

8.5 HISTORY

A general problem of the definition and measurement of information is the dependence on the context involved. Information is nothing intrinsic nor has it a measure of its own. Information is essentially relative; the reference of its measurement being some minimum or maximum with which the pattern can be compared. The reference minima and maxima themselves have unambiguous definitions in the simplest cases only. The minimum that algorithmic complexity refers to is the shortest algorithm that reproduces the pattern. The maximum to which negentropy is relative is maximum entropy. Both algorithmic complexity and negentropy are measures of syntactic information. Measurement of semantic information—if it is possible at all—involves cognition. On the level of cognition, the relation between reference complexity and measured complexity becomes recursive. In cognition, the information squeezed out of a stream of data depends on the structural information to which previous stages of the stream have been compressed. Cognition thus depends on memory. Because memory feeds on data processing on its own, the measure of information in cognition is second-order statistics.

The measure of pragmatic information consists of a combination of surprise and confirmation. Data flows consisting only of patterns never experienced before as well as data flows consisting of nothing but known patterns contain no useful information. The maximum of useful information lies somewhere between these extremes. The concept of memory implied in the notion of pragmatic information is dynamic. Surprise has the effect that memory changes, whereas confirmation reaffirms existing memory. In its higher forms, pragmatic information thus depends not only on memory but also on memory management. Without control of memory change, behaviour instructed by pragmatic information remains limited to simple adaptation. In cognitive terms, pragmatic information processing is limited to simple recognition as long as memory change is not controlled by semantic forms of discrimination.

Semantic forms of discrimination pertain to the interpretation of data, as *representing* something that is not contained in their pattern. In the computer paradigm, elementary discrimination of this kind is that between a constant and a variable. The interpretation of a data string as a constant means that it is taken as such, literally, while its interpretation as a variable means that it is taken as a pointer. On the level of human memory, another instance of such elementary discrimination comes into play. It pertains to the time represented by some informational content. It is the discrimination between actuality and non-actuality. This difference cannot be reduced to a more primitive one without giving up the notion of a memorable past of present experience.

Having a memorable past means more than being able to record, store and retrieve the informational content of experiences made earlier. It consists in the faculty of generating and maintaining a cognitive map depicting a lifetime and more. The consistency of a cognitive map depicting the history of current experience requires continuous update at its front edge as well as in its entirety. Pastness, presence, and futurity are dynamic properties. Each moment, a moment having been future until then becomes present in order to instantly disappear into the past. Each moment, the totality of moments that are still future move closer to the present. Each moment, the totality of moments already past recede further away from the present. Each moment, thus, a unique past and a unique future become actual. Since past and future exist only in actual recollection or anticipation, respectively, events that are past or future carry two dates: the date of their supposedly *real* occurrence and the date of their *actual* imagination.

As soon as past and future events show up in present awareness, the unambiguous chronological order of memorable and foreseeable happenings is lost. In order to re-establish an unambiguous order, a dating system making use of both the real and the imaginary coordinate must be introduced. The events surfacing in the present must be ordered according to the chronology of their real occurrence and according to the chronology of the re- or pre-actualisation as well. The totality of events making up the cognitive map of a memorable past and foreseeable future are thus ordered not in a linear but in a planar way. The cognitive map our history consists in is not one-dimensional but, at least, two-dimensional. For different lines of reasoning that lead to the same result see (Dobbs, 1972; Franck, 1989; Franck, 1994).

The continuous update of the cognitive map-making which our history consists in is what generates the perception that time goes by. In order to localise the place of the now in the order of the memorable and foreseeable events, one entry of the two-entry dating vectors must be continuously adjusted while the other one must explicitly be kept unchanged. Thus, the field constituting time perception is ceaselessly changing in one dimension while it remains unchanged in the other one. It is ceaselessly changing concerning the coordinate of the place where recollection and anticipation are actually possible. It remains unchanged concerning the coordinate of the place where the events that can be remembered or anticipated really happened.

A simple and illustrative proof of our capacity to handle two or even more dimensions of time is our faculty of episodic recollection. Episodic recollection means recollection not simply of facts or events, but of sequences of events that make up a process. Processes of more than negligible length cannot be remembered instantaneously. Episodic recollection is an activity, hence, that takes up time of its own. An activity consuming actual time in order to reproduce processes running in another time encompasses two processes running in different times at the same time. Even though the constituent processes run in different time, the activity of episodic recollection does not lose its identity. An activity, however, that consists of processes running in different times at the same time amounts to patent nonsense—unless the time it occupies allows for more than a single degree of freedom. An additional degree of temporal freedom is tantamount to another dimension of time.

If we go further and try to recollect an episodic recollection itself, we can see that our capacity to handle higher dimensions of time is not limited logically but only energetically. It is very hard to nest processes imaginatively. Imagining a process that is embedded into another process demands effort and training. It is not limited to the first stage, however. Even the thought by which we make clear to ourselves what episodic recollection means consumes time. The thought of an activity, however, that consists in nesting an imagined episode in a process of active imagination is a phenomenon encompassing already three processes, each of which is running in different time. Of course, this thought represents twofold nesting only formally. Probably nobody is capable of the mental acrobatics needed to perform the nesting in full detail.

8.6 CONCLUSION

We are back to the question of how it is conceivable that time goes by. We see that this question has two parts. The first part concerns the very existence of the now and the cause of its travelling. The answer to this first part cannot be given at present. It has to await further progresses in the physical and life sciences. The second part of the question concerns the translation of the experience of time's flowing into consistent thought. The answer to this second part of the question consists in showing which seemingly self-understood habit of thought must be given up in order to make the specious present and the speed with which time flows conceivable. This habit of thought has been identified by now. It is the seemingly self-understood supposition that time has one and only one dimension.

As soon as a second dimension of time is introduced, the phenomena of the specious present and the speed of its travelling lose their perplexity. Since past and future are extended in an 'imaginary' dimension, the specious present may be extended in that dimension as well. As soon as events become possible that carry two dates without losing their identity, processes running in different times at the same time consequently become possible. With processes running in different times at the same time, the speed with which time passes becomes definable. The speed with which time passes becomes a regular kind of speed if it is measured in length of imaginary remembered time divided by real time taken up by the remembrance, using suitable units for measurement on both axes. In fact, the impression that time passes at a certain speed appears where a comparison is made between a sequence of remembered events and the sequence of the acts of performing the remembrance.

Disregarding a whole dimension is a very powerful means of abstraction. If it is correct that time, as we subjectively experience it, occupies more than one dimension, it becomes clear why the natural sciences are so successful in abstracting from subjectivity. In this case, the world is consistently and precisely reduced to physical reality when time is reduced to clock time. Without accounting for a higher dimension of time, unresolvable problems should arise, however, when conceptualising phenomena such as subjectivity, actuality and history. Accordingly, the logical foundation and methodological clarification of the social sciences should remain as dubious as it is at present, unless a higher dimension of time is explicitly introduced. Even the self-account of science should run into serious trouble if the dimensionality of temporal time is not appropriately considered. Science is reductionist in its methods, to be sure. The theory of science cannot abstract from temporality, however, as soon as some notion of the accumulation of knowledge, of paradigmatic shifts or of scientific revolutions becomes unavoidable. If temporality exhibits a higher dimension, the conflict between the logic of discovery and the psychology of research, the dispute between the theory and the history of sciences should persist until the time gestalt of the advancement of knowledge is reconstructed adequately.

Founding the historical sciences is not the only way of testing the hypothesis that historical time has more than a single dimension. There are more immediate tests as well. First, there is the famous proof of the 'unreality of time' by McTaggart (1908). In this proof, he pretends to demonstrate that temporal time cannot exist for logical reasons. He shows that it is impossible to account for the spontaneous change of the truth values of predicates like 'is present' or 'is future' within a conceptual framework of strictly one-dimensional time. In this framework, he insists, temporal time cannot be defined consistently. On the basis that (1) time is one-dimensional and that (2) inconsistent concepts cannot denote something real, he 'proves' that temporal time is an illusion. Contrary to various claims to the opposite, this proof has not been refuted conclusively until today (see Franck, 1989, 1994). It leaves, thus, the question open whether the changing truth-values become controllable in a framework accounting for a higher dimension of time.

A second way of testing the hypothesis that temporal has more than a single dimension consists in measuring the fractal dimension of phenomena such as musical understanding. Musical understanding asks for representing the time 'gestalt' of the whole piece of music in actual time. The time gestalt of a sonata or a fugue exhibits a cascade of self-simular structures the fractal dimension of which can be measured by analysing the score of the phonographic record (Hsü, 1993; Vrobel, 1997). As soon as this fractal dimension proves to be higher than 1, the actual understanding implies the representation of a time gestalt that cannot be accounted for in simple clock time.

Finally, there is a quasi-empirical test of the hypothesis that can be performed by programming computers. In the case that historical time deploys another dimension, any piece of software enabling a machine to deal consistently with historical objects or to use conclusively the grammar of tense should contain a data model accounting for the higher dimension (Franck, 1990). It should not be possible to program a machine to do as if it had a recollectable past and a foreseeable future without implementing a temporal perspective in the data model that makes use of more than a single degree of freedom. In order to disprove the hypothesis, nothing more is needed than programming these capabilities without implementing a planar possibility locus of time.

ACKNOWLEDGEMENTS

For helpful comments and valuable suggestions, I am indebted to Karl Svozil. I also want to express my thanks to Silvia Plaza for stylistic support.

REFERENCES

Carnap, R., 1963, Intellectual autobiography. In *The Philosophy of Rudolf Carnap*, edited by Schilpp, P.A. (La Salle, Ill: Open Court).

Dobbs, H.A.C., 1972, The dimensions of the sensible present. In *The Study of Time*, edited by Fraser, J.T., Haber, F.C. and Müller, G.H. (Berlin: Springer-Verlag).

Einstein, A. and Besso, M., 1972, *Correspondence 1903–1955*, (Paris: Hermann).

Franck, G., 1989, Das Paradox der Zeit und die Dimensionszahl der Temporalität. *Zeitschrift für Philosophische Forschung*, **43** (3).

Franck, G., 1990, Virtual time—Can subjective time be programmed? In *Ars Electronica*, edited by Hattinger, G., Russel, M., Schöpf, Ch. and Weibel, P. (Linz), Vol. 2.

Franck, G., 1992, *Raumökonomie, Stadtentwicklung und Umweltpolitik*, (Stuttgart: Kohlhammer).

Franck, G., 1994, Physical time and intrinsic temporality. In *Inside Versus Outside. Endo- and Exo-Concepts of Observation and Knowledge in Physics, Philosophy, and Cognitive Science*, edited by Atmanspacher, H. and Dalenoort, G.J. (Berlin: Springer-Verlag).

Georgescu-Roegen, N., 1971, *The Entropy Law and the Economic Process*, (Cambridge: Harvard University Press).

Gödel, K., 1949, A remark about the relationship between relativistic theory and idealistic philosophy. In *Albert Einstein, Philosopher–Scientist*, edited by Schilpp, P.A. (La Salle, Ill: Open Court).
Hägerstrand, T., 1970, What about people in regional science? In *Papers and Proceedings of the Regional Science Association*, **24**.
Hsü, K.J., 1993, Fractal geometry of music: From bird songs to Bach. In *Applications of Fractals and Chaos*, edited by Crilly, A.J., Earnshaw, R.A. and Jones, H. (New York: Springer-Verlag).
James, W., 1890, *The Principles of Psychology*, (New York: Henry Holt).
Loasby, B.J., 1976, *Choice, Complexity, and Ignorance*, (Cambridge: Cambridge University Press).
Lorenz, H.-W., 1993, *Nonlinear Dynamical Economics and Chaotic Motion*, (Berlin: Springer-Verlag).
Mach, E., 1883, *The Science of Mechanics: A Critical and Historical Account of its Development*, tr. by McCormack, T.J. (Chicago: Open Court).
McTaggart, J.E., 1908, The unreality of time. In *Mind*, New Series, no. 68.
McTaggart, J.E., 1927, The Nature of Existence, 2 Vols., edited by Broad, C.D. (Cambridge: Cambridge University Press).
Newton, Sir I., 1687, *Principia*, tr. by Cajori, F., 1966, (Berkeley, University of California Press), Vol. 1, p. 6.
Nicolis, G. and Prigogine, I., 1989, *Exploring Complexity. An Introduction*, (New York: Freeman).
Pöppel, E., 1989, Taxonomy of the subjective: An evolutionary perspective. In *Neuropsychology of Visual Perception*, edited by Brown, J.W. (Hillsdale, NJ: Erlbaum).
Price, H., 1996, *Time's Arrow and Archimedes' Point: New Directions for the Physics of Time*, (Oxford: Oxford University Press).
Russell, B., 1903, *The Principles of Mathematics*, (Cambridge: Cambridge University Press).
Salamander, R., 1982, *Zeitliche Mehrdimensionalität als Grundbedingung des Sinnverstehens*, (Bern: Peter Lang).
Shackle, G.L.S., 1958, *Time in Economics*, (Westport, Conn: Greenwood Press).
Shackle, G.L.S., 1990, *Time, Expectations and Uncertainty in Economics: Selected Essays*, edited by Ford, J.L. (Aldershot: Edward Elgar).
Snodgrass, R.T., 1992, Temporal databases. In *Theories and Methods of Spatio-Temporal Reasoning in Geographic Space*, Lecture Notes in Computer Science 639, edited by Frank, A.U., Campari, I. and Formentini, U. (Berlin: Springer-Verlag).
Vickers, D., 1994, *Economics and the Antagonism of Time*, (Ann Arbor: University of Michigan Press).
Vrobel, S., 1997, *Fractal Time*, (Houston: The Institute for Advanced Interdisciplinary Research).

PART THREE

Databases for Temporal GIS

INTRODUCTION

This third part of the book turns gradually from the abstract conceptual questions raised by the initial, more philosophical chapters to the practical implementations of databases for temporal GIS. This part moves from the conceptual to the technical, pointing out why the solutions to the questions are so difficult with today's technology. It becomes quickly evident that database research treats much more complex questions than simple 'storage and retrieval' of data. Database research connects aspects of conceptual modelling (Brodie *et al.*, 1984) with actual implementation and shows how all these considerations are overshadowed by performance issues. It is also concerned how to structure data clearly and still achieve an acceptable performance, even if very large data collections must be dealt with, as is customary for GIS.

The first chapter by Michael Worboys explores the various possibilities. It continues the questions started by Smith and Casati in the previous part, and discusses the nature of objects, changes and events. It also considers what constitutes a coherent set of assumptions and reviews the terminology usual in computer science, especially extensions of the Entity–Relationship Model (Chen, 1976) and object-oriented approaches (Unified Modeling Language, Eriksson and Penker, 1998). The contrast here is more with the terminology than the contents of the philosophy based discussion in the previous part, but it also leads to a transition of very abstract ideas to the concrete problems of managing information about spatial socio-economic units.

Worboys starts a classification of change, linking it to the image schemata listed by Johnson (1987) as related to force and change. This point was already raised in the first Part 'Setting the Stage' by Frank (Chapter 2) and by Cheylan (Chapter 3) when considering the change of objects and is addressed with more detail in the following chapter in this part by Damir Medak. He investigates the possible changes to objects; their separation in parts or the merging of previously independent objects to a single new one are fundamental changes and directly related to the ontological commitments proposed by Smith in the second part. The chapter attempts a formal ontology of objects of the type described by Smith, using an algebraic method to define the semantics of the concepts as clearly as possible. It selects the simplest notion of time—simple ordering of events—and it attempts only an ontological account and does not include the epistemological aspects discussed by Franck. Similarly, the objects only have an identity, which allows them to be traced throughout history, but no spatial or other attributes or changes of these are considered. Medak concentrates on the events that lead to the creation of new objects or the disappearance of existing ones:

- create and destroy, with kill, reincarnate and evolve,
- identify and spawn,
- aggregation and disaggregation,
- fusion and fission.

The chapter suggests that these operations are fundamental and cover all relevant cases. They can be called lifestyles, as certain objects do follow some of them and others are not applicable: for liquids, for example, fusion and fission are applicable, but not aggregation and disaggregation. Ownership parcels of land follow the same lifestyle as liquids, but for other spatial socio-economic units the aggregation and disaggregation lifestyle—the lifestyle of assembled objects—is more appropriate.

This formalisation reflects some of the issues pointed out by Franck: for example, the flow of time is reflected in a succession of snapshots—in this intentionally simple case there is no need to measure time beyond an order of events. Due to the formalisation, the important difference between aggregates—where the process of aggregation is reversible—and liquids—where the fusion of two liquids cannot be undone—, becomes clear and can be carried forward to spatial socio-economic units. What results is that administrative units, especially if considered for statistical purposes, do behave like aggregates, as they can be disaggregated. Ownership parcels as registered in a cadastre, however, are a complex of physical parts (the land) and legal rights (for example, mortgages). The law does typically not provide for easy dissolution of rights once merged (for example, divorce, the undoing of a marriage, is very difficult and does not lead to the initial state before marriage). The law sees rights—metaphorically—as liquids and treats parcels, because they are bundles of rights, as liquids. By the same token, if administrative units are seen as holders of rights, then mergers cannot simply be undone (for example, the treaty establishing the European Union does not even have provisions for its dissolution!).

The third chapter in this part considers changes from a purely database perspective. Therese Libourel follows the customary separation of information about the world in schema information (describing the methods used to form concepts of the world), and the actual data (describing the individuals). Correspondingly, she classifies changes in 'changes in the schema' and also data changes. Changing values are what databases are about and the temporal aspects of changes can be handled in temporal databases. However, how to treat schema changes is an eminent, new problem, which requires a theoretical and a practical solution for temporal GIS urgently. One must not assume that only the values describing data are changing, but also that the concepts applied are changing as well—perhaps less often—but nevertheless sometimes (see later Chapter 22 by Ryssevik). For example, a change in an administrative law very often implies a change in the conceptualisation of the objects (Chapter 18 by Stubkjær): additional properties become relevant, the coding of properties changes or a class is subdivided into several more specialised ones. Such changes affect the framework in which the stored data are interpreted and have therefore more consequences for a database. Libourel differentiates between cases where the stored data evolve to the new schema and solutions that preserve previous states of the database schema. In both cases some links between the semantics in the initial schema and in the later schema are necessary—they may either be 'on the fly conversion' when the data are accessed, or a systematic change of all the data.

To further the discussion, the fourth chapter by Emmanuel Stephanakis and Timos Sellis in this part focuses on the implementation of a temporal GIS. It contrasts the requirements of an application based on spatial socio-economic units—the spread of a fire—with queries that could be asked of a database of ship positions. From these requirements the lack of operations in a DBMS to treat temporal data becomes a gaping hole. For spatial data no consistent and widely accepted algebra is available, even though standards are emerging, both for a space (raster) and spatial objects (vector) data model (OGC, 1996). The chapter starts with Tomlin's map algebra (Tomlin, 1983) and extends it with constructs and operations to deal with the temporal aspects. A temporal overlay operation and temporal distance operations are defined. The distinction between events and intervals, previously discussed by Worboys, is now used to characterise temporal database schema languages. One distinguishes the time a fact becomes valid and the (generally different) time, the new fact is entered into the database (so-called transaction time). The implementation of data storage may time stamp the whole database (the solutions used for the conceptual discussion by Medak but for practical implementation not recommended) or may time stamp any change in an object, or may only time stamp each value describing a property of an object. From a logical point of view these representations are equivalent, but not for the actual performance of a query and also for the amount of storage required.

REFERENCES

Brodie, M.L., Mylopoulos, J. and Schmidt, J., 1984, *On Conceptual Modelling: Perspectives from Artificial Intelligence, Databases, and Programming Languages*, (Springer-Verlag).

Chen, P.P.-S., 1976, The Entity–Relationship Model—Toward a unified view of data. *TODS*, **1** (1), pp. 9–36.

Eriksson, H.-P. and Penker, M., 1998, *UML Toolkit*, (New York: John Wiley).

Johnson, M., 1987, *The Body in the Mind: The Bodily Basis of Meaning, Imagination, and Reason*, (Chicago: University of Chicago Press).

OGC, 1996, *The OpenGIS Abstract Specification: An Object Model for Interoperable Geoprocessing*, (Wayland, MA: Open GIS Consortium).

Tomlin, C.D., 1983, A map algebra. In *Proceedings of the Harvard Computer Graphics Conference*, (Cambridge, MA).

CHAPTER NINE

Modelling Changes and Events in Dynamic Spatial Systems with Reference to Socio-Economic Units

Michael F. Worboys

9.1 INTRODUCTION

The large majority of current systems for handling geospatial information are static, concentrating on a single temporal snapshot, usually the current state. Changes in the application domain are tracked in the system by performing updates and erasing information on the past. In recent years there has evolved a body of research, both in the general database community (Snodgrass, 1992) and in the spatial database community (Al-Taha *et al.*, 1993) for adding temporal dimensions. That research addresses the issue of 'time in the system', where the challenge is to provide computational models that enable past, current and future states of the application domain (valid time) and the system (transaction time) to be handled in the temporal database.

Work presented in this chapter, however, is concerned with a different aspect of temporal systems, referred to as 'the system in time', where we are concerned to handle in a dynamic system a model of the real world as it changes in both the spatial and temporal dimensions. Such dynamic, spatial systems will have a wide range of applications, including transportation networks and environmental monitoring. They have been termed *responsive GIS* (Williams, 1995), with the following characteristics:

a. large amounts of spatially referenced data are required to be readily available;
b. sizeable proportion of the data is regularly updated from external data sources;
c. some of the data are noisy, conflicting and incomplete;
d. rapid decision-making is to be supported.

The purpose of this chapter is to explore modelling constructs that may be useful in constructing dynamic or responsive spatial systems. At first sight, it seems that the notion of time is key here, since by adding temporality to our systems (rather in the same way that we added space) it would appear that we then provide the required functionality. However, time is merely a framework in which change is possible. As expressed by Shoham and Goyal (1988), "The passage of time is important only because *changes* are possible with time. ... The concept of time would become meaningless in a world where no changes were possible." The principal thesis that is explored in this chapter is that it is not *time* that is the key to conceptual modelling of dynamic systems, but *change* and related constructs such as event and process.

The object-oriented approach has provided a large amount of power to modelling and implementing complex spatial systems, so might naturally be thought to be applied in the context of dynamic spatial systems. However, we immediately run into some deep philosophical questions, like "Is an event an object?" for which a positive answer would open up the entire object-oriented approach. The chapter attempts to carefully consider the fundamental constructs of change, event and process, placing the discussion in a spatial context, and concludes by using some of the ideas to propose a base model for dynamic spatial systems. Our principal application is the main theme of the book, namely socio-economic units (SEUs). The chapter includes a classification of change as is applied to SEUs. We begin with a brief review of the object-oriented approach to geo-spatial phenomena.

9.2 MODELLING GEOSPATIAL OBJECTS

In this section, for the sake of completeness, we summarize the basic concepts of the object-oriented approach from the conceptual modelling perspective. Objects model abstract and concrete things in the world. Objects have *attributes*, describing aspects of their state (for example, name of a person) and *operators*, describing ways in which they may behave (for example, birth, marriage, death of a person). Objects with the same structure of attributes and operators may be gathered together in an *object class*, of which each member is an *instance* (for example, object: me, class: person). Each object has a unique *identity*, which is immutable, even if its state and behaviour changes.

Object classes may be structured into hierarchies in two different ways. Firstly, an object class *A* may be a *subclass* of class *B* if every member of *A* is a member of *B*. Thus the class *woman* is a subclass of *person*. Note that a subclass may not only inherit state and behaviour from its superclass, but also have its own individual state and behaviour. The hierarchy of classes structured in this way is called the *inheritance* hierarchy. A second relationship between classes is that of *composition*. For example, class *car* is composed of classes *body*, *wheels*, *engine*, etc. Such a hierarchy of classes may be termed a *composition* hierarchy.

The object-oriented approach is a natural modelling method for geospatial information, because the underlying spatial entities have a complex structure and composition. Spatial object classes have been modelled by several researchers (for example, Egenhofer and Frank, 1987, 1992; Worboys, 1992, 1994).

Socio-economic units are discussed elsewhere in the book. For our purposes, they have the following properties:

a. SEUs can be modelled as objects,
b. an SEU has an attribute that is an aggregate of people,
c. an SEU has a spatial location attribute that is of class *region*,
d. SEUs may change over time.

The location attributes of a collection of SEUs, at an instance of time, combine to form a 'coverage' of a spatial region.

9.3 SPATIAL CHANGE

9.3.1 What is Change?

In the object model, the focus is the collection of object classes. Although objects in the classes may change, the changes themselves are not explicitly handled but are implicit in the variations in the properties of the objects. A fine but important distinction to be made is between change and the results or effects of change.

We may contrast two distinct definitions of change:

Change—Definition 1

An object o changes if and only if there exists a property P of o and distinct times t and t' such that o has property P at t and o does not have property P at t'.

Change—Definition 2

A change occurs if and only if there exists a proposition Π and distinct times t and t' such that Π is true at t but false at t'.

The first definition of change can be traced back to the ancient Greek philosophers (for example, Aristotle, *Physics*, Book 1, Chapter 5). In this case, change is a verb that applies to objects. The focus is on the objects. The second definition (Russell, 1903) makes the notion of change explicit as a noun; it commits to change as an explicit occurrence in the world, without prior reference to objects.

Pursuing definition 1 a little further, there are two distinct ways in which an object may change. Consider a land parcel; it may change ownership or its boundary. In the case of change of boundary, the land parcel *alters* in the sense that its attributes change. With change of ownership, the object itself does not alter, but its relationship to other objects changes. So it seems that we should extend definition 1 to at least allow changes to relationships between objects.

9.3.2 Classification of Change

There have been several attempts to classify change (see, for example, Shoham, 1987). Mark Johnson (1987) defines image schemata of change in terms of 'force' structures as follows:

a. compulsion: stability, diffusion;
b. blockage: disappearance, split;
c. counterforce: disappearance;
d. diversion: transformation, split;
e. removal of restraint: appearance, stability, transformation, union;
f. enablement (potential forces).

A commonly occurring construct associated with change is that of *production*, where one or more objects contribute to the creation of a new set of objects or transmit properties to an existing collection of objects. There are several possibilities:

- Production and death, where the parent and child objects are of the same type. An example is the merging of several land parcels to form a new parcel. The new parcel is a composite of the original parcels. In this case the original parcels cease to exist.
- Production and death, where the parent and child objects are of different types. For example, a building collapses, producing a pile of bricks. The pile of bricks is a decomposition of the building into component parts, of different type from the original building.
- Production and continuation, where the parent and child objects are of the same type. For example, two parents (re)produce a child. Note that there is limited use of the composition construct here, since the parents cannot be thought of as components of the child.
- Production and continuation, where the parent and child objects are of different types. For a first example, consider states associating into a union. The union is a composition of the states. Both states and union continue in existence. The

states continue to have a relationship with the union. Thus, if a state ceases to exist, so the union changes or may itself cease to exist. The second example is a ship having a spillage and creating a pool of oil. The pool of oil is a new type created from a component of the composite object–ship. Eventually, the pool of oil ceases to have a relationship with the ship. The ship may sink but the pool is still present. Example 3 is people constructing a building. In this case, people are agents in the construction event, but have little connection themselves with the final type. The building eventually has little relationship to the constructors; it continues to exist even if the people cease to exist.

9.4 EVENTS AND PROCESSES

9.4.1 Events

An event is maybe the key notion in the modelling of dynamic systems. A dictionary definition is 'a happening, occurrence or episode'. In terms of change, an event has been taken to be a change or composite of changes (it may be absence of change, for example, "the lawn stays wet", "I remained still for five minutes"). Some writers define an event to be instantaneous while others allow it to have a duration. As with objects, we should take care to distinguish event type from event occurrence, for example, horse race is the type and the 1996 Grand National is the occurrence.

There have been several attempts to extend traditional modelling methods, such as the entity–relationship (ER) approach, to handle events. The entity–relationship approach has proved to be enormously successful in modelling fairly simple static systems, but has the limitation that only entities, attributes and relationships are provided as primitive constructs. Thus ER will only be successful if events can be considered as entities or relationships. The attempt to introduce temporal relationships to stand for events results in methods such as ERT (van Assche *et al.*, 1988), ERAE (Dubois *et al.*, 1986), EVORM (Proper *et al.*, 1995) and dynamic modelling (Rumbaugh *et al.*, 1991).

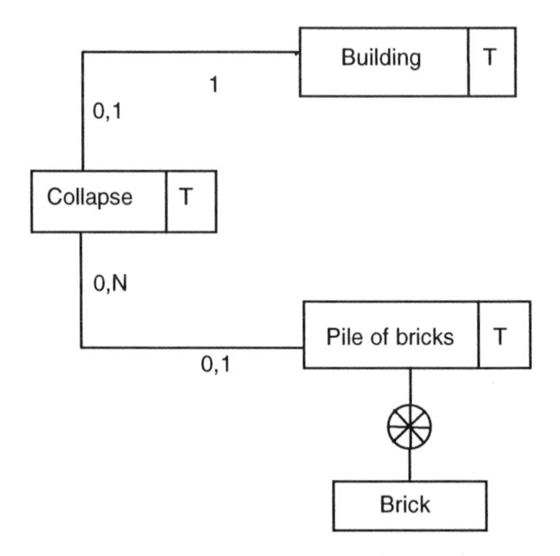

Figure 9.1 ERT model of the collapse of a building producing a pile of bricks

ERT (van Assche *et al.*, 1988) is an extended version of ER that includes inheritance and composition constructs, as well as the capability to timestamp entities and relationships. It does not support schema evolution. An example of its notation is shown in Figure 9.1, modelling the change above, where a building collapses, producing a pile of bricks.

ERAE (Dubois *et al.*, 1986) has constructs entity, relationship, attribute and event. The modelling approach is predicated on the Newtonian view that space is a container for objects and time is a container for occurrences. An event is defined as 'an instantaneous happening of interest'. Each event has an associated function returning a time, while each entity has an associated existence predicate. Figure 9.2 shows the collapse of the building modelled using ERAE.

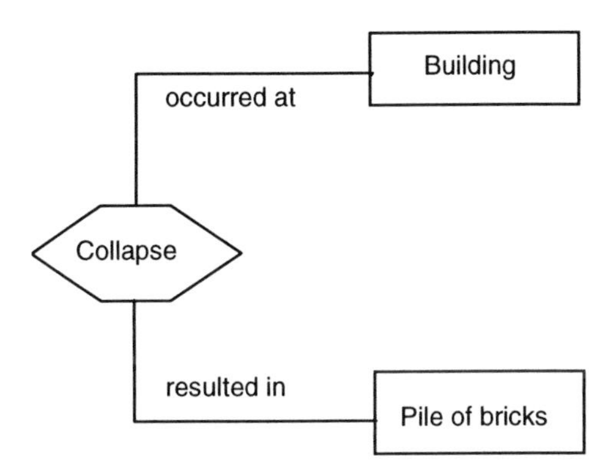

Figure 9.2 ERAE model of the collapse of a building producing a pile of bricks

The EVolving Objects Relationships and Methods (EVORM) approach (Proper *et al.*, 1995) allows models of evolving application domains. The schema itself is allowed to change. Based on ER, its primitives are: evolvable elements (object types, relationship types, operations...), actions (occurring as events), and application model histories.

In the dynamic modelling of Rumbaugh *et al.*, (1991), an event is viewed as an instantaneous one-way transmission of information from one object to another. A sequence of events is defined to be a *scenario*. State and event are seen as dual concepts, where a state separates events and an event separates states. The relationship between states and events in a model is represented by a transition diagram. For Rumbaugh, dynamic modelling consists in identifying scenarios of events and constructing a transition diagram.

9.4.2 Events and the Object-Oriented Approach

This section considers how events fit with the object-oriented approach. There could be a very close coupling if we were able to treat events as objects. Objects are items that:

a. possess properties,
b. stand in relation to one another,
c. undergo changes which constitute events,
d. possess determinate and objective identity conditions,
e. take their place in predicates.

The distinction between events/changes and things/objects is problematic. It seems that events and objects belong to distinct categories. Events *occur*, but objects do not occur. As above (ERAE), space is a container for objects but time is a container for occurrences. Events/changes are reportable. The whole of an object is usually present at any time in its existence, while usually only part of an event is present at any one time. An alternative view (Broad, 1959) is that 'a thing... is simply a long event'. Conversely, many philosophers have reduced objects to a series of events.

Closely related is the question of event identity. Two intensional definitions have been provided by Davidson (1980):

1. Events are identical when they have exactly the same causes and effects.
2. Events are identical when they occupy the same space and time.

An extensional definition is the following, due to Kim (1969), that events are identical when they consist in the same objects having the same properties at the same times. Unfortunately, this makes no provision for changes in objects or relationships between objects. We should note the changes in the objects (or objects before and after the event). We should also consider changes in the relationships between the objects and in the inheritance and composition hierarchies. Therefore, an event (type) may be defined by its time-varying collection of participating objects (types) and their time-varying attributes (types) and relationships.

9.4.3 The Concept of Process

A process is a series of changes with some sort of unity or unifying principle. Process is to change/event as syndrome is to symptom. In order to model process, it is useful to view a process as a composite event. Processes are a setting in which events occur and are related to each other. The next section discusses this setting in more detail.

9.5 MODELLING DYNAMIC SYSTEMS: AN OBJECT-ORIENTED MODEL OF PROCESS, EVENT AND CHANGE

Any object-oriented model of process, event and change needs to frame these concepts in an object-oriented setting. Thus, for the concept of event, for example, are the following constructs appropriate?

a. identity,
b. attributes,
c. relationships,
d. inheritance hierarchy,
e. composition hierarchy.

Event identity has been considered in the previous section. It makes sense for events to have attributes, for example, "our house move (event) took two days (attribute)". The granularity of an event will be one of its attributes. Similarly relationships between events can obtain, for example, "the first war (event 1) ended 21 years before (relationship) the second war (event 2)". An inheritance hierarchy of events may be constructed, for example, the event type 'war' is a subtype of the event type 'European war'. Event composition is a natural construct, where temporally composite events become processes if there is an underlying thread. Other kinds of composition are also possible, for example, spatial composition.

We explore the possibility of relationships between events in more detail. A broad subdivision for relationship types is temporal, spatial and causal/control. Under the heading of temporal come relationships such as concurrency, the Allen relationships, and the notion of temporal relativism (Knight and Ma, 1994). Events may be spatially related, for example, "their weddings took place one mile from each other", although this example stretches the notion of event and raises the question whether 'pure' events have spatial attributes/relationships. Causal/control relationships, such as independence and synchronization, are important. Causality is a powerful notion in this context. A useful definition is that of 'causally before' (Reiter and Gong, 1995). Event e is causally before event f if

a. e happened before f
b. e could possibly have affected the occurrence of f

In object-oriented terms, what could it mean for one event to cause another? Suppose that events have methods **send** and **receive**, allowing them to communicate with each other. Define the one-step causality relation between events e and f as the smallest relation satisfying the properties:

a. if events e and f occur simultaneously at the same process, then $e \Rightarrow f$
b. for any m **send**$(m) \Rightarrow$ **receive**(m)

A question that we leave this section asking is, what is the relationship between agents and events?

9.6 A RESPONSIVE SYSTEM FOR SOCIO-ECONOMIC UNITS

A system that can handle information about SEUs must be both responsive to change (dynamic) and be able to handle current and previous states (temporal). Change may impact both individual SEUs and the whole coverage, including the hierarchy in which the SEUs may be structured. Some of the salient properties of change in this context are:

- stability of the results of change;
- rate of change;
- reversibility of the effects of change;
- continuity vs. discreteness of the events and processes underlying the change;
- magnitude of change.

Changes to individual SEU may be changes to their geospatial attributes (topological and metric) or to non-geospatial attributes. In order to give some idea of the range of changes possible, we list some topological geospatial changes below:

- *changes to interior of SEU*: cohesiveness, internal structure;
- *changes to boundary of SEU*: geometry, permeability, fuzziness, one-way vs. two-way, recognizability, salience, dissonance;

Examples of changes to SEU coverages include:

- creation, modification, and deletion of individual SEUs in the collection;
- changes in the geometry of the coverage.

Examples of changes to a hierarchy of SEUs include:

- introduction and elimination of levels of the hierarchy;
- movement of SEUs between levels;
- merging and splitting of levels;
- changes of linkages between levels.

9.7 CONCLUSIONS

This chapter has considered the notions of change, event and process and compared them with more familiar modelling constructs such as object, entity, identity, attribute and relationship. The main thesis is that to model dynamic systems, we must include extra concepts from those present in temporal information systems. In some ways this is similar to the distinction between temporal data management vs. temporal reasoning. There is a need for temporal data management in the handling of inventories of objects as they evolve through time. But this is only part of the story. Temporal reasoning includes:

- prediction;
- explanation (we know: how things were, rules of change; we want: explanation of why things are as they are);
- planning;
- finding the rules of change.

In order to be able to tackle these higher-order tasks, a richer model of change must be used. This model must then be comprehensive enough to be able to handle some of the aspects of change of socio-economic units discussed here.

REFERENCES

Al-Taha, K.K., Snodgrass, R.T. and Soo, M.D., 1993, Bibliography on spatiotemporal databases. *ACMS SIGMOD Record*, **22**, pp. 59–67, and *International Journal of Geographical Information Systems*, **8**, pp. 95–103.

Assche, F. van, Layzell, P.J., Loucopoulos, P. and Spelltinex, G., 1988, Information systems: a rule-based approach. *Journal of Knowledge Based Systems*, **1**.

Broad, C.D., 1959, *Scientific Thought*, (Paterson, NJ: Littlefield).

Davidson, D., 1980, *Essays on Actions and Events*, (Oxford: Clarendon Press).

Dubois, E., Hagelstein, J., Lahou, E., Ponsaert, P., Rifaut, A. and Williams, F., 1986, The ERAE model: A case study. In *Information System Design Methodologies: Improving the Practice*, edited by Sol, H.G., Olle, T.W. and Verrijn-Stuart, A.A. (Elsevier, North-Holland).

Egenhofer, M.J. and Frank, A.U., 1987, Object-oriented databases: Database requirements for GIS. In *Proceedings of the International GIS Symposium: The Research Agenda*, (Washington, DC: U.S. Government Printing Office), **2**, pp. 189–211.

Egenhofer, M.J. and Frank, A.U., 1992, Object-oriented modeling for GIS. *Journal of the Urban and Regional Information Systems Association*, **4**, pp. 3–19.

Johnson, M., 1987, *The Body in the Mind*, (Chicago: University of Chicago Press).

Knight, B. and Ma, J., 1994, A temporal database model supporting relative and absolute time. *Computer Journal*, **37**, pp. 588–597.

Proper, H.A. and van der Weide, T.P., 1995, A general theory for evolving application models. *IEEE Transactions on Knowledge and Data Engineering*, **7**, pp. 984–996.

Reiter, M. and Gong, L., 1995, Securing causal relationships in distributed systems. *Computer Journal*, **38**, pp. 633–642.

Rumbaugh, J., Blaha, M., Premerlani, W., Eddy, F. and Lorensen, W., 1991, *Object-Oriented Modeling and Design*, (Englewood Cliffs, NJ: Prentice-Hall).

Russell, B., 1903, *Principia Mathematica*, (Cambridge: Cambridge University Press).

Shoham, Y., 1987, *Reasoning about Change*, (Boston: MIT Press).

Shoham, Y. and Goyal, N., 1988, Temporal reasoning in artificial intelligence. In *Exploring Artificial Intelligence*, edited by Shrobe, H. (Los Altos, CA: Morgan Kaufmann), pp. 419–438.

Snodgrass, R.T., 1992, Temporal databases. In *Theories and Methods of Spatio-Temporal Reasoning in Geographic Space, International Conference GIS,* Pisa, edited by Frank, A.U., Campari, I. and Formentini, U. (Berlin: Springer-Verlag), Lecture Notes in Computer Science 639, pp. 22–64.

Williams, G.J., 1995, Templates for spatial reasoning in responsive geographical information systems. *International Journal of Geographical Information Systems*, **2**, pp. 117–131.

Worboys, M.F., 1992, A generic model for planar geographic objects. *International Journal of Geographical Information Systems*, **6**, pp. 353–372.

Worboys, M.F., 1994, Object-oriented approaches to geo-referenced information. *International Journal of Geographical Information Systems*, **4**, pp. 385–399.

CHAPTER TEN

Lifestyles

Damir Medak

10.1 INTRODUCTION

Recent developments in GIS show the lack of tools for managing temporal information. The research of the temporal domain in GIS has been intensified during the last 10 years with the growing need for spatial-temporal databases (Egenhofer and Golledge, 1998). Different applications lead to different interpretation of temporal components. The application of spatial-temporal databases on GIS and other disciplines is hindered by the lack of a well-understood ontology of time.

The ontology of time is just a part of the greater problem that affects Artificial Intelligence research since its foundation: the search for an ontology of common-sense knowledge. The simple collection of rules enabling a robot to act as an intelligent thing in everyday environment outruns by far both the hardware capacities and scientific resources available (Hobbs and Moore, 1985). The physical world appears too complicated to be comprehended and researchers turned to their particular simplifications—the so-called 'naïve' sciences emerged. It was Pat Hayes (1985) who described simple concepts of physics using first-order logic for formalization. A similar approach was recently proposed for geography and GIS area by Egenhofer and Mark (1995).

Research about temporal databases concentrates on the structural and representational issues of time-domain (Langran, 1989). Many temporal query languages that have been proposed are summarized in (Snodgrass, 1992). Since relational database management systems were dominant, respective temporal languages were developed as extensions of standard query language (SQL). The majority of the research is concerned about metrics of time, its interval vs. point representation, i.e., quantitative time. Nevertheless, qualitative approaches emerge in both spatial and temporal GIS. Ordered time scales are analyzed and formalized in (Frank, 1994), while visual descriptive language for qualitative representation of change is proposed by Hornsby and Egenhofer (1997). Galton (1995) focuses on qualitative theory of movement. However, formal modelling of change has not been undertaken.

This chapter presents a formal model for dealing with essential change viewed as a succession of elementary temporal constructs that interfere with the identities of objects. This type of change has nothing to do with the concept of motion, which is about changing the positions. The essential change should be understood as a change in the fundamental property of objects of having an identity. How and when an object changes its identity are fundamental questions here.

The problem is explained informally in Section 10.2. The functional programming language GOFER (Jones, 1994) is used for formalization. The cornerstone of the model is the class-based approach enabling us to build both powerful and easily adaptable prototypes. Section 10.3 explains the formalization techniques. Section 10.4 shows the formalization of all necessary elements grouped into typical lifestyles. The practical examples from the SEU domain are analyzed in Section 10.5. Section 10.6 presents the conclusions drawn from this Chapter.

10.2 FRAMEWORK FOR LIFESTYLES

In the following the characteristics of identities in the proposed view of the world are described. Next, temporal constructs are enumerated and explained. Finally, their grouping into lifestyles is presented.

10.2.1 World of Discourse

The modelling of the physical reality is a very complex task for a database designer. A particular set of entities of interest must be separated from the real world. Such a set is called *the universe of discourse* (or mini-world). Everything what is outside this universe of discourse does not interfere with any actions performed inside it. This view is known as the *closed world assumption* (Reiter, 1994).

To design a database system that captures change, the temporal component must be introduced into the world of discourse. The world has a *clock,* which determines the order of states in the world. This is the simplest notion of time—ordered time, but sufficient to capture the most basic notions in a temporal database. It is assumed that the complete world history is known and sequentially entered into the system. The discussion of valid time vs. transaction time is widely treated in the literature and not of any concern here.

The elements of the world are *objects* of different kinds, which all have just one but essential property in common: they have an identity, such that a changed state of an object at a later time can be linked to the state of the same object at an earlier time. This is achieved with a unique *identifier* for each object, which remains fixed during its lifetime. This feature enables the system to distinguish one object from the other. The identifier must satisfy three conditions: *uniqueness* (no duplicates are allowed), *immutability* (cannot be changed within the system), and *non-reusability* (identities already issued to an object must not be given again).

Most changes in properties of objects or their spatial location, their geometry, etc. do not affect identifiers. The identity of the object before and after such *continuous change* is the same (represented as having the same identifier).

The second type, *catastrophic* or *essential change*, which affects the identity of the objects, deserves to be studied in detail because it captures a very low level of universality in the behaviour of all objects. The most primitive operations for handling this type of change are called *temporal constructs*.

10.2.2 Temporal Constructs

When and how objects change their identities is the fundamental question. The work of many authors was summarized in (Al-Taha and Barrera, 1994) and the list of applicable temporal constructs was proposed. These are shown in Figure 10.1.

The first two operations (create and destroy) seem natural and connected: the object is created and some time later destroyed. While the act of *creation* is indisputable, the concept of *destroying* needs an explanation. The goal of a temporal database is to keep track of the past states of the world. Even if something is destroyed, the information about its previous existence remains in the system and can be retrieved. An example would be the life of a person, which ends with death, but the record of his existence in the past is kept.

The missing part in Figure 10.1 would be a construct for removing the object from the system including all information about its past states, including that it ever existed. In fact, this operation does not fit in Figure 10.1, since it affects all world states while all other constructs lead from one state of the world to the next one. Applying this construct (called *eradicate*) leads to the loss of information, because the operation is irreversible in its nature. This function should be restricted. It may be necessary occasionally, for example, to rectify errors.

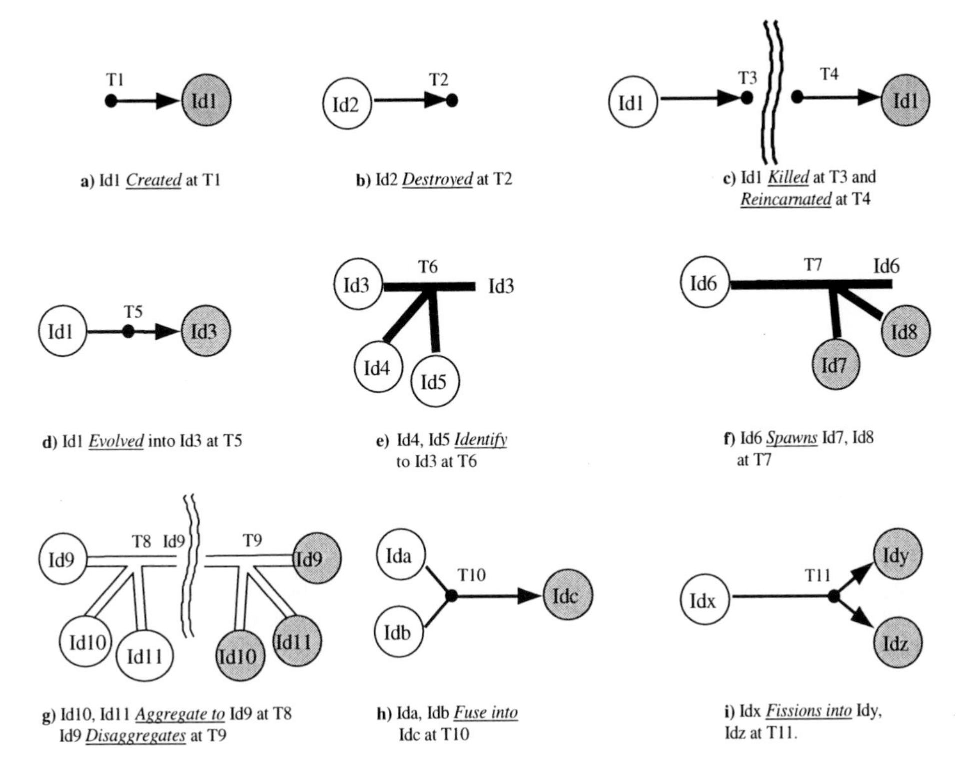

a) Id1 *Created* at T1
b) Id2 *Destroyed* at T2
c) Id1 *Killed* at T3 and *Reincarnated* at T4
d) Id1 *Evolved* into Id3 at T5
e) Id4, Id5 *Identify* to Id3 at T6
f) Id6 *Spawns* Id7, Id8 at T7
g) Id10, Id11 *Aggregate to* Id9 at T8 Id9 *Disaggregates* at T9
h) Ida, Idb *Fuse into* Idc at T10
i) Idx *Fissions into* Idy, Idz at T11.

Figure 10.1 Temporal constructs of identities (from Al-Taha and Barrera, 1994)

The pair of operations *kill* and *reincarnate* represent the change of identity to inactive with the possibility to be activated again. This is not applicable to the biological life of human beings, but can be seen in their professional career: if one changes the job for a while and then comes back to the old enterprise again, it would be a reincarnation, at least from the point of view of the employer.

Evolving can be seen as a combination of *create* and *destroy*: one object leaves and another comes. But, the new object keeps the information about its ancestor. This is necessary for tracking the history, e.g., when a country changes its identity. It is important to keep track of the fact that one country evolved into the other one, but it is not just a change of some attributes. The Austrian Empire disappeared and the (first) Republic of Austria emerged. A similar example would be a tree, which is cut down and thus 'evolves' to become a piece of lumber.

The primary role of *identify* is to keep the identity of one object which is merged with other objects from different sources and referring to the same objects.

The left-hand part of Figure 10.2 shows the previous state of two parcels with identifiers 21 and 22. It is common practice in land-registry offices that the larger parcel retains its identifier after unification with the smaller one.

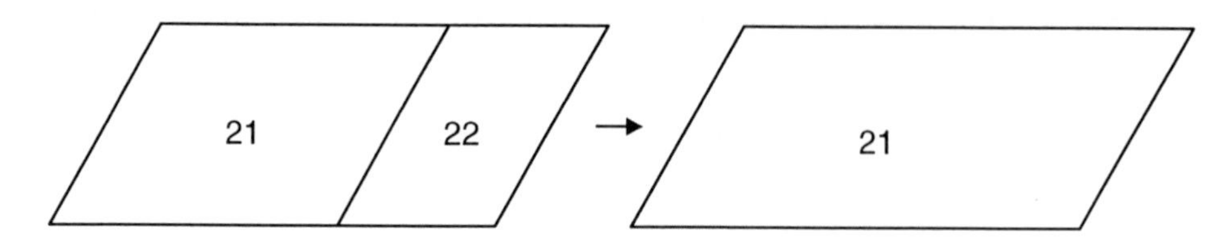

Figure 10.2 Temporal construct *identify* in cadastral practice

One could also see the emerging object as a new object (the fuse operator). The current cadastral practice is justified by savings of new numbers. The parcels are split in the same manner. The parcel continues in its larger piece and spawns some additional parcels, see Figure 10.3.

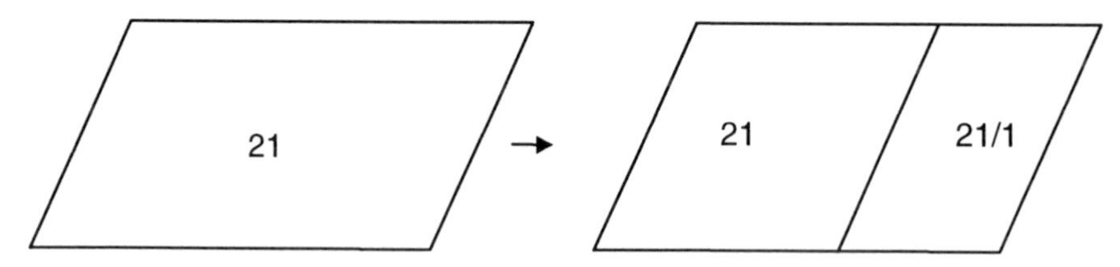

Figure 10.3 Temporal construct *spawn* in cadastral practice

The concept of *aggregation* relies on part/whole relation, which is both well explored and intuitive. The standard example is a car having its chassis number (identifier), engine (which has its own number), steering wheel, tyres, etc. The aggregation is a natural approach when dealing with solid artifacts, especially with manufactured goods. A car retains its identity, even if some parts are removed (disaggregation) and replaced with others (aggregation). Since aggregated objects keep their identity, the aggregation is a reversible operation. Legally, this is not the case for the merge of two parcels.

A *fusion* is not reversible, because fused objects cease to exist, producing new objects. However, a *fission* may be reversed by following fusion. The decision if the object produced through such a reunification gets the old identity (in virtue of reincarnation of the former object) or becomes a new object, depends on the user. The concept of *fission/fusion* captures the behaviour of liquids, which are more difficult to formalize than solids (Hayes, 1985). A glass of wine and a glass of water can be mixed (fusion), but it is impossible to separate the wine and the water again.

10.2.3 Groupings of Lifestyles

The objects in real world can be separated in different groups depending on the purpose of such distribution. The simple clustering regarding the behaviour of objects in view of temporal constructs is shown in Figure 10.4.

It is obvious that every object must be created in a database if a user wants to perform any further operation on it or even to be aware of its existence. Thus, the outer rectangle is the most general for all objects. Since some objects may not be killed (and later reincarnated), the inner rectangle represents a subset of the general case.

Figure 10.4 points out the clusters: possible combinations of temporal constructs applicable to the specific groups of objects. If an object can be killed, then it can also be reincarnated—these two operations are linked. Some objects follow the aggregate/

disaggregate lifestyle—manufactured goods—and others the fusion/fission lifestyle—typically liquids. It is assumed that mixed lifestyles do not exist.

Special attention will be paid later to two clusters bounded elliptically: *fission/fusion* and *aggregate/disaggregate*. Namely, these lifestyles are good abstractions for the behaviour of spatial socio-economic units.

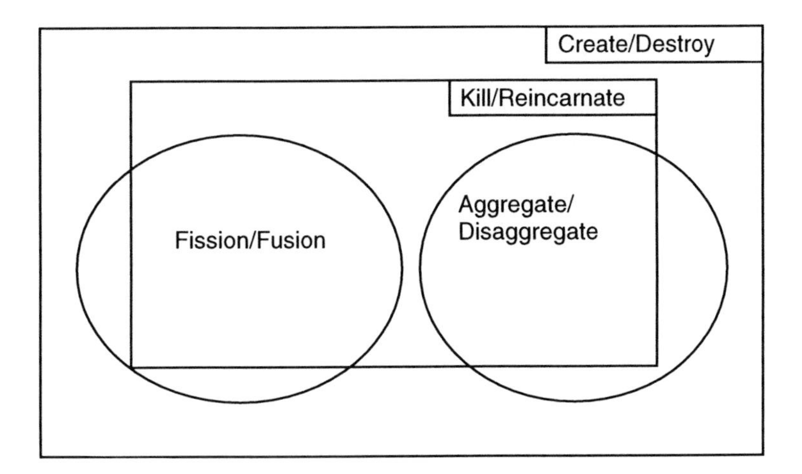

Figure 10.4 Clustering of temporal constructs

10.3 METHOD OF FORMALIZATION

The language used for the formalization is GOFER (Jones, 1994), a functional programming environment suitable for writing executable prototypes. It enables a researcher to easily check her own ideas expressed as algebraic specification for syntactical errors (Frank, Kuhn *et al.*, 1997). GOFER is a strictly typed language: a type is assigned to every object (functions are objects, too) and a type-consistency check is performed automatically.

10.3.1 Notation in GOFER

There are only a few symbols and keywords differing in GOFER from ordinary programming languages like Pascal or C. Such symbols used in this chapter are shortly explained here to make further reading easier.

::	is type of	&&	and	[]	list
.	composition	||	or	:	list constructor
=>	inherits	/=	not equal	--	comment

10.3.2 Functions, Classes and Data Types

The most fundamental building blocks in GOFER are *functions*. The signature of a function, which gives the types of its arguments, is written as:

```
functionName :: argumentType(s) -> resultType
```

Multiple arguments are separated with "->". A function can have more than one argument, but only one result.

Operations can be grouped in *classes*, which characterize all entities with similar behaviour. An example for a very basic class is the class *Eq* (equality):

```
class Eq a where
      (==), (/=) :: a -> a -> Bool
      x /= y = not (x == y)
```

It is stated only that there are two functions taking two arguments of the same type yielding a Boolean value as the result. The class parameter (in this case a) stands for arbitrary datatype. The signatures for two operations (equal and not equal) are given. The last statement indicates that non-equality is inferred from equality. It is an axiom, since it is valid for every implementation of the class.

The universe of values is organized in collections, called *datatypes*. Datatypes are divided in basic (Int, Float, Char, Bool) and compound datatypes. A coordinate in the plane can be abstracted as two integers (basic datatypes), wrapped with the tag Coord into compound datatype Coord:

```
data Coord = Coord Int Int
```

An implementation of the class for a datatype makes the operations from the class signature available for the datatype specified. For example, the instantiation of Eq for a datatype Coord would be:

```
instance Eq Coord where
      Coord a b == Coord c d = a == c && b == d
```

The value of the inequality function (/=) is inherited from the class rule and must not be defined in the implementation. More detailed explanation of the classes-oriented functional programming can be found in (Frank, Kuhn *et al.*, 1997).

10.3.3 Higher Order Functions

Higher order functions are functions that take functions as arguments.

The idea may be, for example, to build a function that is able to reach any object from the list according to specific criteria and to apply a primitive function onto it. There are only a few generic higher order functions for a complex datatype. Therefore, the benefits of their universality are very high.

The standard example for a higher order function is *map* function. Let *square* and *cube* be unary functions with the signature:

```
square, cube :: t -> t
```

Now, in order to perform these functions on the elements in the list, we have to use the higher order function *map,* which takes any function with the signature (t->t) and applies it to all elements of the list.

```
map :: (t -> t) -> [t] -> [t]
```

The list of integers can be squared or cubed knowing only how to evaluate the functions square and cube in the simplest case (square x = x * x, cube x = x * x * x).

```
map square [1,2,3]   yields  [1,4,9]
map cube [1,2,3]     yields  [1,8,27]
```

10.4 FORMALIZATION OF LIFESTYLES

Temporal constructs described in Section 10.2 are formalized here, beginning with the formalization of identifiers, which are key components of objects. Objects are the building blocks of the world. The world is a collection of objects. It is just a snapshot and the list of consecutive worlds is the temporal database holding all the information of interest.

10.4.1 Identifiers

A simple example is the class *IDs* where all operations on objects with identity are 'signed', that is, the types of their arguments are fixed. The generic GOFER class Eq is introduced as the context to remind us that equality operations must be available for any representations of IDs.

```
class Eq i => IDs i where
      nextID :: i -> i
      sameID :: i -> i -> Bool
      sameID i j = i == j
```

To fully define the behaviour of all objects it would be necessary to add an axiom 'notSameID i (next i)', which we cannot express in a constructive sense. The implementation may take different forms depending on users' needs, which is very convenient. A simple example assumes the identifiers as integers:

```
type ID = Int
instance IDs ID where
      nextID i = i + 1
```

If somebody wanted to define ID differently, she needs only to change its datatype and the instances (implementation), while the class header (behaviour) remains the same. For example, much more complex IDs (as typical in databases) are possible.

10.4.2 Objects

The identity of objects is defined by virtue of two operations on the identifiable members of the class and the implementation on the highest level of abstraction. The next step is to see how to extend the concept of identity to more feasible units in our formalized world. The class *Objects,* which describes the behaviour of prototypical objects, is introduced:

```
class Objects o where
      putID :: ID -> o -> o
      getID :: o -> ID
      sameObj :: o -> o -> Bool
      isID :: ID -> o -> Bool
      sameObj o1 o2 = sameID (getID o1) (getID o2)
      isID i o = sameID i (getID o)
```

There are only two fundamental operations on object class at this level of abstraction: *putID* and *getID*. The mechanism for assigning an ID to an object (*putID*) is a constructor, while the function *getID* is an observer. The comparisons operations (*sameObj* and *isID*) are the comparisons of objects' identifiers, and thus implementation independent. The simplest implementation is

```
data SimpleObject = SimpleObj ID
instance Objects SimpleObject where
      putID i (SimpleObj j) = SimpleObj i
      getID (SimpleObj i) = i
```

The use of the *putID* operation must be restricted to a principled use. Therefore, all objects are wrapped into the same collection—*a world.*

10.4.3 World

Observing only one object at a time is not sufficient to describe lifestyles. A world is a collection of objects with some defined actions. These actions ensure that:

- every object has an ID,
- the IDs of any two objects are different,
- objects can be identified by their IDs.

Each such world is a snapshot of the state of the world; if the world evolves, a new snapshot is produced. This allows temporal reasoning, where the succession of world snapshots acts as a time line.

```
data World t = W ID [t]
```

The sort (type) of objects has not yet been defined. The datatype *World* is parameterized with *t*, allowing *any* kind of object.

10.4.4 Create, Destroy

The class *Worlds* is introduced with fundamental operations allowing creation and destruction of objects.

```
class Objects t => Worlds w t where
      createObj  :: t -> w t -> w t
      destroyObj :: ID -> w t -> w t
      eradicate  :: ID -> w t -> w t
      perform    :: (t -> t) -> ID -> w t -> w t
      exists     :: ID -> w t -> Bool
      getObj     :: ID -> w t -> t
```

Instances of the class *Worlds* for the *World* datatype:

```
instance Worlds World t where
      createObj t (W i s) = W i' (t':s) where
            i' = nextID i
            t' = putID i' t
      destroyObj i (W j os) = W j os' where
            os' = filter ((/= i) . getID) os
      perform f i (W i' s) = W i' [if isID i x then f x
            else x|x<-s]
      exists i (W j s) = elem i (map getID s)
      getObj i (W j os) = head (filter ((== i).getID) os)
```

The operation *createObj* adds a new object with value to the world and assigns a unique ID to it; *destroyObj* deletes the object with the given ID from the next version of the world. The higher order function *perform* applies a unary function to the appropriate objects in the world. The observer function *exists* tests if an object with a given ID is available. Finally, *getObj* retrieves an object from the world.

With this code a running example of a world can be built and tested if its behaviour agrees with our intentions. Tests reveal if this model follows the abstract rules stated.

The result seems to be a modest achievement, but there are a few conclusions that can be drawn here. At first, it is possible to create the world, which may grow by adding new objects. Secondly, all objects have *unique* identities within the context—the world. These cannot be changed within the system (*immutability of identity*). For the world keeps track of the last ID, the identifiers already used cannot be given again (*non-reusability*). Therefore, all three basic characteristics of the identity specified in (Al-Taha and Barrera, 1994) are fulfilled.

10.4.5 Kill, Reincarnate

The purpose of temporal databases to track the history of objects strongly recommends the operations sketched in Figure 10.1c: *kill* and *reincarnate.* The *kill* operation does not remove the object from the model. It just changes the state of the object to dead. The object can be *reincarnated.* This is modelled as a class *Killable,* which comprehends both operations necessary, and an inspecting function *alive* to determine the current state of the object:

```
class Killable k where
      kill :: k -> k
      reincarnate :: k -> k
      alive :: k -> Bool
```

Simple implementation may look as:

```
data Eternal = Eternal ID Alive -- object continues for ever
type Alive = Bool
instance Objects Eternal where
      putID i (Eternal j b) = Eternal i b
      getID (Eternal i b) = i
instance Killable Alive where
      kill True = False
      kill False = error "already killed, cannot kill second
            time!"
      reincarnate True = error "alive, cannot reincarnate
            twice!"
      reincarnate False = True
      alive b = b
instance Killable Eternal where
      kill (Eternal i b) = Eternal i (kill b)
      reincarnate (Eternal i b) = Eternal i (reincarnate b)
      alive (Eternal i b) = alive b
```

Note how the polymorph operations propagate through different levels: *kill* is defined on the lowest level (for the generic Boolean type of *Live*) and then simply used in the instantiation for *Eternal* datatype.

```
test1 = alive (getObj 1 (perform kill 1 w2))
```

A test reveals that an object is not alive after it has been killed.

10.4.6 Eradicate

The implementation of *eradicate* as permanently removing an object from the consideration for all times, past and present, is a more complex matter. Assuming that the flow of time is represented as a growing list of snapshots, then to *eradicate* an object is equal to *destroy* it in all object views.

```
data WorldList t = WL [World t] -- the succession of snapshots
instance Worlds World t => Worlds WorldList t where
      eradicate i (WL ws) = WL ws' where
            ws' = map (destroyObj i) ws
```

Although it looks natural to have the power to remove an object from a model (in the case that an object has been accidentally created), *eradicate* is a very dangerous operation, for one could accidentally remove a valid object (Clifford and Croker, 1988) or an object which is referenced by another object (i.e., what happens with the son if the parent is removed? A standard problem for any regime trying to falsify history.) Therefore this operation should be restricted for special cases only.

10.4.7 Evolve

The *evolution* of an object is changing its identity. Since the identity is immutable, i.e., the user cannot change it directly, such a construct includes several operations: *destroy* the old object, *create* a new one while establishing a link between the previous and the actual object. The concept of predecessors of an object is necessary for such temporal relations and can be defined as new class *Preds*:

```
class Objects p => Preds p where
      putPreds :: [ID] -> p -> p
      getPreds :: p -> [ID]
class Evolves e where
      evolve :: ID -> e -> e

instance (Preds t, Worlds World t) => Evolves (World t) where
      evolve j (W i os) = perform g j (W i' os) where
            g  = putID i' . putPreds [j]
            i' = nextID i
```

Assemblies/Parts

The major part of our everyday world are (or can be viewed as) assemblies or aggregates—things composed of other things. The set of examples for this lifestyle is very rich, mostly human artifacts—where pieces are combined to construct a complex mechanism. A car (consisting of a chassis, wheels, engine, steering wheel, battery, headlights, etc.) is a standard example for manufactured goods in general.

The main paradigm for this lifestyle is 'part-of' property. It is possible to model a whole-part situation in two ways: to maintain a list of IDs for the composite object (explicit aggregate) or to tag every partial object with the ID of the aggregate (implicit aggregate).

The following formalization maintains both ends of the part-of relation, trading the simplification of queries for the data redundancy.

```
class Aggregates a where
     aggregate, disaggregate :: ID -> ID -> a -> a
     addPart, addAggr :: ID -> a -> a
     isPartOf :: ID -> a -> Bool

instance (Worlds World a, Aggregates a) => Aggregates (World a)
   where
     aggregate i j w = perform (addPart j) i ((perform
          (addAggr i) j w))
     disaggregate i j w = perform (removePart j) i w1 where
          w1 = perform (removeAggr i) j w
```

This specification is used in Section 10.5.1 for describing the lifestyle of administrative units.

10.4.8 Liquid-Like Objects

In his research in naive physics, Hayes (1985) analyzed various states of liquids and partly formalized them using first-order logic. Liquids have a lifestyle different from aggregates, as fusion is not reversible, aggregation is. Hayes pointed out two abstractions: liquids as containers and liquids as objects.

Having in mind a picture of many glasses with water, the first abstraction of a world of liquids could be that of *containers*. They may or may not have limited *capacity*, but they must contain some *amount* of liquid, allowing zero-amounts. The essential feature of our world is the conservation of mass; once poured in, the amount of liquid cannot just disappear.

```
class Containers c where
    container          :: Float -> c
    empty, full        :: c -> Bool
    content,capacity   :: c -> Float
    takeOut, putIn     :: Float -> c -> c
    pour               :: Float -> ID -> ID -> c -> c
```

The implementation for the datatype *Container*:

```
data Container = C ID Float Float -- capacity, content
instance Objects Container where
    putID j (C i e f) = C j e f
    getID (C i e f) = i
instance Containers Container where
    container cap = C 0 cap 0.0
    empty (C i cap cont) = cont == 0.0
    full (C i cap cont) = cap == cont
    content (C i cap cont) = cont
    capacity (C i cap cont) = cap
    takeOut q (C i cap cont) = if q>cont then error
                 "not enough in cont"
```

```
          else C i cap (cont-q)
    putIn q (C i cap cont) = if q>(cap-cont) then error
               "would overflow"
          else C i cap (cont + q)

instance (Worlds World c, Containers c) => Containers (World c)
    where
    pour q i j w = perform (putIn q) j (perform (takeOut q) i w)
```

The container lifestyle is stable; the identity of the container is preserved from the creation onwards. However, it is an ontology of containers, not of liquids.

Concentrating on liquids, we treat quantities of liquids as objects. Once mixed with each other, there is no way of establishing the previous state. This irreversible merging of two objects, differing from reversible aggregation of solid objects, is called *fusion*. The process of splitting one liquid object into two other objects *(fission)* is reversible. Both processes establish new entities and dispose of existing ones. Therefore, appropriate operations are grouped under class *Fluids*.

```
class Fluids f where
      fluidObj    :: Float -> f
      content2    :: f -> Float
      takeOut2    :: Float -> f -> f
      putIn2      :: Float -> f -> f
      split       :: Float -> ID -> f -> f
      fuse        :: ID -> ID -> f -> f
```

There are complex rules required to answer the 'right to reincarnation'. Take the following situation: the liquid from a bottle is split in two glasses and then the two glasses are poured back into the bottle. This succession of fission and fusion should result in the initial state, but tracking the identifiers and predecessors is very complex if one requires that succession of fission and fusion should cancel each other, see Figure 10.5.

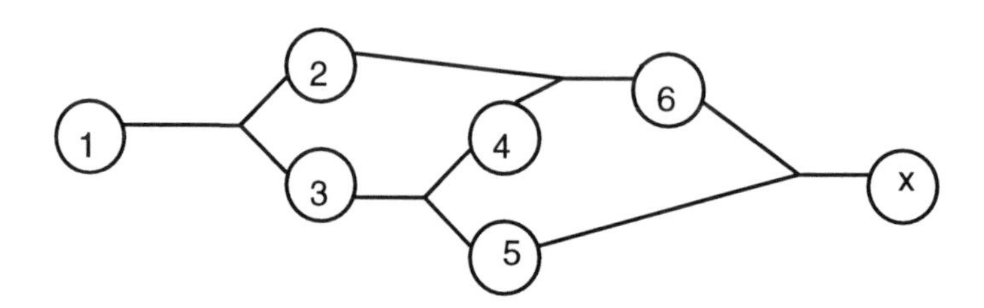

Figure 10.5 Reversibility of fission/fusion operations.

The following instantiation of *split* and *fuse* shows how this lifestyle can be formalized without specifying the type of objects:

```
instance (Worlds World f, Fluids f, Preds f, Killable f) =>
          Fluids (World f) where
     split q i w = perform kill i w1 where
          w1 = createObj o1 (createObj o2 w)
          o1 = putPreds ps1 (fluidObj q)
          o2 = putPreds ps1 (takeOut q (getObj i w))
          ps1 = [i] ++ getPreds (getObj i w)
```

```
fuse i j (W i1 ts) = perform kill i (perform kill j w1)
     where
     w1 = createObj o1 (W i1 ts)
     o1 = putPreds ps1 (fluidObj (qi + qj))
     ps1 = [i,j]++getPreds(getObj i w)++
           getPreds (getObj j w)
     qi = content2 (getObj i (W i1 ts))
     qj = content2 (getObj j (W i1 ts))
```

10.5 EXAMPLES IN SEU DOMAIN

The apparatus so created can be used to point out a surprising difference in the behaviour of SEUs. Some—for example, administrative units for statistical purposes—behave like aggregates, others—for example, parcels—behave for legal reasons like liquids.

10.5.1 Administrative Units

Administrative units are understood here as the elements of the political or administrative sub-divisions. Beside the partition of the state territory, which differs from country to country, political unions (for example, European Union) or regions (for example, Scandinavia) are examples. The elements (units) remain discernable—their identity and name are saved.

Obviously, this class of SEUs corresponds very well with the aggregate lifestyle formalized in Section 10.4.8. A small example demonstrates how easy this behaviour can be implemented using the previously established operations, which guarantee the desired properties.

```
type AllParts = [ID]
type PartOf = [ID]
type UnitName = String
data UnitClass = Parish | County | Province | State
data AdminUnit = AU ID AllParts PartOf UnitClass UnitName
instance Objects AdminUnit where
     putID i (AU j as ps c n) = AU i as ps c n
     getID (AU i as ps c n) = i
instance Aggregates AdminUnit where
     addPart i (AU j as ps c n) = AU j (i:as) ps c n
     addAggr i (AU j as ps c n) = AU j as (i:ps) c n
     isPartOf i (AU j as ps c n) = elem i ps
```

10.5.2 Cadastre Parcels

Cadastre parcels are the basic elements for land ownership. They have different lifestyles than administrative units because they lose their identity when split or merged. A parcel created from merging two parcels is a new entity from which the previous ones cannot be reconstructed (merging a parcel merges the encumbrances, e.g., the mortgages; splitting a parcel requires to separate them—not a reversible process). A cadastre saves previous data and marks it as invalid for future inspection of former states.

```
type Area = Float
type Pred = [ID]
type Exist = Bool
data Parcel = P ID Exist Area Pred
instance Objects Parcel where
     putID i (P j b a p) = P i b a p
     getID (P i b a p) = i
instance Killable Parcel where
     kill (P i b a p) = P i (kill b) a p
     reincarnate (P i b a p) = P i (reincarnate b) a p
instance Preds Parcel where
     getPreds (P i cont b ps) = ps
     putPreds is (P i cont b ps) = P i cont b (is++ps)
instance Fluids Parcel where
     fluidObj a = if a > 0.0 then P 0 True a []
                  else error "cannot create empty object"
     content2 (P id b cont p) = cont
     takeOut2 q (P id b cont p)
          | q > cont = error "not enough content"
          | otherwise        = (fluidObj (cont - q))
```

All other operations necessary are already defined in Section 10.4.9. It means that the core of the class behaviour (*split* and *fuse*) is inherited from the previous specification, allowing code-reusability.

10.5.3 Comparison

The aggregate lifestyle of administrative units stresses their structure as physical objects and in particular as containers for other objects. Ownership parcels should have the same lifestyle, but a careful analysis of the legal rules for merging and splitting points out that a merge is not reversible.

The legal justification stems from the rights which go with a parcel (for example, mortgages), which are conceived such that in general merging rights is easy but irreversible (the best known example is marriage). An ownership parcel being a complex form of physical land and abstract legal rights, the 'liquid' lifestyle of rights dominates the aggregate lifestyle of land pieces.

Dissolution of administrative units, when considered as owners of rights, is not a reversion of the merge. For example, the contract for the European Union has no provisions for 'undoing' the union.

10.6 CONCLUSIONS

In this chapter we showed that formalization is the perfect approach to separate the details of the behaviour of SEUs. The complex design of temporal databases is based on a finite number of simple operations (temporal constructs). Class-based approach allowed better insight in the most fundamental properties of temporal constructs. The most important temporal features are modelled without explicitly introducing the notion of time.

The design process is simplified when we assume that specific object classes have similar lifestyles. A lifestyle comprises the set of common temporal operations a class of

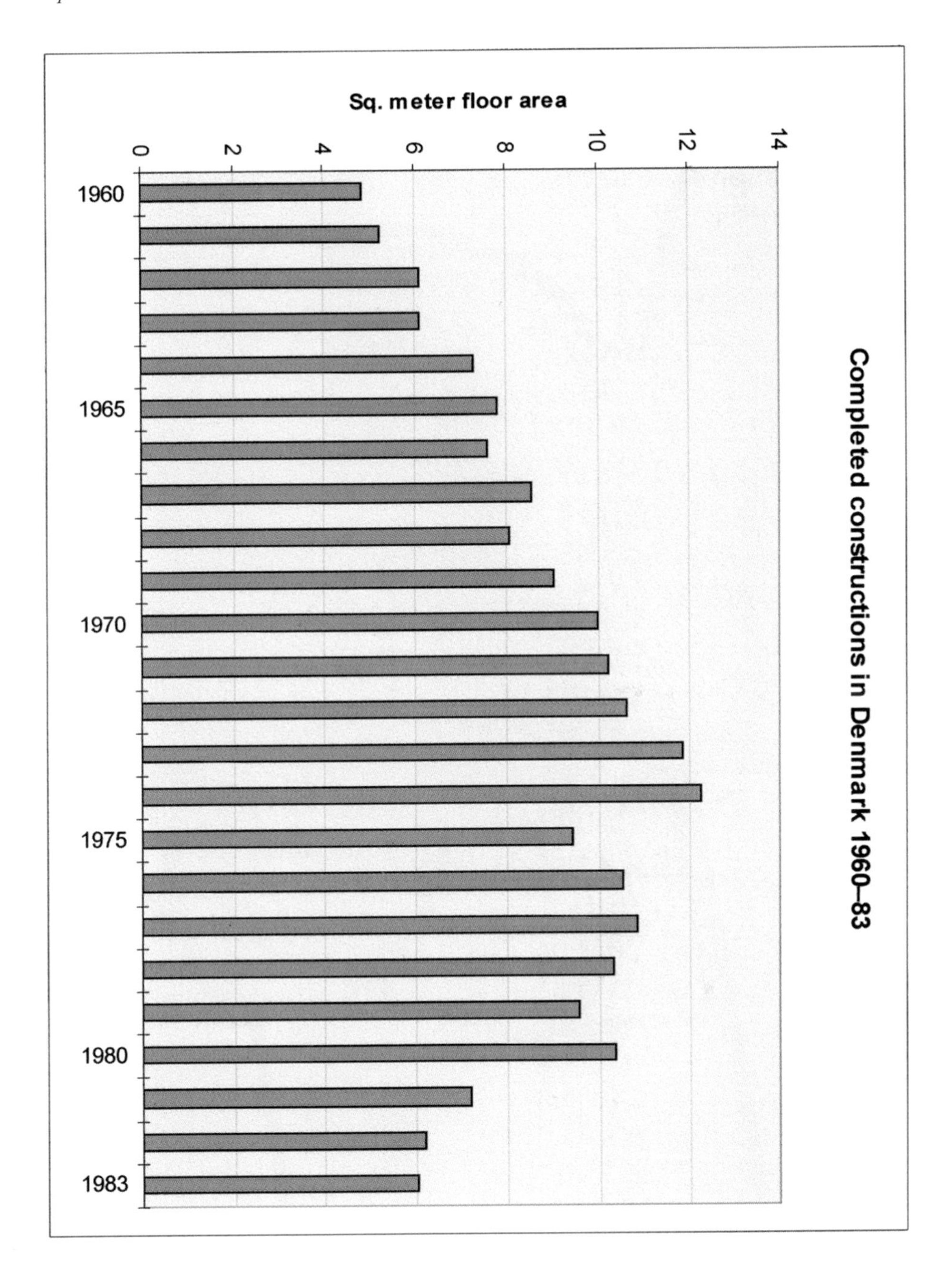

Figure 18.1 Construction activity (Fuldført byggeri ialt) in Denmark 1960-1983 (1.000 sq. meter floor area); Source: Danmarks Statistik

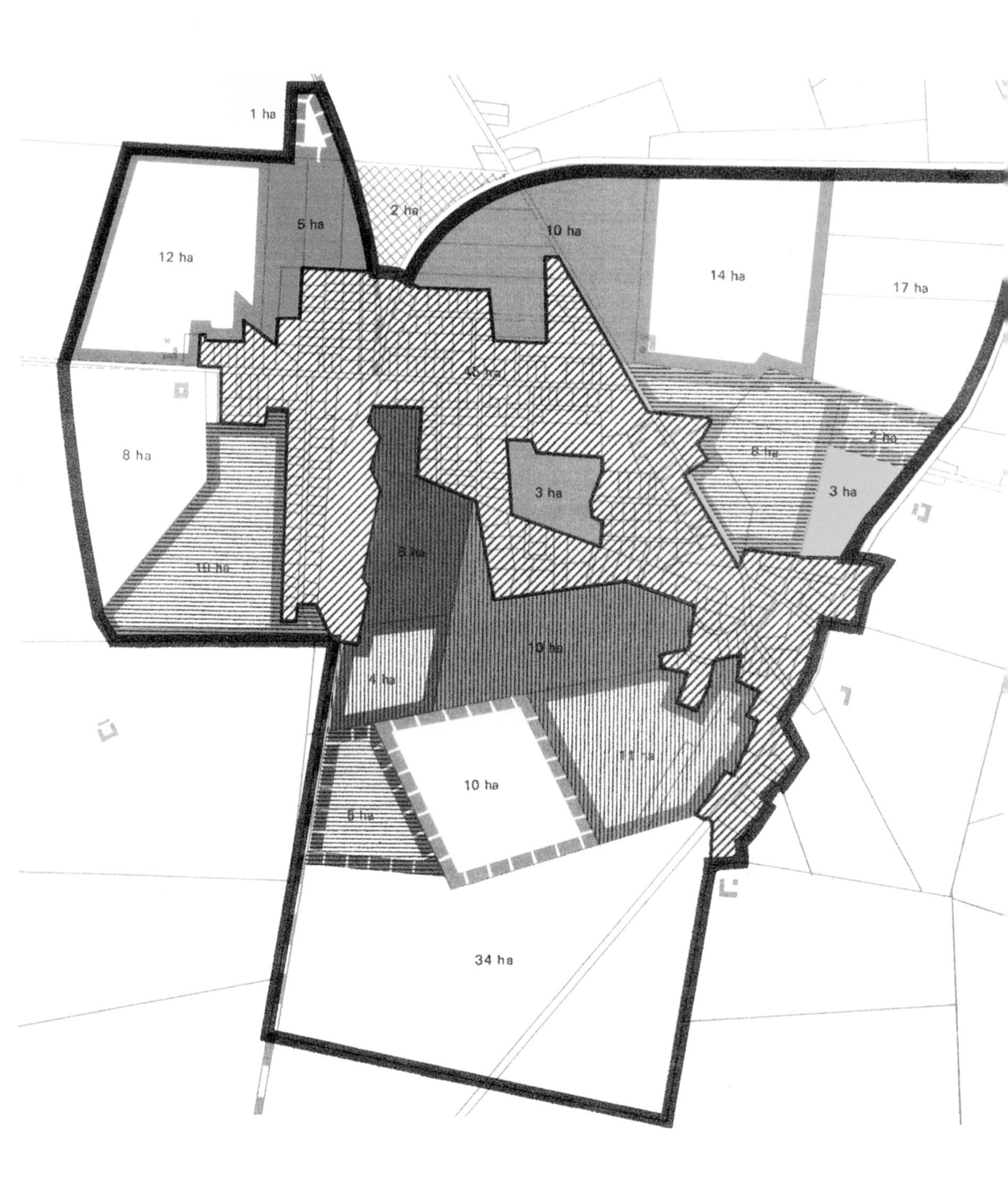

Figure 18.2 Illustrative map of scale 1:10.000 regarding the Development Inquiry as of 1 January 1970 of the Greater Copenhagen Council. The heavy line depicts the zone boundary. For colour codes, etc., see the text and Figure 18.3 (Egnsplanrådet, 1971, Extract from annex E)

	byggemodne arealer pr. 1. 1. 1970				arealer, der forventes byggemodnet i perioden 1. 1. 1970 – 31. 12. 1972				arealer, der forventes byggemodnet i perioden 1. 1. 1973 – 31. 12. 1975			
anvendelse / ejerforhold	haveboliger	etageboliger	industri og håndværk	off. formål og lign.	haveboliger	etageboliger	industri og håndværk	off. formål og lign.	haveboliger	etageboliger	industri og håndværk	off. formål og lign.
privat eje												
ha (u. decimaler)												
sociale boligsel-skabers eje												
ha (u. decimaler)												
offentlig eje												
ha (u. decimaler)												
i alt ha												

Figure 18.3 Account of developed areas as of 1.1.1970, and areas to be developed 1970-72, and 1973-75 specified with respect to land use and ownership. For translations, see the text. (Egnsplanrådet, 1971, Extract from annex D)

Figure 18.4 Aerial photo of the Western fringe of Copenhagen, about 20 km from the city centre (up North)
Source: Base map. Copyright: Kampsax Geoplan. Original scale 1:25.000.
Zone boundary: County of Copenhagen. Communication of 13.03.98

objects can undergo. These lifestyles are successfully mapped onto prototypical spatial socio-economic units. When assemblies are a logical parallel to hierarchical administrative units, it is surprising that cadastre parcels correspond to the world of liquids. The cadastre parcels may be merged or divided, but these operations are not reversible. It is their legal nature, which disables reversibility of the processes; the legal rights are easy to unite, but very complicated to split.

REFERENCES

Al-Taha, K. and Barrera, R., 1994, Identities through time. In *International Workshop on Requirements for Integrated Geographic Information Systems*, New Orleans, Louisiana, pp. 1–12.

Clifford, J. and Croker, A., 1988, Objects in time. *Database Engineering*, **7**, pp. 189–196.

Egenhofer, M.J. and Golledge, R.G., Eds, 1998, *Spatial and Temporal Reasoning in Geographic Information Systems*. Spatial Information Systems Series (New York: Oxford University Press).

Egenhofer, M.J. and Mark, D.M., 1995, Naive geography. In *Spatial Information Theory—A Theoretical Basis for GIS, International Conference COSIT '95*, Lecture Notes in Computer Science 988, edited by Frank, A.U. and Kuhn, W. (Berlin: Springer-Verlag), pp. 1–15.

Frank, A.U., 1994, Qualitative temporal reasoning in GIS—ordered time scales. In *Proceedings of 6th International Symposium on Spatial Data Handling*, Edinburgh, UK, (IGU Commission on GIS), pp. 410–430.

Frank, A.U., Kuhn, W., *et al.*, Eds, 1997, *Gofer as used at GeoInfo (TU Vienna)*. GeoInfo Series, Vol. 12 (Vienna: Dept. of Geoinformation, Technical University Vienna).

Galton, A., 1995, Towards a qualitative theory of movement. In *Spatial Information Theory—A Theoretical Basis for GIS, International Conference COSIT '95*, Lecture Notes in Computer Science 988, edited by Frank, A.U. and Kuhn, W. (Berlin: Springer-Verlag), pp. 377–396.

Hayes, P.J., 1985, Naive physics I: Ontology for liquids. In *Formal Theories of the Commonsense World*, edited by Hobbs, J.R. and Moore, R.C. (Norwood, NJ: Ablex Publishing), pp. 71–107.

Hobbs, J.R. and Moore, R.C., Eds, 1985. *Formal Theories of the Commonsense World*. Ablex Series in Artificial Intelligence (Norwood, NJ: Ablex Publishing).

Hornsby, K. and Egenhofer, M.J., 1997, Qualitative representation of change. In *Spatial Information Theory—A Theoretical Basis for GIS, International Conference COSIT '97*, Lecture Notes in Computer Science 1329, edited by Hirtle, S.C. and Frank, A.U. (Berlin: Springer-Verlag), pp. 15–33.

Jones, M.P., 1994, Gofer—Functional Programming Environment, Report, Dept. of Geoinformation, Technical University Vienna.

Langran, G., 1989, A review of temporal database research and its use in GIS applications. *International Journal of Geographical Information Systems*, **3**, pp. 215–232.

Reiter, R., 1994, On specifying database updates. *The Journal of Logic Programming*, **19** (20).

Snodgrass, R.T., 1992, Temporal Databases. In *Theories and Methods of Spatio-Temporal Reasoning in Geographic Space*, Lecture Notes in Computer Science 639, edited by Frank, A.U., Campari, I. and Formentini, U. (Berlin: Springer-Verlag), pp. 22–64.

CHAPTER ELEVEN

How Do Databases Perform Change?

Thérèse Libourel

11.1 INTRODUCTION

For the last decade, the world of Data Base Management Systems (DBMSs) has followed various technological and conceptual advances. The main challenge has been to open DBMSs to many application fields (biology, geographical information, CAD, etc.) and thus to face new types of data and processes. Within these new fields, the evolution concept is closely linked to the understanding and the description of the observed field. In this chapter, we introduce how DBMSs approach and realise the mechanisms of evolution.

In the first part, we quickly present the basic terminology used in the world of databases. In the second part, we define the concept of evolution in databases. Finally, some existing solutions are introduced, which can be used to represent change of spatial socio-economic units.

11.2 SOME DEFINITIONS

In Figure 11.1 we show the usual process for the creation of databases.

The conceptualisation step is based on a model (relational model, entity/relation model, object model, etc.) and gives a unique conceptual schema describing the semantics related to the part of study. Then data are stored in one or several databases, building up a group of values in correspondence with the schema and its constraints.

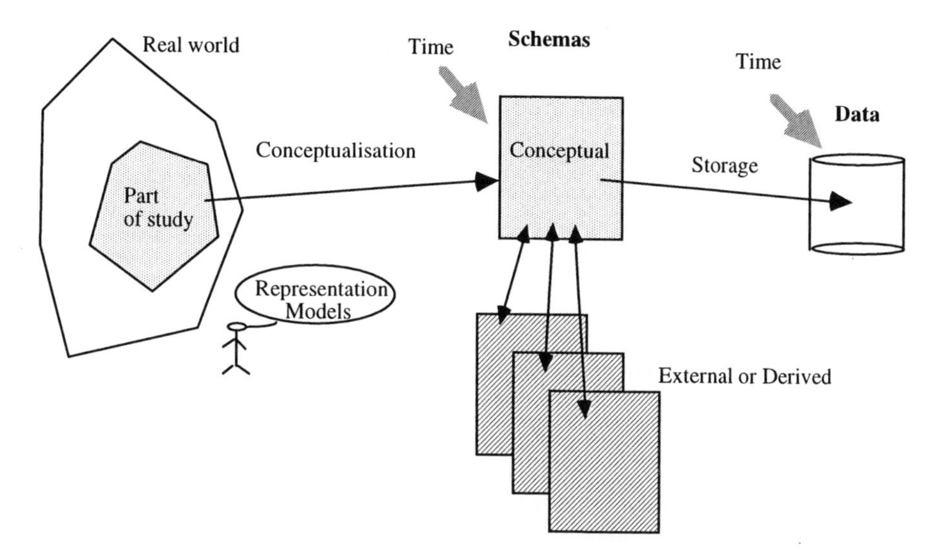

Figure 11.1 Steps for the creation of databases

Different perceptions of the same conceptual schema (semantic relativism) can lead to some external or derived schemas. These derived schemas present the semantics of 'virtual' or 'imaginary' objects (Abiteboul and Bonner, 1991), which, according to the authors, "allows the programmer to adapt the database reality to the needs of the user".

For spatial and temporal entities, some spatial and temporal features have to be added to the usual thematic characteristics in the model. Though not connected to the subject, we can, however, notice the imperfection of the available standard models for DBMSs.

11.3 THE COVERING OF THE 'WORLD EVOLUTION' IN THE FIELD OF DATABASES?

To present the 'world evolution' within DBMSs, according to Figure 11.1, we propose the following partition of problems occurring in change.

11.3.1 Schema Evolution or Schema Mutation

The real world shown either evolves (through extensions, modifications, deletion) or its perceptions change. The schema level has to take these evolutions into account.

Modifications may be performed within the chosen model *(schema evolution)* (see Figure 11.2). However, the designer may decide to change the overall semantics of this approach and then go from one model to another (for instance, from the relational model to the object one). This is called *schema mutation* (see Figure 11.2).

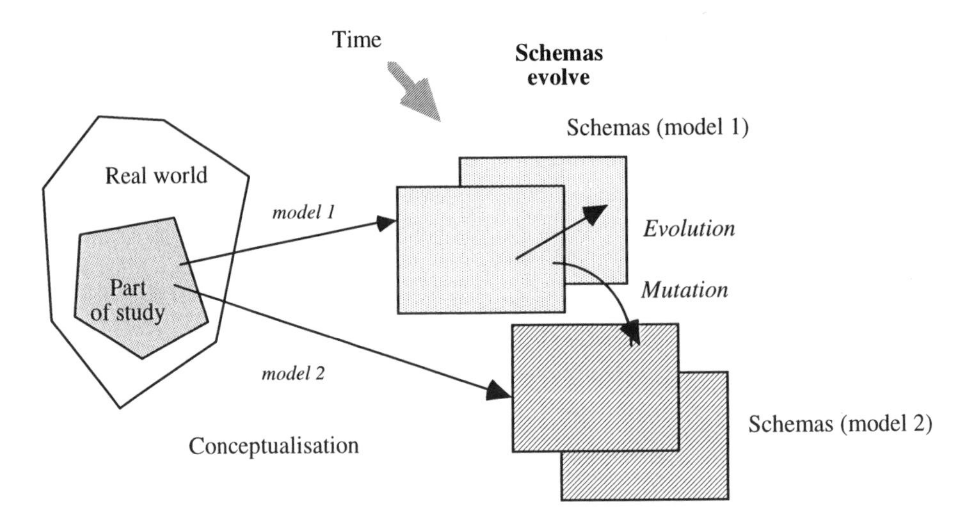

Figure 11.2 Evolution and mutation schemas

11.3.2 Data Evolution

The real world shown at the schema level involves entities and relations between these entities that are changing along with the time. The representation at the schema level remains valid and constant, but changes in data have to be taken into account (see Figure 11.3).

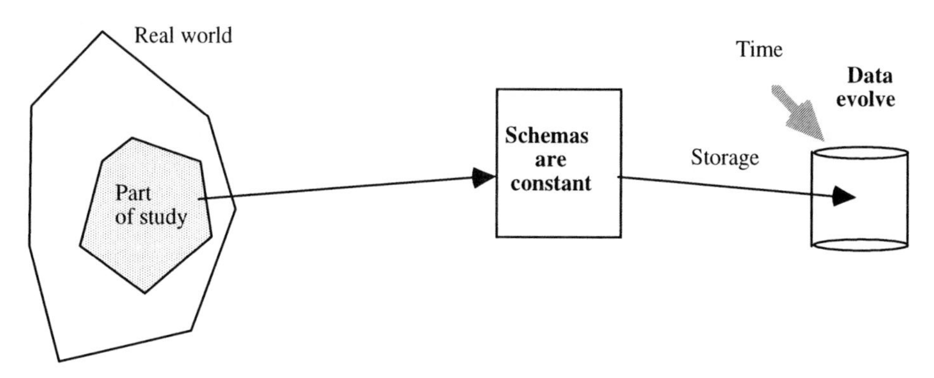

Figure 11.3 Data evolution

This concerns presently more or less complex *updatings*:

- When *time is implicit* and there is no need to keep trace of evolutions, the database gives solely a snapshot. Updatings are made either by changing values or by migrating objects.
- When *time is explicit*, temporal types (instant, interval, temporal set) and temporal association must be defined. Moreover, temporal values require an interpretation (valid time, transactional time and user time) (Snodgrass and Ahn, 1986). Keeping trace of the evolution may lead to two kinds of updatings: historical updatings (valid time) or bitemporal updatings (valid and transactional time) (Snodgrass and Ahn, 1985).

Data may follow two kinds of evolution:

a. *instantaneous change* from one state to another, like the merging of two land parcels, and
b. *continuous change* like the growth of a city.

Computer systems are not able to take into account continuous phenomena. Therefore, updates are stored in a discrete way. The final report depends on the way to handle interpolation between those various updates (Yeh, 1995; Ryssevik, this volume; Worboys, this volume).

11.3.3 Schema and Data Evolution

Evolutions may occur at the schema level as well as at the data level, and various constraints of the system have to be ensured.

This obvious problem raises more difficulties according to the role given to time:

- When *time is implicit* and there is no need to keep trace of evolutions at the schema level, data have to follow the variations of their representation so that the consistence of the database is preserved. This is called *data adaptation*.
- When *time is explicit*, either we may keep a review of consistent versions (schema and related data), or in a more complex way the schema absorbs and maintains the old and new definitions. In the latter case, the user may query transparently the various steps from the past to the present. This means that the system manages equivalencies between past and present structures and behaviours.

Now, we present for spatial socio-economics units what is defined as relevant evolutions and which solutions are available in today's DBMSs.

11.4 EVOLUTION WITHIN DATABASES VS. EVOLUTION OF SOCIO-ECONOMIC UNITS

11.4.1 Requirements

For general purposes, a database user may handle the evolution in three main cases (Libourel, 1994):

- *Evolution with loss of the past*: a new system is built each time a significant event occurs. Usually this is the case when the schema level changes and previous data are lost.
- *Evolution with partial loss of the past*: in order to deal with new requirements a new schema is built. Usually this is the case when the schema level changes and existing data are partially lost (we try to adapt them into the new schema).
- *Evolution with maintenance of the past*: all changes that occur to the data or schema level must be stored so that we can access any state at any time.

11.4.2 Case Study

A first model with the OMT (Object Modelling Technique) formalism (Rumbaugh *et al.*, 1991) is proposed in Figure 11.4. (The formalism used in OMT for object diagram is close to the ER formalism and includes classes and relations symbols. Among the relations we note also aggregation and generalisation/specialisation.)

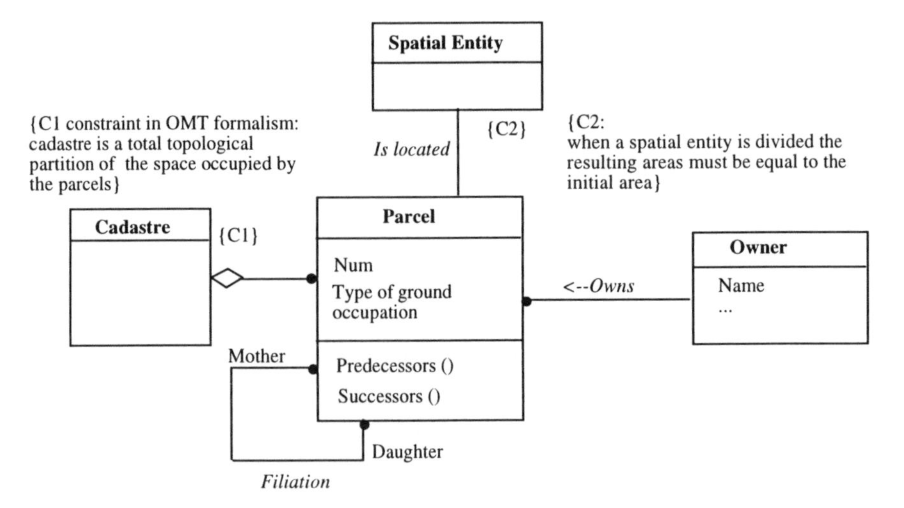

Figure 11.4 Simplify modellisation of the cadastre

In this representation, time has been intentionally forgotten. We suppose that only the state of the cadastre at a given date has to be known (punctual time representation). The dynamic of the socio-economics units requires an explicit management of time:

- *At the level of attribute for an entity or an association.* For instance, to the attribute 'type of ground occupation', we associate the instants of apparition, life span, etc. Acquisition time may also be added to the association *owns.*
- *At the level of entity and association.* For instance, spatial entities like parcels and the cadastre itself may be associated to temporal intervals of their validity.

Our cadastral model is generic and could be used in other applications. Parcel could be a state and cadastre the whole country, in this case the owner could be, for example, the Member of Parliament (MP).

11.4.3 Schema Evolution or Schema Mutation

The designer observes that the schema is not any more in correspondence with the modellised reality. Therefore he designs a new schema with a possible altered semantics, no matter whether the depending values are lost. This kind of evolution may be qualified as *'evolution with loss of the past'.*

Taking our case study, first the model takes into account some variations, for instance, we can add the date of registry of owners (see Figure 11.5).

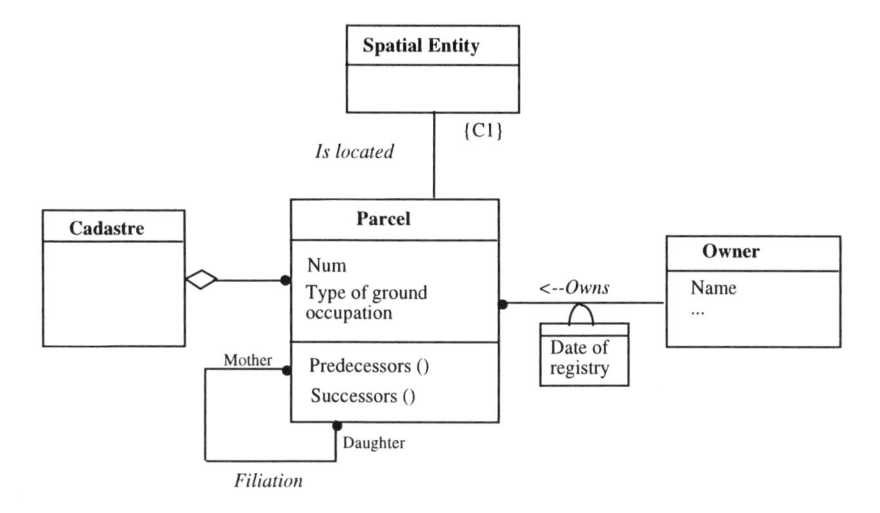

Figure 11.5 First evolution of schema modellisation of cadastre

More completely, the model is altered, modifying the semantics, to introduce time explicitly. Each parcel corresponds now to a spatio-temporal entity. Its characteristics are inherited from a spatial entity and a temporal entity. The attribute 'Type of ground occupation_H' points out that the previous states of 'Type of ground occupation' remain stored. The associations 'Owns_H' and 'Filiation_H' show a time feature as well. Hence, an evolution of the representation results (see Figure 11.6).

Several publications are related to this kind of evolution (Skarra and Zdonik, 1986; Penney and Stein, 1987; Kim and Chou, 1988; Nguyen and Rieu, 1989; Kim, 1990; Libourel, 1992; Dicky *et al.*, 1995, 1996), which is presently mainly a *conception* issue.

Solutions available propose some tools for modifying step by step or globally the part of study. The flexibility of the system depends on the available operations for modifications and the proposed help with respect to the relevance of these operations.

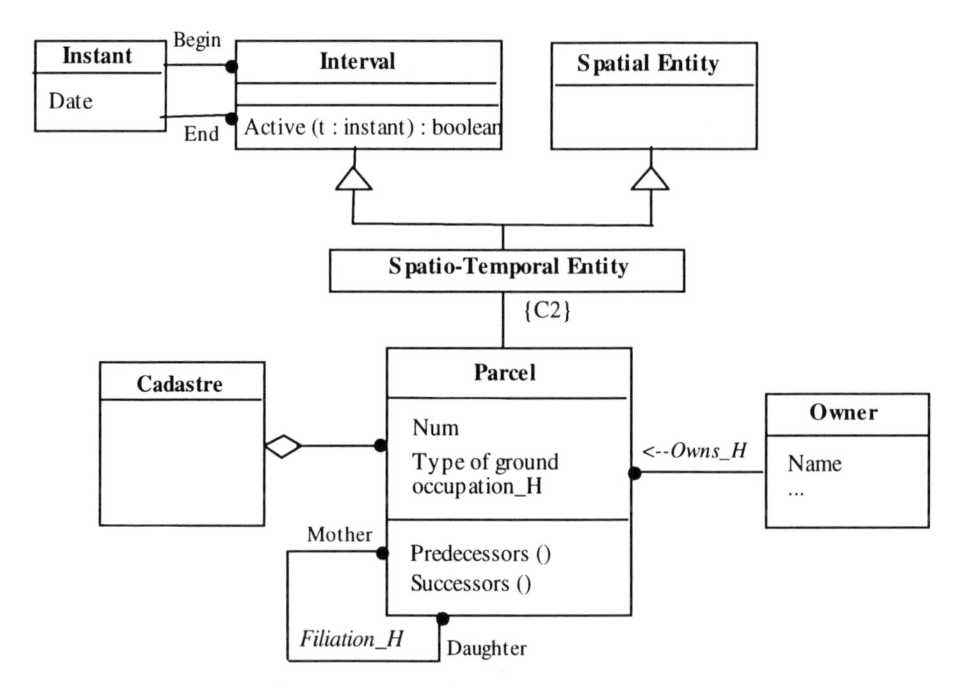

Figure 11.6 Spatio-temporal modellisation of the cadastre

OMT Model

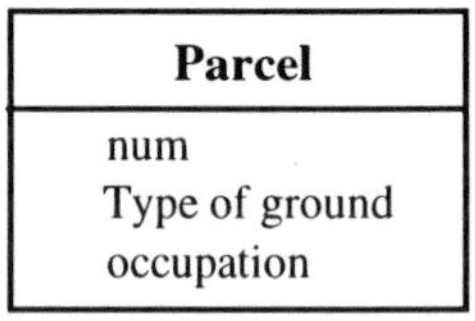

Parcel
num Type of ground occupation

Data Dictionary

Table Parcel

Attribute	Null?	Domain
IDparcel	N	ID
num	N	Integer
Type of ground occupation	Y	String

Primary Key IDparcel

SQL

```
CREATE TABLE Parcel
(IDparcel ID not null,
 Num Integer not null,
 Type of ground occupation char(35),
PRIMARY KEY (IDparcel));
```

Figure 11.7 Example of data dictionary and its implementation

Relational DBMSs introduce the notion of a *meta-level* of data representation. This meta-level consists in a meta-base, which is a data dictionary (i.e., a description of relational schemas stored in the database).

The meta-level allows describing schema evolutions. A relative flexibility is thus provided concerning semantic evolutions as main changes occur in the dictionary (Kerherve, 1986).

The allowed operations on this dictionary are:

a. creation and destruction of tables,
b. alteration of a table definition by modifying its attributes or adding some missing ones,
c. creation and modification of keys in tables.

For instance, the SQL sentence 'ALTER TABLE Parcel ADD Owner char(20)' adds the attribute Owner in the table Parcel. The main existing systems dealing with such evolutions are: Ingres, DB2, Oracle, Sybase, Informix, etc.

In Object-Oriented paradigm, the conceptual schema is a graph whose nodes are classes and links ('is-a', 'part-of' or 'reference' relation) (Cattell, 1994). Some *Object-Oriented* DBMSs (Gemstone (1986), Orion (1987), O2 (first prototype in 1989, commercially offered in 1991), Iris (1986), Ontos (Ex Vbase 1987), etc.) give a methodology for processing variations on a database conceptual schema.

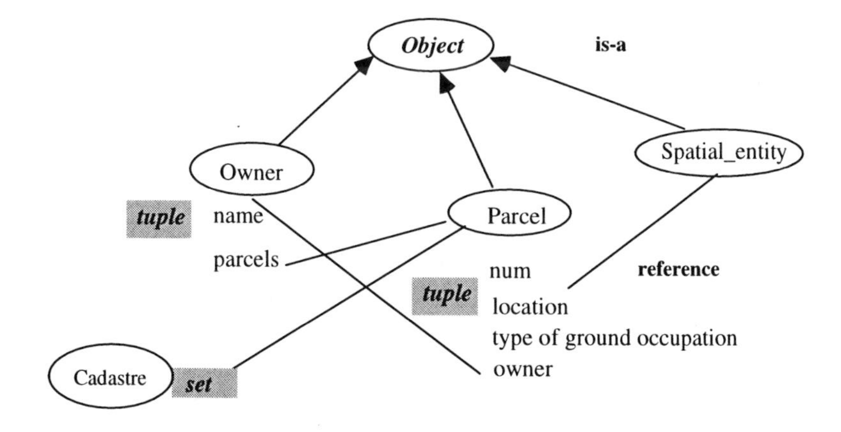

Figure 11.8 Object schema of cadastre

The taxonomy of evolutions generally adopted is:

- modifications of a class;
  - structure modification (create, delete or update attributes)
  - behaviour modification (create, delete or update operations)
- modification of the inheritance graph;
  - class addition
  - class deletion
  - addition of an 'is a' link
  - deletion of an 'is a' link

The evolution operations are monitored by a set of rules and constant properties relative to the schema entities that have to be maintained during the evolution.

This first kind of evolution is not more important for spatio-temporal entities than for any other. It allows just to bring the representation closer to the studied problem or to generate rapidly a prototype of the semantics of these representations.

However, a more interesting problem arises in the case of an *'evolution allowing loss of the past'* at the schema level, but *'conserving the past'* of data. How are old data read in the new model? The solution is to provide a more or less complicated mechanism for *converting* data. For instance, we get some cadastral data where the representation of spatial entities relative to the parcels is type changing (from the vectorial 'spaghetti' model to the topological model). Other kinds of evolution deal, in one way or other, with *'the maintenance of the past'* (Skarra and Zdonik, 1986; Monk and Sommerville, 1993; Benatallah, 1996).

11.4.4 Data Evolution

The *schema remains constant* when data move with time. For standard DBMSs, when *time is implicit,* the issue is a classical *updating*. Relational DBMSs propose some operations such as insertion, updating and deletion of tuples. Some *triggers* ensure the control of constraints at the schema level or between the schema level and the data level for updatings. These are carried out through *transactions* ensuring the consistence of the database.

In a similar way, Object-Oriented DBMSs provide operations for creation, modification and deletion of objects or instances. Some systems use a mechanism of automatic '*classification*' of the objects (Nguyen and Rieu, 1989; Carré *et al.*, 1990) that allows the migration of objects while they are evolving, with an inheritance mechanism.

Using the example in Figure 11.4, we can notice that operations called by Frank *life operations* (this volume), such as creation, disappearance, subdivision, insertion and deletion of a parcel require the check of the constraints {C1} or {C2}. The changes of owner of a parcel make up an operation of modification without checking the constraints {C1} and {C2}. However, if we wish to *conserve the past,* we have to keep trace of the evolutions during the modification or the migration of values. For this purpose, the *time dimension* has to be explicitly added.

Temporal databases, which at the present time are well developed only at the relational level, are designed to make a review of the successive events data are subject to (Clifford and Warren, 1983; Adiba and Bui-Quang, 1986; Adiba *et al.*, 1987). *Version management* requires for each modification of a value, to keep the former state of the concerned data. The notion of *historical data,* more precisely, has to take semantics into account, both for the modification of data values and the notion of periodicity desired by the designer. Various researchers have proposed an extension of the relational algebra to temporal operators. (A summary is presented by McKenzie and Snodgrass (1991) and Tansel *et al.* (1993).)

Regarding spatio-temporal entities, the temporal databases approach may be useful for interpreting the notion of movements or of life operation (when the object has the same identification from its birth to its death). We can also notice that using dated versions helps to reproduce movement or life only when some mechanisms of interpolation are provided. (Yeh, 1995).

The problems raised by the notion of life operations with the possibility of transformations, leading to filiations of identifiers, have not been studied for DBMSs. This is typically the kind of problem appearing in the model in Figure 11.6 as a consequence of some operations of fusion, fission and reorganisation that may occur for a parcel of the cadastre. In relational DBMSs identification is managed with keys, when in OODBMSs each object has its own and unique object identifier (oid) which is independent of its state. Therefore it is possible to manage object filiation.

11.4.5 Data-Schema Evolution

These evolutions with the *maintenance of the past* are the most complicated to deal with. A first kind of solution admits that the *schema evolves*, but to each evolution a *version* corresponds and the data of each version are juxtaposed in the system. However, at a given instant j, we access only to a consistent state corresponding to a global stamped schema and data version (Sj, Ij).

Some systems with objects—Orion (Kim and Chou, 1988; Kim, 1990), Encore (Skarra and Zdonik, 1986)—propose a strategy for maintaining *versions*. Regarding Geographical Information some GIS allow the management of versions (Smallworld). Parallel to this option (version management), the repercussions of each schema evolution towards each instance lead to a choice of strategy. We choose a propagation that can be either *immediate or early* (cf. Gemstone (Penney and Stein, 1987)) or *delayed or lazy*; the modification is effective only when the concerned object is used (cf. Orion).

Another kind of approach (Cellary and Jomier, 1990; Gançarski and Jomier, 1994) is to work on the notion of *versioned database (VDB)* conserving, for each stamped version, the new appeared objects, the values variations of the modified objects or the references of the 'stable' objects.

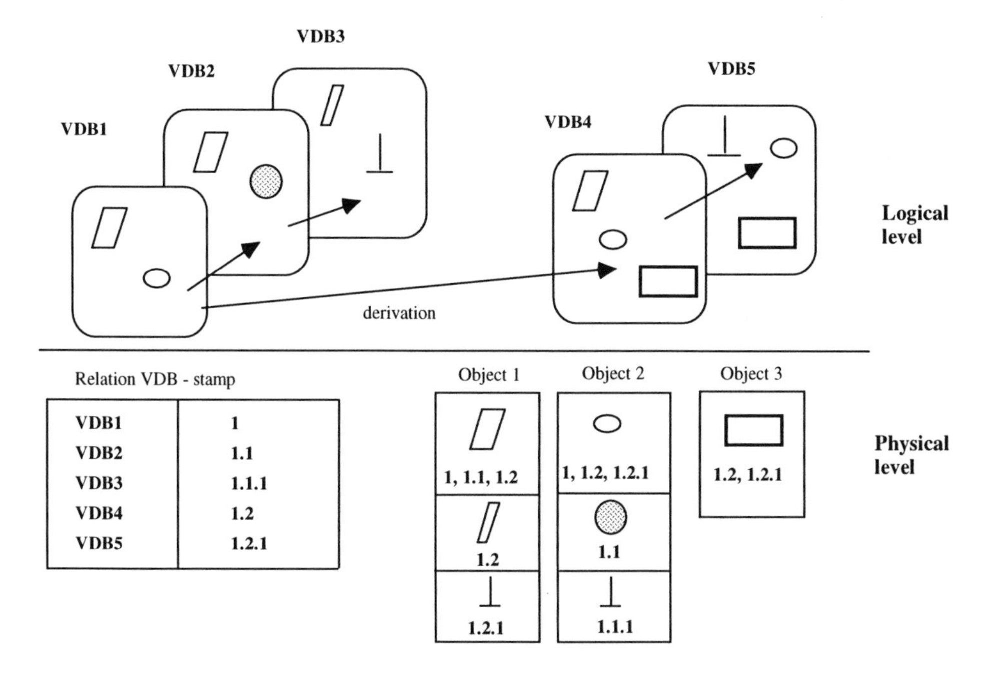

Figure 11.9 Database version

At the beginning, the work context (or VDB1) shows two objects (a parallelogram and a ring). Database users may have different perceptions of the world (in our case, two perceptions that are VDB1+VDB2+VDB3 or VDB1+VDB4+VDB5). At the physical level, each VDB is stamped with a unique stamp that explains the derivation tree (see the relation between VDB and stamp in Figure 11.9). Each object is stored separately with its properties and evolutions. Object 1 is created in VDB1 and remains unchanged in VDB2 and VDB4. We notice that in VDB3 the size of the parallelogram changes but other characteristics remain constant. For the physical description of this object, only evolutions are registered, other characteristics are 'inherited' from the previous state. In VDB5, object 1 is deleted at the logical level, but it remains at the physical level. Its state references are nil.

The existing solutions for *versions management* do not treat the time dimension in an implicit way through the stamping of versions, which are more often carried out through the transactional time. Temporal databases also deal with version management, but they are models where time is introduced explicitly. One of the most famous systems is Postgres (Stonebraker, 1985). Performance of the various operations considered is important. To keep the base consistent requires expensive strategies in terms of run time and storage.

The schema *absorbs and maintains permanently* the old and new definitions. The ideal solution consists of presenting the mechanisms of selectivity and reversibility. The user chooses the context of his/her work, accesses to the definitions and the applications of this context. For instance (see Figure 11.10), the system is completed with the management of farms, parcels aggregated and farmers, and we need to make a review of those farms as a complement of the review of the concerned parcels. The implied data may be built on existing data or may be added. The user may expect from the system that these various views can be asked for at any time.

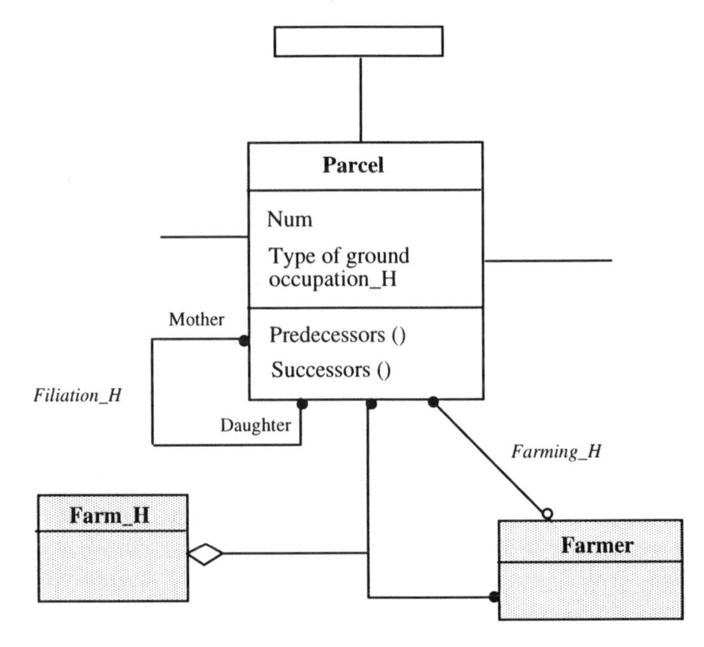

Figure 11.10 Global evolution

11.5 PERSPECTIVES

The evolution of DBMSs is still far from completely integrating the spatio-temporal needs. Systems providing the new schema and object evolution functionalities are generally built from atemporal models. In this chapter, we have admitted that applications evolve, which is presently a problem in itself. The object-oriented approach seems to be promising as dynamism is considered already in the model. It raises a complementary crucial problem: the principle of unique object identification. Network breakthroughs, moreover, require some dispatching mechanisms and means for data access. Most of the database systems (Oracle, Sybase, Ingres, etc.) provide the possibility to *distribute data* on various sites. The management of distributed transactions uses data or referential dictionaries (repositories). The management of identifiers corresponding to fragmented objects is therefore at the heart of the problem. Most of the database systems provide as well data sharing and delocalised updating: *synchronous and asynchronous replication*. It remains the main challenge how the various 'times' (valid and transactional) will be processed in these new environments so that representation is allowed and the management of spatio-temporal entities.

ACKNOWLEDGEMENTS

The author thanks J.-P. Cheylan, E. Dellerba, D. Gautier, C. Mende and L. Spery for many useful comments on several issues in this chapter.

REFERENCES

Abiteboul, S. and Bonner, A., 1991, Objects and views. In *Proceedings of ACM/SIGMOD Conference*, Denver, CO, pp. 238–247.

Adiba, M. and Bui-Quang, N., 1986, Historical multimedia databases. In *Proceedings of the 12th VLDB Conference*, Kyoto, Japan, pp. 63–70.

Adiba, M., Bui-Quang, N. and Collet, C., 1987, Aspects temporels, historiques et dynamiques des Bases de Données. *TSI*, **6**, pp. 457–478.

Benatallah, B., 1996, Un compromis: Modification et versionnement du schéma. In *Proceedings of BDA '96*, Cassis, France, pp. 373–396.

Carré, B., Dekker, L. and Geib, J.M., 1990, Multiple and evolutive representation in the ROME language. In *Proceedings of TOOLS2*, Paris, pp. 101–109.

Cattell, R.G.G. *et al.*, 1994, *Object Data Management*, (Reading, MA: Addison-Wesley).

Cellary, W. and Jomier, G., 1990, Consistency of versions in object-oriented databases. In *Proceedings of the 16th VLDB Conference*, Brisbane, Australia.

Clifford, J. and Warren, D.S., 1983, Formal semantics for time in database. *ACM-TODS*, **8**, pp. 214–254.

Dicky, H., Dony, C., Huchard, M. and Libourel, T., 1995, ARES: Adding a class and REStructuring inheritances hierarchies. In *Proceedings of BDA '95*, Nancy, France.

Dicky, H., Dony, C., Huchard, M. and Libourel, T., 1996, On automatic class insertion with overloading. In *Proceedings of OOPSLA '96,* San Jose, CA.

Gançarski, S. and Jomier, G., 1994, Managing entity versions within their context: A formal approach. In *Proceedings of DEXA '94.*

Kerherve, B., 1986, *Vues relationnelle.* Thèse Paris.

Kim, W. and Chou, H.T., 1988, Versions of schema for object-oriented databases. In *Proceedings of the 14th VLDB Conference*, Los Angeles, CA, pp. 148–159.

Kim, W., 1990, *Object-Oriented Databases: An Introduction*, (MIT Press).

Libourel, T., 1992, *Introduction de relations pour exprimer l'évolutivité dans un système d'objets.* Thèse Université Montpellier II.

Libourel, T., 1994, Evolutivité dans les SGBD et les systèmes d'objets. In *Congrès Systèmes Flexibles AFCET* Versailles.

McKenzie, L.E. Jr. and Snodgrass, R.T., 1991, Evaluation of relational algebras incorporating the time dimension in databases. *ACM Computing Surveys*, **23**, pp. 501–543.

Monk, S. and Sommerville, I., 1993, Schema evolution in OODBs using class versioning. *SIGMOD Record*, **22** (3).

Nguyen, G.T. and Rieu, D., 1989, Schema evolution in object-oriented database systems. *Data & Knowledge Engineering*, North Holland, **4** (1).

Penney, J. and Stein, J., 1987, Class modification in the Gemstone object-oriented DBMS. In *Proceedings of OOPSLA Conference*, Orlando, FL, pp. 111–117.

Rumbaugh, J., Blaha, M., Eddy, F., Premerlani, W. and Lorensen, W., 1991, *Object-Oriented Modeling and Design*, (Englewood Cliffs, NJ: Prentice Hall).

Santos, C., Delobel, C. and Abiteboul, S., 1994, Virtual schemas and bases. In *Proceedings of Extensive Data Base Technology.*

Skarra, A.H. and Zdonik, S.B., 1986, The management of changing types in an object-oriented database. In *Proceedings of OOPSLA Conference*, Portland, pp. 483–495.

Snodgrass, R.T. and Ahn, I., 1985, Taxonomy of time in database. In *Proceedings of ACM/SIGMOD '85 Conference*, **14**, pp. 236–246.

Snodgrass, R.T. and Ahn, I., 1986, Temporal databases. *IEEE Computer*, **19**, pp. 35–42.

Stonebraker, M., 1985, The design of the Postgres System. In *Proceedings of the 11th VLDB Conference*, Stockholm, Sweden, pp. 405–417.

Tansel, A.U., Clifford, J., Gadia, S., Jajodia, S., Segev, A. and Snodgrass, R.T., Eds, 1993, *Temporal Databases Theory, Design and Implementation*, (Benjamin Cummings).

Yeh, T.S., 1995, Modélisation de la variabilité des entités dans un système d'information géographique. Thèse Université Paris 6.

CHAPTER TWELVE

Towards the Design of a DBMS Repository for Temporal GIS

Emmanuel Stefanakis and Timos Sellis

12.1 INTRODUCTION

Geographic Information Systems are computer-based systems designed to support the capture, management, manipulation, analysis, modelling and display of spatially referenced data at different points in time (Aronoff, 1989). Today, GIS are widely used in many government, business and private activities, which fall into three major categories (Maguire *et al.*, 1991):

1. *socio-economic applications* (for example, urban and regional planning, cadastral registration, archaeology, natural resources management, market analysis, etc.);
2. *environmental applications* (for example, forestry, fire and epidemic control, etc.);
3. *management applications* (for example, organization of pipeline networks and other services, such as electricity and telephone, real-time navigation for vessels, planes and cars, etc.). The role of GIS in these applications is to provide the users and decision-makers with effective tools for solving the complex and usually ill- or semi-structured spatial problems (Hopkins, 1984), while providing an adequate level of performance.

GIS are enormously complicated software systems. A GIS is supported by an operating system, and depends on the presence of a graphic package for input and output, routines supplied by the programming language in which the GIS is written, and numerous other software products (Goodchild, 1991). Many of the more powerful contemporary GIS are designed to rely on a Database Management System (DBMS), relieving developers of many of the common data housekeeping functions. However, traditional DBMSs have only dealt with alphanumeric domains and have proved not to be suitable for non-standard applications, such as GIS, which are characterized by more complex domains.

The design and implementation of a DBMS repository for the application domain of GIS is an active area of research. Several architectural approaches have been proposed in the past to implement commercial or prototype systems and satisfy the urgent needs for operational geographic data handling (Aronoff, 1989; Stefanakis and Sellis, 1996). However, current GIS packages—though powerful toolboxes, most with hundreds of functions—suffer from several limitations, which render them inefficient tools for decision-making. It has been widely recognised (Al-Taha, 1992; Laurini and Thompson, 1992; Leung and Leung, 1993; Fischer, 1994; Stefanakis and Sellis, 1996; Stefanakis, 1997) that current commercial systems:

- do not accommodate the temporal dimension of geographic data;
- are based on an inappropriate logical foundation which is unable to handle the imprecision of data;
- provide a limited number of built-in analytical and modelling functionalities; and
- have an inadequate level of intelligence.

This study focuses on the first feature. The other limitations have been examined elsewhere (Stefanakis and Sellis, 1996; 1997; 1998; Stefanakis, 1997). Nowadays, the methods used in commercial GIS packages for both representation and analysis of geographic data are inadequate because they do not directly address the matter of time. On the other hand, most geographic applications involve data that change over time (Langran, 1989). Provided that the primary goal for any information system, including GIS, is to model accurately the real world, it is obvious that a dynamic world can only be captured and represented by a DBMS repository that handles changing information.

The scope of this study is to examine some issues towards the development of a DBMS repository for *temporal GIS*, able to model a dynamic world by capturing and presenting the world's changes over time. The discussion is organized as follows: Section 12.2 provides a brief presentation of two representative real-world examples, which highlight the requirements for modelling and analysis of spatio-temporal data. Section 12.3 presents the current state-of-the-art in the area of spatio-temporal databases. Two issues are addressed next: a *representational* and a *reasoning* issue. Specifically, Section 12.4 shows how the temporal dimension of geographic entities may be incorporated into a general spatial data model first introduced by Tomlin (1990) and briefly presented in Section 12.3; while Section 12.5 shows how the basic data interpretation operations for geographic data handling (also presented in Section 12.3) may be extended to support spatio-temporal reasoning. Representative operations, such as the temporal overlay, the temporal distance and the temporal selection, are examined in more detail and accompanied by simplified real-world examples. Finally, Section 12.6 concludes the discussion by summarizing the contributions of the chapter and giving hints for future research in the area of temporal GIS.

12.2 SPATIO-TEMPORAL APPLICATIONS AND REQUIREMENTS

Geographic data have a set of properties that make them distinctly different from the common alphanumeric lists and tables appearing in conventional information systems developed for business applications. Specifically, geographic entities have at least five dimensions (Stefanakis and Sellis, 1997), namely:

a. thematic,
b. spatial,
c. temporal,
d. quality, and
e. multimedia,

along which attributes may be measured. In addition, their analysis involves a large variety of operations which are highly application-dependent. Current GIS accommodate the spatial and thematic dimensions of geographic entities and provide a basic set of operations for general-purpose geographic data handling.

A DBMS repository for temporal GIS is supposed to support one additional dimension of geographic entities, the temporal dimension, as well as operations for temporal and spatio-temporal analysis. In the following, two real-world applications, where the temporal dimension of geographic data is involved, are presented, while several requirements for modelling and analysis of spatio-temporal entities are highlighted. The first concerns moving point objects and the second areal units that change over time.

Case 1: Vessel Trajectories

In maritime world (Dooling, 1994) there is a need for modelling and analysis of vessel trajectories. This problem can be viewed as the representation and reasoning on point objects whose positions change over time, i.e., moving point objects. Figure 12.1a illustrates a simplified example of two vessel trajectories (*A*, *B*) in the territorial waters (*TW*) of a country. Assuming navigation on a plane surface, each trajectory consists of a set of triplets (X_i, Y_i, T_i), where (X_i, Y_i) indicate the position of the vessel at time T_i. Some typical queries regarding this situation are:

V1. Does the vessel pass through point (*X*,*Y*)?
V2. When does the vessel pass through point (*X*,*Y*)?
V3. Where is the vessel at time T_k?
V4. Does the vessel cross the national boundary (*NB*) on the 3rd of July?
V5. Which is the shortest distance of the vessel from the seashore (*SSh*)?
V6. Do two trajectories (*A*, *B*) intersect?
V7. Do the two vessels collide? (a distance threshold is involved in this query)
V8. Which is the shortest distance between two trajectories (*A*, *B*)?
V9. Which is the shortest distance between the two vessels?
V10. What is the vessel velocity at time T_k?
V11. Which is the length of the vessel trajectory between two locations?
V12. Which is the duration of the trip between two locations?

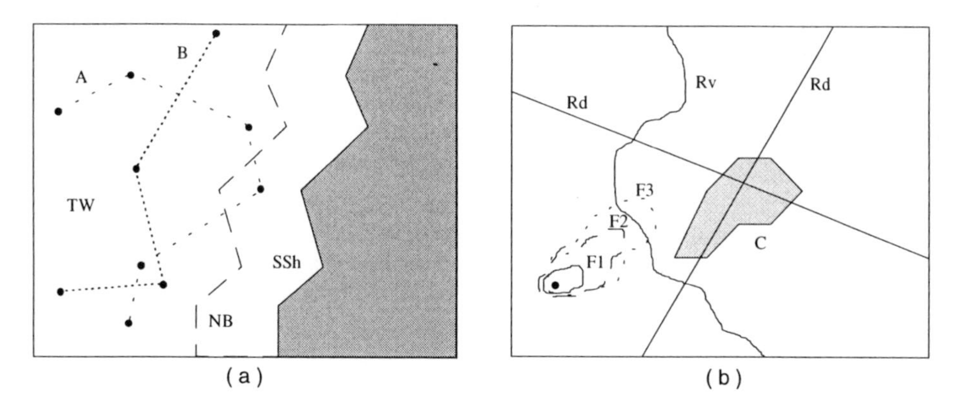

Figure 12.1 Two real world examples with spatio-temporal entities

Case 2: Forest Fire Spread

In environmental applications, there is a need to represent objects whose position, shape and size change over time. Focusing on forest management (Lymberopoulos *et al.*, 1996), the representation and analysis of areal units that change over time, such as the burning area in case of fire, is a common example. Figure 12.1b illustrates three snapshots (F_1, F_2, F_3) of the burning area in a forest. Assuming a fire on a plane surface (a projection of the earth surface), the burning area consists of a set of pairs (GEO_i, T_i), where GEO_i describes the geometry of the burning area at time T_i. In general GEO_i is a composite polygon (areal unit), possibly with holes (islands). Some typical queries regarding this situation are:

F1. Does the fire pass through point (*X*,*Y*)?
F2. When does the fire pass through point (*X*,*Y*)?

F3. Which area is burning at time T_k?
F4. Which is the burned area at time T_k?
F5. Does the fire cross the road (*Rd*) or the river (*Rv*)?
F6. Which is the shortest distance of the fire from the city (*C*)?
F7. Do two fire-fronts meet (in case of more than one fire)?
F8. Which is the fire-front velocity at time T_k?

Examining the two applications above, several of the requirements for modelling and analysis of spatio-temporal data are highlighted. Specifically, the data model should provide constructs for the representation of change over time. Most geographic entities are not *static*, as considered in conventional GIS, but *dynamic*. That is, both their position (e.g., vessel) and shape (e.g., burning area) change over time. Hence, the data model should be able to represent (Peuquet and Wentz, 1994):

a. the location and shape of individual entities through time;
b. the state of a geographic area (i.e., collection of entities) at different temporal instants; and
c. the distribution of geographic entities at various spatial and temporal scales (resolutions).

Regarding the operations for spatio-temporal reasoning, they should be able to retrieve, manipulate and analyze geographic entities based on both their spatial and temporal relationships through a flexible and easy-to-learn interface (Peuquet and Wentz, 1994). These operations fall into two categories depending on whether they form a sequence of individual spatial and temporal operations or not.

Most of the queries given above can be considered as queries of the first category. These queries are called *spatio-temporal selections* and constitute a sequence of individual spatial and temporal operations. For instance, query *V4*: "Does the vessel cross the national boundary on the 3rd of July?" consists of:

- a temporal selection, which derives the vessel trajectory (a set of successive vessel locations (X_i,Y_i) forming a polyline) during July the 3rd; and
- a spatial operation (an intersection test between two polylines), which returns true if the vessel trajectory intersects the polyline of the national boundary and false otherwise.

Similarly, query *F8*: "Which is the fire-front velocity at time T_k?" consists of:

- a temporal selection of fire-front locations GEO_{k-1}, GEO_k, GEO_{k+1} at temporal instants T_{k-1}, T_k and T_{k+1} respectively; and
- a spatial operation, which computes the distances between the three fire-fronts and computes the velocity of interest (using as parameters the values T_{k-1}, T_k and T_{k+1}).

On the other hand, there are queries that cannot be split into individual spatial and temporal operations. Those operations are called *spatio-temporal joins* and compute the intersection of spatial and temporal dimensions assigned to many geographic entities of the database. Example queries of this category are *V7*, *V9* and *F7*. For instance, query *V7*: "Do the two vessels collide?" consists of an intersection (join) of all triplets (X_i, Y_i, T_i), describing the trajectory of the first vessel (or the line segments defined by all pairs of successive triplets) with all triplets (X_j, Y_j, T_j) describing the trajectory of the second vessel (or the corresponding line segments). Similarly, query *F7*: "Do two fire fronts meet?" consists of an intersection (join) of all pairs (GEO_i, T_i) describing the burning area of the first fire with all pairs (GEO_j, T_j) describing the burning area of the second fire.

12.3 STATE OF THE ART

12.3.1 Spatial Data Modelling and Operations

A DBMS repository for the application domain of GIS should provide two alternative views for representing the spatial component of geographic information (Smith *et al.*, 1987; Gatrell, 1991, Güting 1994). Those views are: a) *objects in space* (or single spatial objects); and b) *space* (or spatially related collections of spatial objects).

Spatial objects are defined as a set of locations together with a set of properties characterizing those locations. The basic abstractions of these objects in two-dimensional space are: *point*, *line*, and *region*. On the other hand, the *space* is defined as a set of objects to which attributes (properties) may be attached, together with a set of relationships defined on these objects. The basic abstractions for spatially related collections of spatial objects are: *partitions*, *networks*, *nested partitions*, and *digital terrain (elevation) models*.

This study adopts a model first introduced by Tomlin (1990), which treats in an elegant manner both single and related collections of spatial objects, and can be used to define spatial operations independently on the fundamental models available in literature (Aronoff, 1989): the *vector* model and the *raster* model.

Geographic information can be viewed as a hierarchy of data (Tomlin, 1990; Samet and Aref, 1995; Stefanakis and Sellis, 1996). At the highest level, there is a library of *maps* (more commonly referred to as *layers*), all of which are in registration (i.e., they have a common coordinate system, see Figure 12.2a). Each layer corresponds to a specific theme of interest and is partitioned into *zones* (regions), where the zones are sets of *individual locations* with a common *attribute value* (Figure 12.2b). Examples of layers are the land-use layer, which is divided into land-use zones (for example, wetland, river, desert, city, park and agricultural zones) and the road network layer, which contains the roads that pass through the portion of space that is covered by the layer.

There is no standard algebra defined on geographic data. This means that there is no standardized set of base operations for geographic data handling. Hence, the set of operations available in geographic information systems vary from one system to another and heavily depend on the application domain. However, their fundamental capabilities can be expressed in terms of four types of operations (Tomlin, 1990): a) programming, b) data preparation, c) data presentation, and d) data interpretation operations. *Data interpretation operations* are those that transform data into information and as such they comprise the heart of any system for handling geographic data. Consequently, the following discussion focuses on them.

Data interpretation operations available in GIS characterize (Aronoff, 1989; Tomlin, 1990; Samet and Aref, 1995; Stefanakis and Sellis, 1996): a) *individual locations*, b) *locations within neighbourhoods*, and c) *locations within zones*; and constitute respectively the three classes of operations:

- *Local operations*: they include those that compute a new value for each individual location on a layer as a function of existing data explicitly associated with that location (Figure 12.2d);
- *Focal operations*: they compute new values for every individual location as a function of its *neighbourhood* (Figure 12.2e). A neighbourhood is defined as any set of one or more locations that bear a specified distance and/or topological or directional relationship to a particular location (or set of locations in general), the *neighbourhood focus*; and

- *Zonal operations*: they include those that compute a new value for each individual location as a function of existing values associated with a zone containing that location (Figure 12.2f).

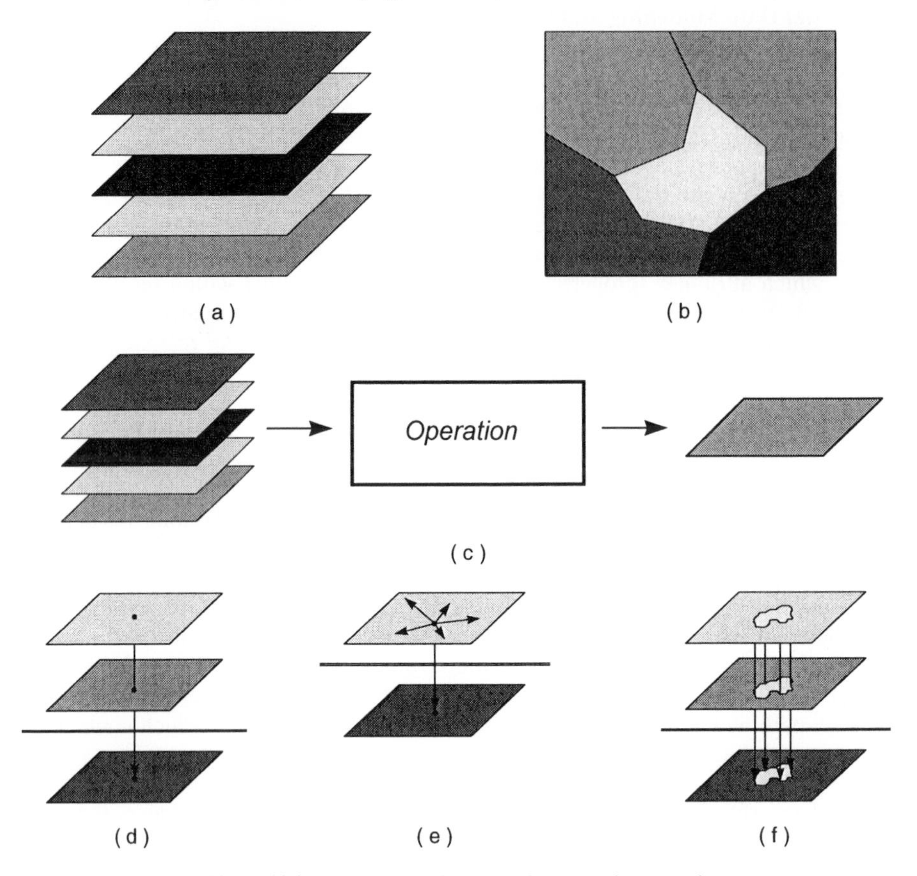

Figure 12.2 The data model and data interpretation operations

All data interpretation is done on a layer-by-layer basis. That is, each operation accepts one or more existing layers as input (the operands) and generates a new layer as output (the product), which can be used as operand into subsequent operations (Figure 12.2c).

Table 12.1 summarizes the basic classes of data interpretation operations accompanied by representative examples (Aronoff, 1989; Tomlin, 1990; Stefanakis and Sellis, 1996; 1998). Notice that data interpretation operations may be combined to compose one or more procedures (a *procedure* is any finite sequence of one or more operations that are applied to meaningful data with deliberate intent) and accomplish a composite task posed by the spatial decision-making process. A simplified example for the task of *site selection* for a residential housing development can be found in (Stefanakis and Sellis, 1996). The basic approach to this is to create a set of *constraints*, which restrict the planning activity, and a set of *opportunities*, which are conducive to the activity. The procedure of site selection is based on the sets of constraints and opportunities, and consists of a sequence of operations that extract the best locations for the planning activity. Notice that a more advanced decision-making process can be achieved by incorporating fuzzy logic methodologies into data interpretation operations (Stefanakis *et al.*, 1996; 1997).

Table 12.1 Basic classes of data interpretation operations

Classes of Operations	Examples of Operations
Local Operations	
• Classification & recoding	*re-code, re-compute, re-classify*
• Generalization	*generalize, abstract*
• Overlay (spatial join)	*overlay, superimpose*
Focal Operations	
• Neighbourhood	
– window & point queries	*zoom-in, zoom-out, point-in-polygon*
– topological	*disjoint, meet, equal, contains, inside, covers, overlap*
– direction	*north, north-east, weak-bounded-north, same-level*
– metric (distance) & buffer zones	*near, about, buffer, corridor*
– nearest neighbour	*nearest-neighbour, k-nearest-neighbours*
• Interpolation	
– location properties	*point-linear, (inverse) distance-weighted*
– thiessen polygons	*thiessen-polygons, voronoi-diagrams*
• Surfacial	
– visualization	*contours, TINs*
– location properties	*height, slope, aspect, gradient*
• Connectivity	
– routing & allocation (network)	*optimum-path-finding, optimum-routing, spread, seek*
– intervisibility	*visible, light-of-sight, viewshed, perspective, illumination*
Zonal Operations	
• Mask queries (spatial selection)	*select-from-where, retrieve*
• Measurement	*distance, area, perimeter, volume*

12.3.2 Temporal Data Modelling and Operations

Several attempts have been made in the past to accommodate and incorporate temporal aspects within information systems (Snodgrass, 1992; Bohlen, 1995). Most of these systems handle one or two basic components of time: a) *valid time*, which records the actual occurrence of real-world events; and b) *transaction time*, which records the time when the information about real-world events is entered into the system. Information systems can be subdivided into four types according to the temporal dimensions they support (Snodgrass, 1992): *static*, *historical*, *rollback*, and *bitemporal*. Static systems do not provide time support; historical systems support valid time only; rollback systems support transaction time only; and bitemporal systems provide support to both valid and transaction times. The rest of the chapter focuses on valid time representation. However, the same general issues are involved in the representation of transaction time as well (Snodgrass, 1992).

Temporal information is represented through *time-stamps* (Snodgrass, 1992). There are three basic types of time-stamps:

- *event time-stamps:* they consist of single chronon identifiers;
- *interval time-stamps:* they consist simply of two event time-stamps, one for the starting event of the interval and one for the terminating event of the interval; and

- *spans:* directed durations of time with known length, but no specific starting and ending events (for example, the duration of 'one week' is known to have a length of seven days, but can refer to any block of seven consecutive days).

Temporal relationship	Definition
A before *B*	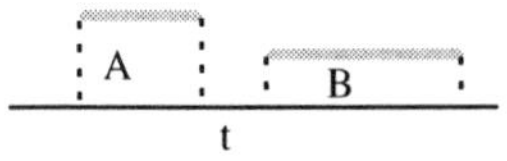 A B t
A equal *B*	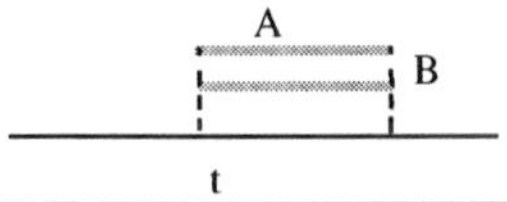 A B t
A meets *B*	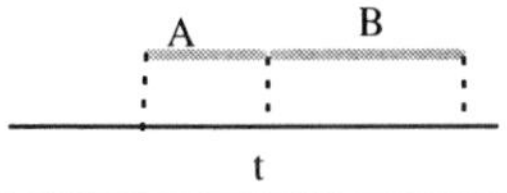 A B t
A during *B*	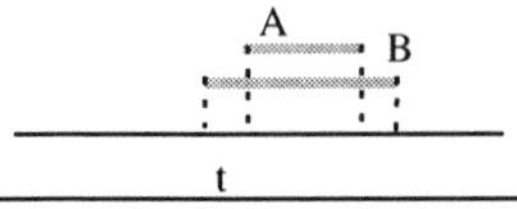 A B t
A overlaps *B*	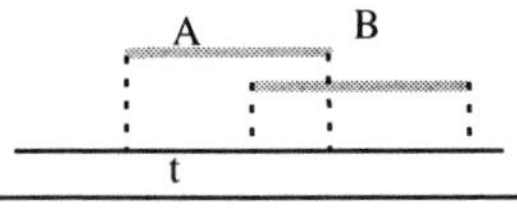 A B t
A ends *B*	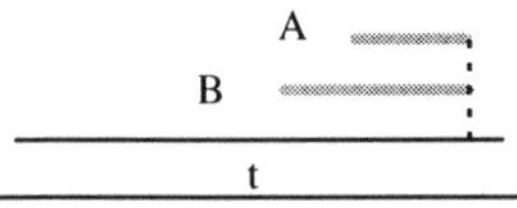 A B t
A starts *B*	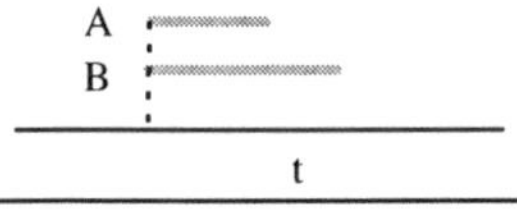 A B t

Figure 12.3 Topological relationships between two interval time-stamps *A* and *B*

These time-stamps may be associated either with entire entities (*temporal entities*) or with individual attributes (*temporal attributes*) characterizing entities.

Although several data models have been enhanced with temporal aspects (e.g., the entity-relationship model in Klopprogge, 1981; semantic data models in Urban and Delcambre, 1986; knowledge-based data models in Dayal and Smith, 1986; and object-oriented models in Rose and Segev, 1991, Wuu and Dayal, 1992), the modelling of temporal information has been predominantly embedded in a relational setting, i.e., time is modelled as an attribute of the relational scheme (Snodgrass and Ahn, 1985; Gadia, 1986; Langran, 1989; Snodgrass, 1992). This is mostly due to the relatively simple structure and semantics of temporal data (in comparison to spatial data) that can, in some way, be represented by the relational model. The extensions to the relational model, to incorporate time, support different types of time-stamps, which are associated with entire tuples or with individual attribute values. An extensive classification of these models to represent both valid and transaction time can be found in (Snodgrass, 1992).

In general, there are many operators over time-stamps that are useful for different applications. The most common operations include *temporal selections* and *temporal joins* (Tansel *et al.*, 1993; Snodgrass, 1995), while more advanced operations, such as *coalescing*, *temporal difference* and *temporal aggregation* have been also examined (Snodgrass, 1992; Jensen *et al.*, 1992).

Temporal selections are the most common operations in spatio-temporal databases (Section 12.3.3). A temporal selection is the retrieval of database entities that satisfy a temporal predicate. The temporal predicate may be a metric or a topological relationship (Allen, 1983) on time-stamps associated either to database entities or to their attributes. Time-stamps can be seen as point (events) or line entities (intervals and snaps) in the one-dimensional space. Hence, the only temporal metric is the *time length* (also referred to as temporal distance), which is a measure of the duration of an interval defined between two events (Peuquet and Wentz, 1994).

On the other hand, topological relationships are defined in terms of relative locations of events and intervals (Allen, 1983; Frank, 1994). Specifically, seven topological relationships hold between two temporal intervals *A* and *B* (Figure 12.3). Notice that an event time-stamp can be seen as an interval time-stamp with the same starting and terminating events.

12.3.3 Spatio-Temporal Data Modelling and Operations

Several spatio-temporal data models have been proposed in the past to record spatial changes over time. Those models adopt the raster or vector model for representing the spatial dimension of geographic entities and extend them to include time.

The only data model available within commercial GIS that can be seen as a spatio-temporal data model is the maintenance of a series of *snapshot images* (Peuquet and Wentz, 1994). Based on the raster model, Langran (1992) assigns to each cell (pixel) a variable-length list denoting successive changes that occurred at the corresponding location; while based on the vector model Langran (1989) introduced the concept of *amendment vectors*, in which any changes subsequent to some initial point in time in the configuration of polygonal or linear features are incrementally recorded as additions to the original features or objects.

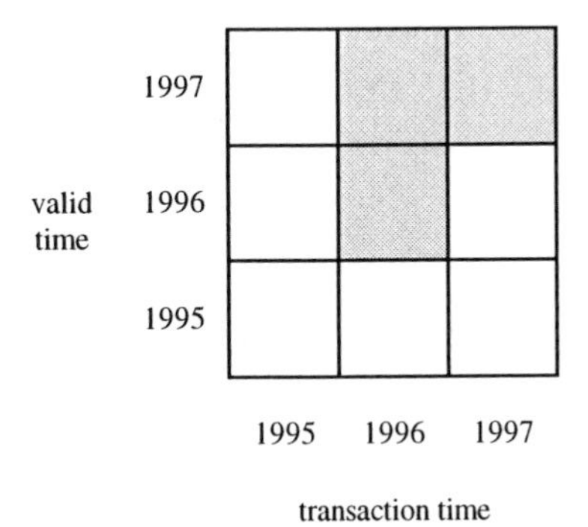

Figure 12.4 Example of a bitemporal element

In order to consider the spatial and temporal domains under a unified model, Worboys (1994a, b) assigns *bitemporal elements* (BTE) to geometric primitives of simplicial complexes (Worboys, 1992), which are adopted to represent the spatial dimension of geographic entities. A BTE (Snodgrass, 1992) is defined as the union of a finite set of Cartesian products of intervals of the form $I_T x I_V$, where I_T is an interval of transaction time and I_V is an interval of valid time. For a BTE T, $(t_T, t_V) \in T$, if and only if, at time t_T there is information in the database that the object bitemporally referenced by T exists at valid time t_V. Figure12.4 illustrates the BTE for a real-world object through the last three years and with a time interval of one year. Furthermore, operations, like union, intersection, difference, spatial projection and temporal projection on spatio-temporal objects are provided by the unified model.

12.4 REPRESENTATION ISSUES: EXTENDING THE DATA MODEL

The purpose of this section is to show how the temporal dimension of geographic features may be incorporated into the general spatial data model introduced by Tomlin (1990) and briefly presented in Section 12.3. In that model *individual locations* constitute the basic entities and *attribute values* regarding various *themes* (layers) are assigned to them.

Each theme (layer) of the region under study is described through a set of attribute values and each individual location is assigned one of these values. The assignment of an attribute value to an individual location is assumed to last forever and no reference to its appearance or duration is made. This assumption is not convenient for most geographic applications, where the attributes characterizing individual locations change over time (for example, changes in crop type of a land).

In order to represent temporal changes there are two possibilities:

1. assignment of *time-stamps* in the themes (layers) associated to the region under study; and
2. attributes) to individual locations of each theme.

The first solution is the most trivial one and is raised in the maintenance of snapshots for the themes associated to the region under study. Figures 12.5a and 12.6a illustrate two examples of this representation for the applications presented in Section 12.2. Notice that this model provides images of the region of interest at different points in time.

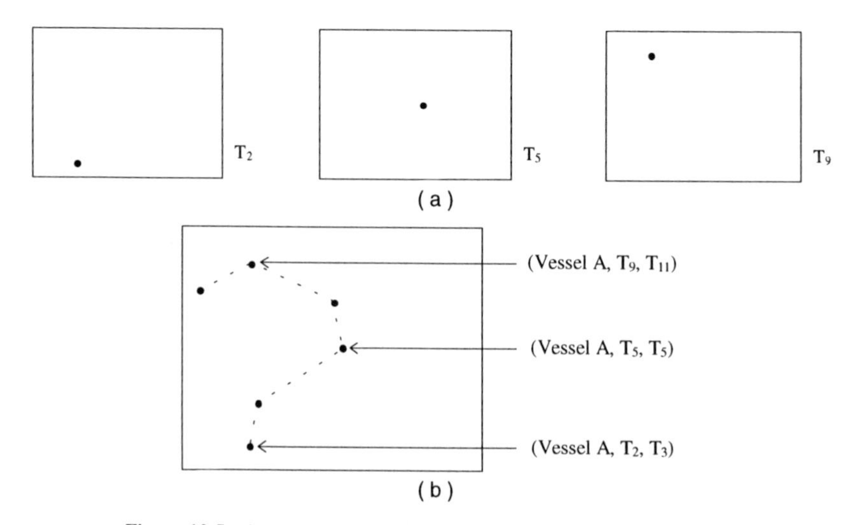

Figure 12.5 The two alternative representations for a moving point object

The second solution incorporates explicitly the temporal dimension of geographic features into the data model. What is proposed is to replace the single attribute value, which is assigned to each individual location of a layer, with composite attribute values with time-stamps. Specifically, each individual location on a layer is assigned a list of *temporal attributes*, that is a set of ordered triplets: (AV, T_{from}, T_{to}), where AV denotes the attribute value characterising that location in the corresponding layer during the interval time-stamp (T_{from}, T_{to}). Figures 12.5b and 12.6b illustrate two examples of this representation for the applications presented in Section 12.2.

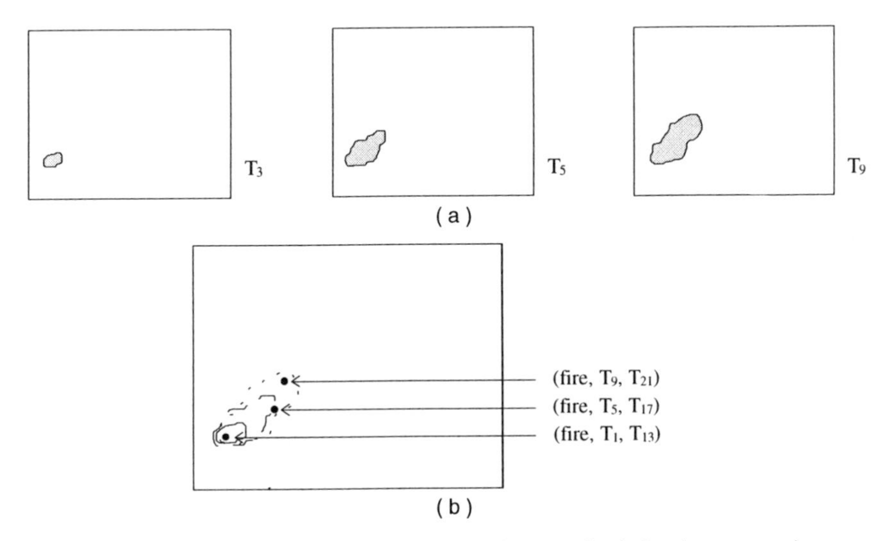

Figure 12.6 The two alternative representations for an areal unit that changes over time

With the latter representation a change to an individual location regarding its attribute values (for example, crop type) is recorded as a new triplet in the list of triplets (Figure 12.7) assigned to that location in the corresponding layer (for example, layer of crop). Hence, minor changes in the area of interest are easily recorded, in contrast to the first solution (i.e., the maintenance of snapshot images), in which major changes are only recorded in order to avoid the generation of new layers.

The concept of zone is also changing in the extended model. Specifically, a *temporal zone* is defined as the set of individual locations on a layer with a common attribute value during an interval time-stamp that characterizes the zone. Notice that this time-stamp does not necessarily coincide with the time-stamps of the individual locations composing the zone and may be degenerated into an event time-stamp (i.e., an instant zone).

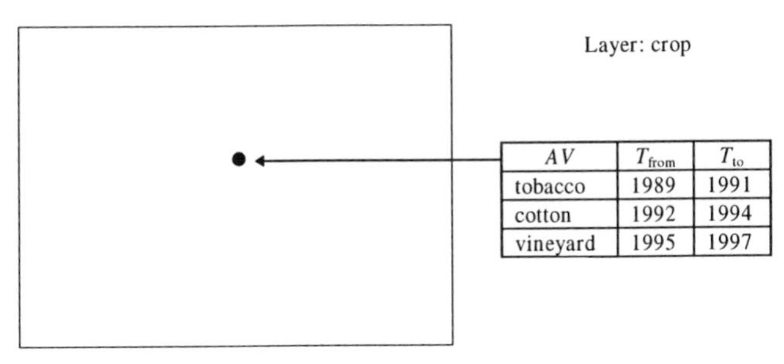

AV	T_{from}	T_{to}
tobacco	1989	1991
cotton	1992	1994
vineyard	1995	1997

Figure 12.7 Representation of change using a list of temporal attributes

In summary, the spatio-temporal model consists of a library of *layers*, all of which are in registration. Each layer corresponds to a specific theme of interest and consists of a set of *individual locations*. Each individual location in a layer is assigned a *list of temporal attributes*, where a temporal attribute is defined as an attribute value with an interval time-stamp. All individual locations on a layer with common attribute values assigned to them during a specified interval time-stamp constitute a *temporal zone*.

12.5 REASONING ISSUES: EXTENDING THE OPERATIONS

Following the incorporation of temporal dimension of geographic entities into the spatial data model, the reasoning issue should be examined. Specifically, the basic data interpretation operations (Section 12.3.1) should be extended in order to handle the lists of temporal attributes assigned to individual locations on a layer. After a redefinition of the basic classes of data interpretation operations, one representative operation for each class is examined, accompanied with examples that commonly appear in spatio-temporal reasoning.

Provided that the thematic information associated with individual locations on a layer changes over time and this change is recorded using temporal attributes, the three classes of data interpretation operations are redefined as follows:

- *Temporal local operations*: they include those that compute a new list of temporal attributes for each individual location on a layer as a function of existing data explicitly associated with that location; for example, temporal overlay operation.
- *Temporal focal operations*: they compute a new list of temporal attributes for each individual location on a layer as a function of its *neighbourhood*. A neighbourhood is defined as any set of one or more locations that bear a specified distance and/or topological or directional relationship to a particular location (or set of locations in general), the *neighbourhood focus*; for example, temporal distance operation; and
- *Temporal zonal operations*: they include those that compute either a new list of temporal attributes or single attribute values for each individual location as a function of existing values associated with a temporal zone containing that location; for example, temporal select operation.

12.5.1 Temporal Overlay Operation

The overlay operation is commonly applied in GIS. It is analogous to join operation in conventional database systems, and is defined as the assignment of new attribute values to individual locations resulting from the combination of two or more layers. The *temporal overlay operation* takes a more general form and is defined as the assignment of a new list of temporal attributes to individual locations resulting from the combination of the lists of temporal attributes associated with that location on two or more layers.

In order to make clear how temporal overlay operation works, let us assume two layers *A* and *B* and an individual location *L* on them, which is assigned the lists of temporal attributes shown in Table 12.2. Assuming that A_i (B_i) occurs before A_{i+1} (B_{i+1}) and there are no gaps between successive interval time-stamps (i.e., $t_{(Ai)to} = t_{(Ai+1)from}$ and $t_{(Bi)to} = t_{(Bi+1)from}$) the temporal attributes can be represented as line segments in the one-dimensional space of time.

Table 12.2 Two lists of temporal attributes

layer A			*layer B*		
AV	T_{from}	T_{to}	*AV*	T_{from}	T_{to}
A1	$t_{(A1)from}$	$t_{(A1)to}$	B1	$t_{(B1)from}$	$t_{(B1)to}$
A2	$t_{(A2)from}$	$t_{(A2)to}$	B2	$t_{(B2)from}$	$t_{(B2)to}$
A3	$t_{(A3)from}$	$t_{(A3)to}$	B3	$t_{(B3)from}$	$t_{(B3)to}$

Figure 12.8 illustrates graphically the situation presented in Table 12.2 as well as the result of the overlay operation. Depending on the *duration* and *temporal distribution* of interval time-stamps the overlay operation results in a new set of interval time-stamps, which are assigned composite attribute values derived from the combination of attribute values of input layers. These composite attribute values with the corresponding interval time-stamps constitute the temporal attributes of the new layer.

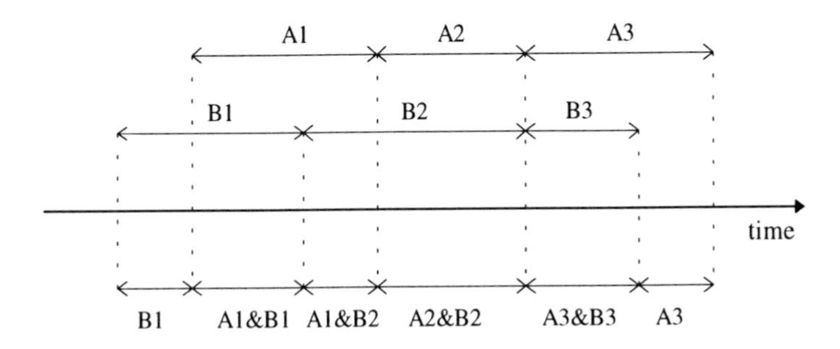

Figure 12.8 Temporal overlay operation

Figures 12.9 and 12.10 illustrate two examples of temporal overlay operation. The first example (Figure 12.9) shows how a layer of *tobacco farmers* is generated by performing a temporal overlay operation with the operands being the layer of *land owners* and the layer of *crop types*. The second example (Figure 12.10) shows how the layer *of wind velocity at vessel A* is derived from the layer of *vessel trajectories* and the layer of *wind velocity*. Obviously, only locations traversed by vessel *A* are assigned lists of temporal attributes in the new layer.

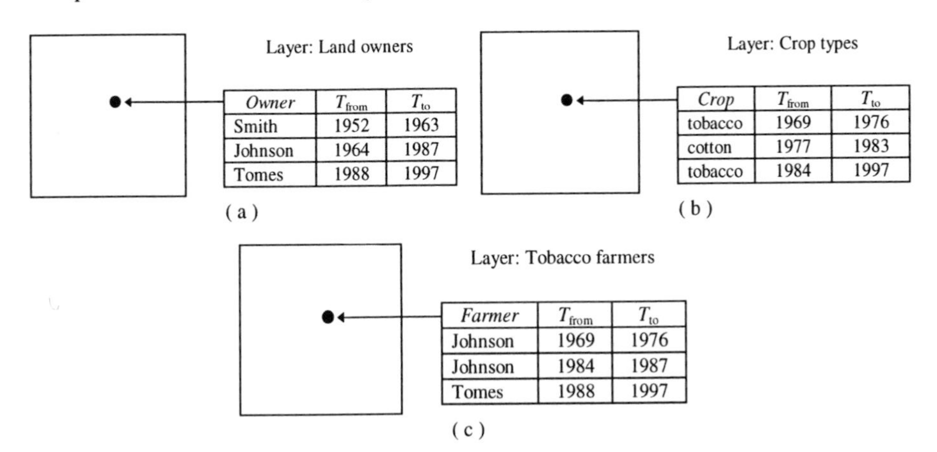

Figure 12.9 Temporal overlay operation: A real-world example for changing areal units

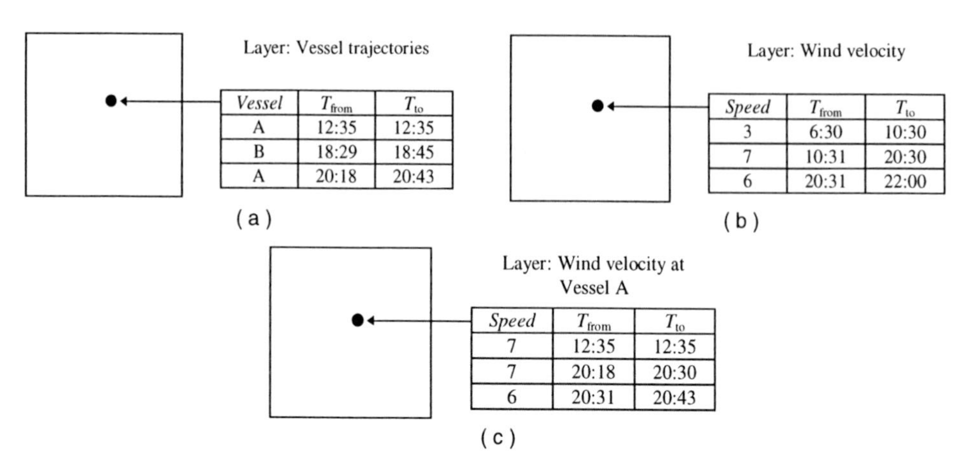

Vessel	T_{from}	T_{to}
A	12:35	12:35
B	18:29	18:45
A	20:18	20:43

Speed	T_{from}	T_{to}
3	6:30	10:30
7	10:31	20:30
6	20:31	22:00

Speed	T_{from}	T_{to}
7	12:35	12:35
7	20:18	20:30
6	20:31	20:43

Figure 12.10 Temporal overlay operation: A real-world example for moving point objects

12.5.2 Temporal Distance Operation

A distance metric is usually required in order to analyze the spatial relationships between entities in GIS. Several metrics are available (Preparata and Samos, 1985; Gatrell, 1991). The choice depends on the application domain and the requirements posed by decision-makers. In this study, the conventional Euclidean distance is considered. For two individual locations i, j the Euclidean distance is given by the formula:

$$d(i,j)=\sqrt{\left(x_i-x_j\right)^2+\left(y_i-y_j\right)^2} \tag{12.1}$$

where (x_i, y_i) and (x_j, y_j) denote the coordinates of the two locations i, j.

What differentiates *temporal distance operation* from conventional distance operation is that, since the entities location and shape change over time, the distance metrics referenced to those entities changes accordingly. Hence, the distance between two dynamic entities or between a static and a dynamic entity always refers to a particular event or interval time-stamp. Following, the problem of assigning to an individual location K a list of temporal attributes depicting its distance from a moving point object is examined.

Assume a moving point object P whose track is represented in a layer A: Each individual location L of layer A traversed by P is assigned a triplet of the form (P, T_{from}, T_{to}), where (T_{from}, T_{to}) is the interval time-stamp during which object P has been at L. In order to compute the list of temporal attributes depicting the distance of individual location K from P, initially, all these intervals time-stamps are retrieved (by performing a temporal select operation; see Section 12.5.3) along with the coordinates of the corresponding locations. Hence a list of new triplets of the form: $(L_i, T_{(i)from}, T_{(i)to})$ is generated. Then, using Eq. 1, a distance measure $D_i = d(K,L_i)$ is assigned to each interval time-stamp (i) of the list and the list of temporal attributes $(D_i, T_{(i)from}, T_{(i)to})$ is finally assigned at location K.

More complex problems, such as the computation of temporal distances between a given location and a changing areal unit, or between two dynamic objects, can be solved in a similar manner. Figure 12.11 illustrates an example of generating a new layer, which depicts the temporal distance of a moving vessel B from a moving vessel of A (i.e., a moving neighbourhood focus). Obviously, only locations traversed by vessel A are assigned lists of temporal attributes in the new layer. Notice that, similarly to temporal distance operation, other focal operations, such as temporal direction (for example, north, east, south-west, etc.) and temporal topological operations (for example, disjoint, meet, overlap, etc.) may be defined.

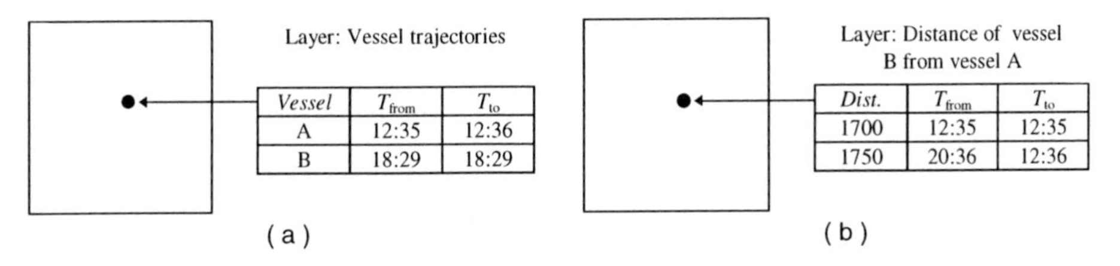

Layer: Vessel trajectories
Vessel
T_{from}
T_{to}
A
12:35
12:36
B
18:29
18:29
(a)
Layer: Distance of vessel B from vessel A
Dist.
T_{from}
T_{to}
1700
12:35
12:35
1750
20:36
12:36
(b)

Figure 12.11 Temporal distance operation: A real-world example for moving point objects

12.5.3 Temporal Select Operation

The scope of *temporal select operation* is to highlight individual locations on a layer based on the list of temporal attributes assigned to them. The temporal predicate involved in the selection may be a metric or topological relationship on interval time-stamps of temporal attributes (Section 12.3.2). Notice that those locations may be filtered by static or temporal zones of other layers as in standard zonal select operation (mask query).

Obviously, the temporal select operation is usually applied in order to generate *snapshot images* (layers) depicting the situation of single entities or of the entire area under study at specified points in time. Those static layers are compatible with the standard model for pure spatial data, presented in Section 12.3.1, and all data interpretation operations for handling the spatial dimension of geographic data can be applied.

12.6 CONCLUSION

The primary goal for any information system, including GIS, is to model accurately the real world. Obviously, a dynamic world can only be captured and represented by a DBMS repository that handles changing information. Provided that most geographic applications involve data that change over time, a DBMS repository for the application domain of GIS should accommodate the temporal dimension of geographic entities. Focusing on this direction the chapter examines several issues regarding the development of a DBMS repository for temporal GIS.

Specifically, the contribution of the chapter can be summarized as follows:

- Through a couple of simplified real-world example applications, with moving point objects or areal units that change over time, several of the requirements regarding the modelling and analysis of spatio-temporal data are highlighted.
- After a brief presentation of the current state-of-the-art in the areas of spatial, temporal, and spatio-temporal databases the issues of representation and reasoning in temporal GIS are addressed.
- The general spatial data model first introduced by Tomlin (1990) is extended to accommodate the temporal dimension associated to geographic entities.
- It is shown how the standard data interpretation operations for handling static geographic data can be extended to handle the new constructs of the extended data model and support spatio-temporal reasoning. Specifically, representative spatio-temporal operations, such as the temporal overlay, temporal distance, and temporal selection operations are examined in more detail. Several simplified real-world examples are given.

Future research in the area includes:

- The design and implementation of a prototype DBMS repository for the application domain of temporal GIS. Such a task includes: a) the design of appropriate data types for representing spatio-temporal data at the physical level; b) the development of operations, query languages and flexible user-interfaces to meet the requirements for spatio-temporal reasoning in various application domains; and c) the design and implementation of effective algorithms, techniques and structures to support the efficient execution of single spatio-temporal operations or composite procedures.
- The incorporation of fuzzy concepts into the spatio-temporal representation and reasoning. The problem of incorporating fuzzy set methodologies into the spatial data model and the basic data interpretation operations for handling static geographic data has been examined in (Stefanakis *et al.,* 1997). Apparently, the extension for dynamic geographic data should be beneficial, since the *imprecision* in both the spatial and temporal dimensions of geographic entities will be addressed and handled.

ACKNOWLEDGEMENTS

This research has been partially supported by a research grant from the General Secretariat of Research and Technology of Greece (YPER '94) and by the European Commission-funded TMR project CHOROCHRONOS.

REFERENCES

Allen, J.F., 1983, Maintaining knowledge about temporal intervals. *Communications of the ACM*, **26**, pp. 832–843.

Al-Taha, K.K., 1992, Temporal Reasoning in Cadastral Systems. Ph.D. thesis, University of Maine, USA.

Aronoff, S., 1989, *Geographic Information Systems: A Management Perspective*, (WDL Publications).

Bohlen, M.H., 1995, Temporal database system implementations. *SIGMOD Record*, **24**, pp. 16–30.

Dayal, U. and Smith, J.M., 1986, PROBE: a knowledge-oriented database management system. In *Knowledge Base Management Systems: Integrating Artificial Intelligence and Database Technologies*, (Springer-Verlag).

Dooling, D., 1994, Navigating close to shore. *IEEE Spectrum*, **31**, pp. 24–31.

Fischer, M.M., 1994, Expert systems and artificial neural networks for spatial analysis and modelling. *Geographical Systems Journal*, **1**, pp. 221–235.

Frank, A.U., 1994, Qualitative temporal reasoning in GIS–ordered time scales. In *Proceedings of the 6th International Symposium on Spatial Data Handling (SDH '94)*, Edinburgh, UK, pp. 410–430.

Gadia, S.K., 1986, Toward a multihomogeneous model for a temporal database. In *Proceedings of the IEEE International Conference on Data Engineering*, Los Angeles, CA, pp. 390–397.

Gatrell, A.C., 1991, Concepts of space and geographical data. In *Geographic Information Systems: Principles and Applications*, edited by Maguire *et al.* (Longman), pp. 119–134.

Goodchild, M.F., 1991, The technological setting of GIS. In *Geographic Information Systems: Principles and Applications*, edited by Maguire *et al.* (Longman), pp. 45–54.

Güting, R.H., 1994, An introduction to spatial database systems. *VLDB Journal: Special Issue on Spatial Database Systems*, **3**, pp. 357–399.

Hopkins, L.D., 1984, Evaluation of methods for exploring ill-defined problems. *Environmental Planning B: Planning and Design*, **11**, pp. 339–348.

Jensen, C.S., Clifford, J., Gadia, S.K., Segev, A. and Snodgrass, R.T., 1992, A glossary of temporal database concepts. *SIGMOD Record*, **21**, pp. 35–43.

Klopprogge, M.R., 1981, TERM: an approach to include the time dimension in the entity-relationship model. In *Proceedings of the 2nd International Conference on the Entity Relationship Approach*, Washington, DC, pp. 477–512.

Langran, G., 1989, A review of temporal database research and its use in GIS applications. *International Journal of Geographical Information Systems*, **3**, pp. 215–232.

Langran, G., 1992, *Time in Geographic Information Systems*, (London: Taylor & Francis).

Laurini, R. and Thompson, D., 1992, *Fundamentals of Spatial Information Systems*, (San Diego: Academic Press).

Leung, Y. and Leung, K.S., 1993, An intelligent expert system shell for knowledge-based GIS: 1. The tools, 2. Some applications. *International Journal of Geographical Information Systems*, **7**, pp. 189–213.

Lymberopoulos, N., Papadopoulos C., Stefanakis, E., Pantalos N. and Lockwood, F., 1996, A GIS-based forest fire management information system. *EARSeL Journal, Advantages in Remote Sensing: Remote Sensing and GIS Applications for Forest Fire Management*, **4**, pp. 68–75.

Maguire, D.J., Goodchild M.F., and Rhind, D.W., Eds, 1991, *Geographic Information Systems: Principles and Applications*, (Longman).

Peuquet, D. and Wentz, E., 1994, An approach for time-based analysis of spatio-temporal data. In *Proceedings of the 6th International Symposium on Spatial Data Handling (SDH '94)*, Edinburgh, UK, pp. 489–504.

Preparata, F.P., and Samos, M.I., 1985, *Computational Geometry*, (Springer-Verlag).

Rose, E. and Segev, A., 1991, TOODM: a temporal object-oriented data model with temporal constraints. In *Proceedings of the 10th International Conference on Entity Relationship Approach*, San Mateo, CA, pp. 205–230.

Samet, H. and Aref, W.G., 1995, Spatial data models and query processing. In *Modern Database Systems*, edited by Kim, W. (ACM Press), pp. 339–360.

Smith, T.R., Menon, S., Star, J.L. and Estes, J.E., 1987, Requirements and principles for the implementation and construction of large-scale GIS. *International Journal of Geographical Information Systems*, **1**, pp. 13–31.

Snodgrass, R.T., 1992, Temporal databases. In *Theories and Methods of Spatio-Temporal Reasoning in Geographic Space, International Conference GIS, Pisa*, Lecture Notes in Computer Science 639, (Berlin: Springer-Verlag), pp. 22–64.

Snodgrass, R.T., 1995, *The TSQL2 Temporal Query Language*, (Kluwer Academic Publishers).

Snodgrass R.T. and Ahn, I., 1985, A taxonomy of time in databases. In *Proceedings of the ACM SIGMOD Conference*, Austin, Texas, pp. 236–246.

Stefanakis, E., 1997, Development of Intelligent GIS. Ph.D. thesis, Dept. of Electrical and Computer Engineering, National Technical University of Athens.

Stefanakis, E. and Sellis, T., 1996, A DBMS repository for the application domain of GIS. In *Proceedings of the 7th Int. Symposium on Spatial Data Handling (SDH '96)*, Delft, The Netherlands, pp. 3B19–3B29.

Stefanakis, E. and Sellis, T., 1997, Towards the design of a DBMS repository for the application domain of GIS: Requirements of users and applications. In *Proceedings of the 18th ICA International Cartographic Conference*, Stockholm, Sweden.

Stefanakis, E. and Sellis, T., 1998, Enhancing Operations with Spatial Access Methods in a DBMS for GIS. *Cartography and Geographic Information Systems*, **25**, pp. 16–32.

Stefanakis, E., Vazirgiannis, M. and Sellis, T., 1996, Incorporating fuzzy logic methodologies into GIS operations. In *Proceedings of the 1st International Conference on Geographic Information Systems in Urban, Regional and Environmental Planning*, pp. 61–68. Also in *Proceedings of the XVIII ISPRS Congress*, Vienna, Austria.

Stefanakis, E., Vazirgiannis, M. and Sellis, T., 1997, Incorporating fuzzy set methodologies in a DBMS repository for the application domain of GIS. Technical Report, Dept. of Electrical and Computer Engineering, National Technical University of Athens, Greece. Submitted for publication.

Tansel, A., Clifford, J., Gadia S., Jajodia S., Seveg A. and Snodgrass R.T., 1993, *Temporal Databases: Theory, Design and Implementation*, (Benjamin-Cummings).

Tomlin, C.D., 1990, *Geographic Information Systems and Cartographic Modeling*, (Prentice Hall).

Urban, S.D. and Delcambre, L.M.L., 1986, An analysis of the structural, dynamic, and temporal aspects of semantic data models. In *Proceedings of the IEEE International Conference on Data Engineering*, Los Angeles, CA, pp. 383–389.

Worboys, M.F., 1992, A generic model for planar geographical objects. *International Journal of Geographical Information Systems*, **6**, pp. 353–372.

Worboys, M.F., 1994a, A unified model for spatial and temporal information. *The Computer Journal*, **37**, pp. 26–34.

Worboys, M.F., 1994b, Unifying the spatial and temporal components of geographical information. In *Proceedings of 6th International Symposium on Spatial Data Handling (SDH '94)*, Edinburgh, UK, pp. 505–517.

Wuu, G. and Dayal, U.A., 1992, A uniform model for temporal object-oriented databases. In *Proceedings of the IEEE International Conference on Data Engineering*, Tempe, Arizona, pp. 584–593.

PART FOUR

Applications

INTRODUCTION

This part opens with an application-oriented paper, which reviews the question Jonathan Raper has posed initially: how and why do we form Spatial Socio-Economic Units? Stuart Aitken asks the question in the concrete setting of practical work in a suburban environment and the planning of a freeway extension. With this application area also comes a concern not so much for abstract space, than for the meaningful space filled with human activity. In this post-modern view using critical social theory, the complex interaction between space and the population is a consequence of social processes. It argues against a separable fundamental nature of either space or time and stresses the central concern for change. This coincides with the argument that space is not an absolute category, but is the product of social interaction and thus different spaces are qualitatively (socially) different. Introducing scale and collections of increasingly larger spaces related to corresponding activities allows the study of how the meaning of space travels around these 'scale' relationships between nested regions.

The chapter then shows how these abstract considerations are put to work in a concrete example of route selection for a freeway. In one case, a tentatively planned route and the acquisition of land in its corridor is related to a later survey of the population of the area of interest. One can observe that people's familiarity with the space around them has a sharp break, which coincides with the land previously acquired by the state and then left undeveloped. The tentative route then becomes a self-fulfilling prophecy and the construction of the freeway does not meet with resistance. But how is this outcome determined by spatial constraints or social process?

The following case study by Denis Gautier shows how advanced methods of spatial analysis can be employed to understand the spatial dynamics of a community, allowing insight into the interaction between the evolution of a biological ecotype and the development of the rural economy. It focuses on the interaction and the change in the natural and human environment—the landscape and its ecology compared with the rural community and the changes it experiences as a result of the economic changes in this century. Two time scales play a part: the annual cycle of plant growing and fruit harvest, and the growth and change of ecological units over multiple years. A similar case was described in great detail in (Cheylan and Lardon, 1993), where diurnal, multiple week cycles and annual cycles of movement of sheep on the Alps overlap. But also multiple spatial scales must be used, from the region to the morphological unit of the landscape. It further describes how visible morphological elements of the landscape change under the influence of human activities or patterns of human activities and their relationship to the economic cycles. After long periods, human activity can significantly change the face of the landscape, and are eventually visible at a large scale (Papagno, 1992).

The particulars of the example is a study of the decrease in the harvesting from chestnut forest in central France, which influences the state of the forest and in consequence becomes a visible change of the landscape—with a considerable lag from the cause. Understanding such complex chains of causes and effects and counteracting them with appropriate measures is crucial for rational planning of the use of space in a changing world.

The method used applies three different levels of analysis: difference between successive states, difference between the exploitation processes in different epochs and finally, the construction of function models located in space. This chapter thus exemplifies some of the demands of the first chapter in this part and shows how modern methods of spatial analysis can be used to address the topics advocated by critical theory. There are a number of practical problems, including questions of the scale used for data collection, which must be overcome to achieve the results presented here, in order to integrate the understanding and the interaction of human and ecological processes at multiple scales.

The study of May Yuan links semantics and space with temporal objects, in this case objects which move, change their form and other characteristics. She repeats Aitken's criticism that the geometric objects managed in a GIS are meaningless (in the sense of 'free of human meaning') containers and she complements them with meaningful spatial objects, movement of which can be traced in time and space. A model that gives equal weight to the semantic, spatial and temporal aspects of an object avoids the disadvantage of the current partial data models used in GIS. She connects these three aspects with the 'life and 'motion' distinction of Frank (Chapter 2) and extends it to queries. A challenging set of spatio-temporal queries about the locations and the dominance of service areas of supermarkets are shown and a simple query language given, with which these queries can be asked. Yuan also sketches the process of query evaluation to demonstrate that these concepts can be implemented.

In the fourth chapter Mauro Salvemini reports from the position of a city planner where the conflict between physical space and the relevance of space for processes is further complicated by the multiple processes that lead concurrently to multiple subdivisions of urban space of widely varying spatial size and form. It is difficult to identify a common building block useful for all purposes and a single level of resolution for all processes cannot be identified. The aggregation of small pieces of land as necessary for one aspect does not necessarily provide an appropriate model of space for another process. Different types of application, from a spatial decision support system to exploratory spatial data analysis pose different and conflicting requirements: ease of analysis or visualization are quite different uses of the same data.

Modern tools, from temporal databases for GIS to data mining, are very attractive and open new possibilities for planners, but do not overcome the fundamental problems in planning: the data available are typically not directly related to the planning decision. At the end this chapter presents an example that demonstrates how the statistical definition of spatial socio-economic units for data collection remains lagging behind the actual changed territory. This is extremely sensible, in areas where unorganized building activities change the gestalt of the territory quickly (for example, squatters, massive urbanization without planning and building permits) and the adaptation of the spatial socio-economic units to capture the new reality always lags behind. The planned use of census data to update the planning database failed due to the differences in the location of census tract boundaries and planning zones.

The last chapter in this section on applications returns to the cadastre. The cadastre is a fundamental instrument in the human relationship to space and has acquired an elaborate set of rules over the hundreds of years of the existence of the modern cadastre. The last chapter by Khaled Al-Taha explores these rules not only to help the designers of GIS who are used to manage cadastral data, but also to make the accumulated knowledge about the social management of land parcels available for the design of GIS in general. He assumes that legal rules clarify the interaction between people and land and that the methods found to resolve conflicts may be useful beyond the limited purview of the cadastre.

Al-Taha investigates the cadastral system in the United States of America. He first gives diagrams for the succession of events that lead to a current cadastral situation (an alternative graphical language has recently been presented by Hornsby and Egenhofer, 1997). He then details the legal rules describing a valid transfer of ownership, with respect to the dates a contract is concluded between the parties and the time it is registered. Registration makes the content of a contract available to others so they can know it: in legal terms, one gives notice. This is a prime example for the valid time and transaction time in temporal databases.

The basic principle is that one must check that the person from whom one buys a property has properly acquired it and not sold it previously—or at least that a previous sales contract is not yet recorded (Frank, 1996). Different states in the USA give various amounts of protection for somebody who acquires a parcel without registration of his contract. Al-Taha formalizes first the rules for a good title and then proceeds to give a formal reasoning system for ownership in first-order predicate calculus. A GIS that allows queries based on these rules has—and this is a technically important point—to select the logic used, the set of rules applied, according to the statute of the state of the property, and must be prepared to use different rules for the same query.

REFERENCES

Cheylan, J.-P. and Lardon, S., 1993, Towards a conceptual data model for the Analysis of spatio-temporal processes: The example of the search for optimal grazing strategies. In *Spatial Information Theory—Theoretical Basis for GIS (European Conference COSIT '93)*, Lecture Notes in Computer Science 716, edited by Frank, A.U. and Campari, I. (Berlin: Springer-Verlag), pp. 158–176.

Frank, A.U., 1996, An object-oriented, formal approach to the design of cadastral systems. In *Proceedings of the 7th International Symposium on Spatial Data Handling, SDH '96*, Delft, The Netherlands, (IGU) pp. 5A.19–5A.35.

Hornsby, K. and Egenhofer, M.J., 1997, Qualitative representation of change. In *Spatial Information Theory—A Theoretical Basis for GIS (Int. Conference COSIT '97)*, Lecture Notes in Computer Science 1329, edited by Hirtle, S.C. and Frank, A.U. (Berlin: Springer-Verlag), pp. 15–33.

Papagno, G., 1992, Seeing time. In *Theories and Methods of Spatio-Temporal Reasoning in Geographic Space*, Lecture Notes in Computer Science 639, edited by Frank, A.U., Campari, I and Formentini, U. (Berlin: Springer-Verlag).

CHAPTER THIRTEEN

Critically Assessing Change: Rethinking Space, Time and Scale

Stuart C. Aitken

> ... in its broadest formulation, society is necessarily constructed spatially, and that fact—the spatial organization of society—makes a difference to how it works (Doreen Massey, 1994, p. 254).

13.1 INTRODUCTION

It is trite to say that ours is a spatial world through which we pass in time. Nonetheless, the apparent naturalness of our everyday time–space geographies often relegates our understanding of space to a mosaic that simply contains social activities. Within this formulation, our activities fill space, and time is a sequential, if not linear, constraint upon those activities. This chapter begins with the suggestion that questions posed by geographers and other social scientists interested in socio-economic units are still, to a large degree, circumscribed by narrow structural definitions of space and time. For example, geographers who ask, "Why do people do *that* there?" are not much more than cartographers who map social patterns. Even questions that seemingly probe the social and cultural contexts of space—"What do people have to do to keep doing *that* there?"—for the most part assume a Cartesian view of space. The overarching premise of this chapter is that we need to reconstitute our questions to penetrate the power of space and time or, put another way, we need to pose our questions in a way that de-stabilizes the 'fundamental nature' of space and time. This premise is based upon the belief that space and time are complex social products and the social is a complex spatial and temporal product. For example, a question such as "What is intrinsic about the way space and time are constituted there that enables people to keep doing *that* and how, in turn, does *that* actively constitute space?" opens up the possibility of different productions of space and time. A central tenet of the chapter, then, is the nonsense of an assumed 'fundamental nature' for either space or time.

Certain advances in critical social theory enable us to pose questions that contest the naturalness of space and time. In an attempt to align this discussion more coherently with the book's focus on 'fluid socio-economic' units, the chapter extends insight we have gained from a critical social theory of space and time to a consideration of the production of regions and scale. The chapter begins with a brief review of some of the ways our understanding of space–time relations has changed. Issues that arise from a critical appraisal of those relations are then highlighted and extended to a consideration of scale and regionalization. In the latter half of the chapter, this understanding of scale and region is used to rethink the production of Geographic Information Systems (GIS) in ways that communicate more fully with the people who have to live their day-to-day lives in fluid socio-economic units.

13.2 TIME, SPACE, CHANGE AND CONTESTATION

> No telos of the final society exists, moreover; society understood as a moving and contradictory process implies that change for the better is always possible and always necessary. (Iris Marion Young, 1990, p. 315)

A focus on person/environment *change* offers considerable insight into the interaction between people and places, and society and space (Aitken and Bjorklund, 1988; Aitken, 1992; Kitchin, 1996). Although most of us come to know our environments as stable and enduring phenomena, in actuality they are dynamic and unstable, comprising constant disturbances. Progressive social and economic change eventually foments some disturbance that heightens awareness of the change. When individuals are confronted with extraordinary transformations and restructuring, their alternatives are to accommodate or live with it, to contest or try to change it, to retreat from it, or to be overcome by it. As such, it can be argued that focusing on change as a 'unit of analysis' is preferable to the static or time-slice accounts of human spatial behaviour and potential that circumscribes much empirical research on socio-economic units (cf. Langran, 1989). The problems with operationalizing and interpreting time-slice accounts of socio-economic units are discussed by May Yuan and Mike Worboys (this volume) to the extent that they need not be elaborated upon here. It is important to point out, however, that focusing on change as the unit of analysis is a different ontology from that suggested by Barry Smith and Roberto Casati (this volume). Their Aristotelian reductionism posits a universe of substance and shadow respectively, which needs in some way to be aligned with an empirically observed reality. To counter this, later in this chapter, it will be argued that we need to engender a deep suspicion of empirical regularities. The focus on person/environment *change* taken here falls squarely within Carola Eschenbach's (this volume) ontological concerns with change in European capitals. It is the *change* itself that is of concern rather than the effects of the change on socio-economic units. At a somewhat naive level, it can be argued that the basic units of measurement in this kind of research are changes in the quality of people's relationships with their surroundings (Stokols, 1988, p. 242).

13.2.1 Focusing on Change

Although focusing upon change can be an important starting point for research, it is important to note that the questions asked of behaviourally-based GIS are constrained by the very nature of those databases (cf. Aitken and Prosser, 1990; Aitken *et al.*, 1993; Askov *et al.*, 1994). In particular, we need to question spatio-temporal GIS that suggest the power of the system lies in its ability to 'manipulate' to 'ensure that data will be useful' (Langran, 1993). This problem with contemporary GIS databases will be returned to at the close of this chapter, but for now is important to consider the primary conceptual benefit of focusing upon change; namely, that it side-steps the incredibly thorny problem of separating individuals from their environments, and time from space.

To begin, it seems reasonable to briefly outline some of the ways geographers consider space and time. Jonathan Raper (in this volume) also considers this, but the perspective here extends his arguments by drawing more fully on some recent critical social theoretic perspectives. Specifically, it is suggested that change cannot be understood unless we think critically about the 'naturalness' of space and time **and** about the social production of scale and region. This suggestion leads to speculation on the 'nonsense' of the space of current GIS because that space is almost always embedded in a strategic and instrumental rationality that has little bearing upon lived experience. This is an assertion that has some justification in current academic concerns for the societal impacts of spatial technologies (Aitken and Michel, 1995; Wiener *et al.*, 1995; Pickles, 1995). In what follows, the arguments against mechanistic approaches to planning are broadened with a consideration of the social production of scale and region, but first some groundwork is laid with a discussion of some problems associated with trying to understand the complex relations between space and time.

13.2.2 Space and Time as 'Natural' Resources

For many social scientists, space and time are conceived of solely in terms of their resource potential. Just as we think of allocating particular portions of space to particular uses, so we allocate particular intervals of time to particular uses. Time and space differ from 'natural resources' because neither can be accumulated or moved, and each may be represented as fluid, at least in a psychological sense. Nonetheless, many social scientists do not contest the assumed 'naturalness' of space and time because space is constituted as something that can be colonized and controlled and, similarly, time is managed and used up as it passes.

A significant shift in our thinking about the resource potential of time and space was highlighted by Torsten Hägerstrand's (1970, 1975) now well-worn suggestion that we think of time-space relations in terms of *constraints* on human potential. In addition, his focus on individual space–time trajectories, although not unproblematic, was an important grounding for subsequent work on the relations between social structures and human agency (Giddens, 1979, 1984; Gregory, 1985). As Raper (this volume) notes, Anthony Giddens' structuration theory contrives a spatial formulation for the connections between social structures and human agency. In addition, some theorists have attempted to reconceptualize people's time–space trajectories in ways that illuminate regional and even national identity (Pred, 1984, 1986). Cheylan and Lardon (1993), for example, conceptualize individual spatio-temporal processes in regional configurations that have some important implications for GIS. In many of these studies, however, Cartesian space is still assumed, and scale and region remain as relatively unproblematic existent realities.

13.2.3 The Feminist Critique of Cartesian Space

In addition to the work on structuration theory and realism identified by Raper (this volume), feminist geographers such as Jacqueline Tivers (1985, 1988) and Isobel Dyck (1990) add an additional constraint of *gender identity* to Hägerstrand's theorems and note that local everyday social transactions are an important part in the construction of larger scale political identities, notably motherhood. This work opens up the possibility of understanding how multiple and contradictory identities are contextualized in space and time (Valentine, 1993; Compare, 1993). It also suggests an avenue of inquiry that focuses on how space and scale are produced in social and political contexts (cf. Keating, 1995; Smith, 1995).

In important ways, feminist critiques of Hägerstrand's work are the basis for contemporary debates on the nature of space, time and power. For example, Gillian Rose (1993) points out that although feminists use time geography to highlight everyday constraints on women, such a focus denies the differences between men's and women's lives precisely because time geography assumes that space is transparent:

> Women know that spaces are not necessarily without constraint; sexual attacks warn them that their bodies are not meant to be in public spaces, and racist and homophobic violence delimits the spaces of black and lesbian and gay communities. Transparent space then mimics the public space of Western empowered men, its violence repressed. This space... claims transparency and universality, and represses any difference from itself. (Rose, 1993, p. 76)

It is important to note that Rose's attack on the assumed transparency and neutrality of space comes after nearly a decade of questioning the general nature of space and time and, in particular, the privileging of time over space in Western thought.

13.2.4 The Constrained Knowledge of Space through Time

With Jean Baudrillard's semiotic appraisal of the American psyche in *America* (1989), Fredric Jameson's (1984, 1992) exploration of 'the cultural logic of late capitalism' and Henri Lefebvre's (1991) project on the reproduction of space, there began a reassertion of the importance of the power of space in social life. In the words of Neil Smith (1993, pp. 96–7) "... space is being rediscovered as a neglected world of potentially novel and unexplored concepts" by literary and cultural theorists, social theorists, feminists and, of course, geographers. Amongst geographers, Edward Soja (1989, 1996) is a particularly supportive account of the reassertion of space over time in the humanities and social sciences. Soja's work represents an explicit effort to re-centre geographical space "... against the grain of an ontological historicism that has privileged the separate constitution of being in time for at least the past century" (1989, p. 61). Taking his lead from Jameson (1984) and Foucault (1977), Soja argues that contemporary culture is increasingly dominated by space and a spatial logic rather than time. This logic is neither linear nor hierarchical, nor is it decipherable in any of our old ways of knowing. As such, Smith (1992, p. 60) points out that by the late 1980s, geographical space emerged as a preferred language for interpreting post-modern social experience. The film texts that Jameson analyzes in the *Geopolitical Aesthetic* (1992), for example, are illustrations of the ways the post-modern world conflates ways of knowing "... with geography and endlessly processes images of an unmappable system" (p. 14). Put simply, our post-modern world is structured and reproduced by images with their own spatial logic and this logic cannot be mapped by traditional positivist or structural methods of understanding. Jameson exposes the post-modern city as a living image that alienates people both politically and symbolically from new urban geographies. These geographies are constituted by rapid change, diversity and conflicts of scale that can, for example, bring together elements of the First and the Third World in the space of a city block. As such, a delimitation of any kind of understandable region in terms of socio-economic units is difficult, and will certainly reflect the biases of the delimiter rather than any kind of 'natural' order. As residents of this post-modern world, we are often constrained by an anachronistic Cartesian logic that does not help us to make sense of these new spaces and scale relations.

13.2.5 Post-Cartesian: New Ways of Knowing Space

To begin unravelling contemporary conceptions of space it is useful to describe how our thinking changes when we stop viewing space as simply a two-dimensional mosaic occupied by social activities. New work in the social sciences and humanities throughout the 1970s suggested that space was neither natural, nor merely a container of activities. With the work of Michel Foucault (1977) amongst others, we began to understand that space was constituted through social relations and material social practices. The space of the home, for example, is constructed around a set of social relations between men, women and children. These relations are hierarchical and depict power structures often based upon a patriarchal system within which women and children are subordinate to men. No one should disturb Dad working in the garage or the study, for example, but the kitchen (traditionally the domain of women) is used, walked through and transgressed in a myriad of ways. As such, space is a social construction that defines sets of power relations. By the late 1980s, this notion of space also began to appear naive and one-sided. Henri Lefebvre (1984, 1991) argued that the "space as a social construction" thesis implies that geographical forms and distributions are constituted as simple outcomes of power relations and material social practices. Lefebvre's ideas began to crystallize with the work of Jameson and Soja into an apprehension that not only is "space a social construction" but the social is also a spatial construction (cf. Keating, 1995). Soja (1985) coined the term 'spatiality' as the set of dialectical processes whereby the spatial becomes the social and the social becomes the spatial. For example, patriarchal relations are not only reflected in space, they are also formed by space. Dad's garage and study, to return to the previous example, are often separated from the rest of the house by space, or at least a door. The kitchen, however, is usually the most accessible room in the house and is often travelled through to get elsewhere. If Mum controls the kitchen, does she also control the domestic sphere, or is 'her place' one of subservience to some form of family norm? The power of space is equally evident at the larger urban scale when we reflect on the gender relations sanctioned by suburbs and the private, autonomous 'neotraditional' communities that are now so prevalent in the United States and parts of Europe (Aitken, 1998).

Not only do we need to understand the ways in which meaning translates between the individual, household, community, urban, regional and national scales, but we must also develop a critical appreciation of the social construction of spatial scale itself. The evolving social theory of space begun with the work of Lefebvre, Jameson and Soja is now joined by a critical appraisal of spatial scale and regionalism (Smith, 1992, 1993; Marston, 1995; Herod, 1996). If we assume that power structures are reflected in, and formed by, space then we must also take account of how they are hierarchically ordered and, in turn, produce regions.

13.3 SCALE: RETHINKING SPACE, TIME AND REGIONALIZATION

Scale, like space, is neither natural nor incontrovertible:

> There is nothing ontologically given about the traditional division between home and locality, urban and regional, national and global scales. The differentiation of geographical scales establishes and is established through the geographical structure of social interactions. (Neil Smith, 1992, p. 73)

Neil Smith points out that the language of difference may very well be articulated through spatial scale because it is the social construction of this hierarchical ordering that creates borders and boundaries between people and places: "... it is (the naturalization of) scale that delimits the prison walls of social geography" (Smith, 1992, p. 76). As Sallie Marston (1995) notes, the connections between scales are not given naturally, but rather they are made. It seems reasonable to assert, then, that there are important connections between the production of scale and the delineation of regions.

13.3.1 Region-Making

In his theory of structuration, Anthony Giddens (1984, p. 119) notes that differentiation of space, which he calls 'regionalization,' is not just about some static locale, but refers rather "... to the zoning of time–space in relation to routinized social practices". In an attempt to re-centre space, Giddens explicitly rejects regionalization as an absolute conception of space. He implies that space is differentiated into regions through social practices. Andrew Herod (1996) suggests that such a view of region-making sees the production of geographical scale as very much interconnected with the social practices of everyday life. He goes on to point out, however, that Giddens does not really "... theorize the *process* of scale *production*, and particularly does not relate it to the processes which seek to differentiate and/or equalize space." Neil Smith (1992, p. 67) is equally derisive of Giddens' project, suggesting that his geography is merely 'a mosaic of locales', and, as such, space is constituted as merely a container of social activities:

> In fact, produced space is not simply a mosaic but within capitalist society it is intensely hierarchical, according to divisions of race and class, gender and ethnicity, differential access to work and services and so on.
> The difference between a mosaic and hierarchical space is that in a mosaic difference has been reduced and reified to a single *spatial* dimension that abstracts from the more dynamic and multifaceted political differentiation of space. (Neil Smith, 1992, p. 67)

13.3.2 Temporal and Geographic Scale-Making

It is, of course, important not to privilege space over time and, as such, we need to recognize the interdependence between the production of geographical and temporal scales. If we are willing to accept the premise that geographic scale delimits regions and the production of scale is a deliberate political act, then we must also in some way understand the relationship between temporal and geographic scales. Some of this understanding comes from focusing upon change as a unit of analysis rather than contriving an artificial separation between time and space. Although scale is a key political process in regionalization, there is also a temporal and spatial differentiation that marks the landscape of regions. On the face of the landscape lie its history and the history of the individuals and groups that are a part of it (cf. Papagno, 1992). Such physical historicity, as well as the functions and groups that live in a region at any given time, creates its spatial differentiation. The region is a network of hierarchically created networks that precipitate out as discretely understood places such as neighbourhoods, parks and so forth. As a physical landscape, the region offers changes and surprises not only in transition from one place to another, but also through time (Young, 1990, p. 318).

Change is geographic but it is also rooted in past everyday details as well as serving as the roots of future everyday details (cf. Pred, 1973, 1984). For example, when an individual's daily path is steered through specific environments as a result of involvement in a particular event such as taking children to day-care, she or he is embedded in a local environment of personal contacts and information in general. A person may consciously or unconsciously employ the mental experiences and practical knowledge acquired in this way to formulate goals and intentions, and she or he can delimit possible long-term projects such as residential relocation. These long-term changes are circumscribed by opportunity and constraint and, as such, delineate a larger spatio-temporal environment within which our day-to-day potential operates (Aitken and Fik, 1988, p. 465). This example demonstrates quite clearly the ways that temporal and geographic scales fold into each other. The contention here is that a research focus may begin with change, but it should also consider the ways in which people make, and are made by, scale. This latter point perhaps needs a little more elaboration.

Andrew Herod (1996) suggests that in our rush to understand place-making we have neglected the production of scale as a delimiter of regions. His point is that, as social scientists, we conceive scale primarily as a convenient, somehow pre-given mechanism for ordering social processes and phenomena. Moreover, when we imagine the regional scale, it is often as a purely mental construct for dividing the world, with one geographer's region as good as the next. Herod argues that traditionally, multi-scale studies of time, space, and change usually do not question how certain regions and scales come about but focus, rather, upon how changing the scale of analysis reveals different insights into particular social and economic processes. Although not without merit, these studies assume that the scales are natural and inviolable. As an alternative, Smith (1992) proposes three tenets that might form the basis of a critical social theory of scale. First, we need to be able to interpolate the 'translation rules' that allow us to understand not only the construction of scale itself but also the ways in which meaning translates between scales. Second, in that scale is used to give meaning to social interactions, we need to understand its metaphorical power. A critical theory of the production of scale would differentiate and integrate the meaning of scale as metaphor for social relations and as a grounded material practice (see also, Smith and Katz, 1993). Third, a critical theory of scale must speak to the construction of difference. Although boundaries are continually forged and reforged in social practice, powerful individuals and groups continue to metaphorically appropriate space to establish the centrality of their own subject positions. Smith (1992, p. 78) argues further that the goal of a politics of spatial justice is not only "... to overcome social domination exercised through the exploitative and oppressive construction of scale," but also "... to reconstruct scale and the rules by which social activity constructs scale." In addition, Sallie Marston (1995) points out that different scales are important at different times and the relative importance of different scales during any period is an empirical question that relates to changing power structures within a region.

In the balance of this chapter, discussion focuses on GIS-based projects that clearly anticipate a more critical appraisal of time, space, region and scale.

13.4 NATURALIZING SCALE: QUESTIONS OF POWER AND COMMUNICATION

13.4.1 Creating Regions of Least Resistance

Questions related to the natural ordering of temporal and geographical scale are highlighted in a project that attempted to assess the impact of a proposed freeway extension on a region in San Diego (Aitken *et al.*, 1993; Askov *et al.*, 1994). The exact route of the expressway had yet to be ascertained when the study was commissioned as part of the social and economic impact assessment required by the United States National

Environmental Policy Act of 1969 (NEPA) and the California Environmental Quality Act of 1970 (CEQA). The research was concerned with creating a geographic basis for analyzing residents' cognition of their local area in the light of possible disruption by the freeway extension. Perceptual and behavioural data garnered from a sample of 464 residents from the surrounding communities were constructed in a GIS to represent the spatial contexts of residents' familiarity with, and experience of, the region. Several forms of spatial dependency analyses suggested a possible routeway for the freeway that would do least damage to the integrity of the region defined in terms of the perceptual and behavioural data. In short, the study showed that an area of familiarity and experiential discontinuity existed along the principle routeway through the region that effectively divided it in two. The transportation planners who commissioned the work were delighted with this result because it suggested a route of 'least resistance' for the expressway.

Several problems arose with this project that relate to interesting conceptualizations of geographic and temporal scales around planning and policy requirements. These problems are posed here because they are exemplary of the kinds of issues that may arise from the convergence of GIS technology and planning. First, the region of study was defined *a priori* by the California Department of Transportation (CALTRANS) to comprise parts of eight relatively distinct communities that had little, if any, political cohesion. In prescribing this region, the primary consideration of the transportation planners was to effect an expressway that connected two existing freeways. Second, although they were unsure of the best location for the expressway, CALTRANS had been buying land adjacent to the principal routeway through the region for some time in preparation for the construction of the expressway. There is nothing in NEPA or CEQA to prevent the acquisition of land for a proposed development as long as the land remains undeveloped. The study was commissioned after much of the land had been vacated and buildings were left derelict along either side of the principle routeway. The point is that if there had been any regional integrity or political cohesion across the principle routeway then it most likely would have disappeared with the creation of the 'wasteland' of vacant lots and derelict buildings. Although the study suggested an area through which the expressway could be constructed with minimal impact on the surrounding communities, there is no way of assessing how the protracted purchase of land surrounding the principle routeway molded the current lack of regional political identity. Clearly, unproblematized temporal and geographic scale issues converged to contrive a corridor of discontinuity in this region as a self-fulfilling prophecy. As such, the snap-shot-in-time, behaviourally-based GIS research was unable to discern such nuances.

13.4.2 Virtual Geographies

With the above San Diego freeway example in mind, some recent work on GIS and planning suggests a need to understand the complexity within which policy decisions are made, and how GIS technology can contextualized those decisions (Aitken and Rushton, 1993; Aitken and Michel, 1995). Left unconsidered in that work is the power of representation and modes of visualization in contriving scale and creating regions. As John Pickles (1995, p. 9) points out, the emergence of spatial digital data, computer graphic representations, and virtual reality creates an intertextuality that directs attention to the multiple fragments, multiple views, and layers that are assembled under new laws of ordering and re-ordering. Pickles looks forward to the development of a global village on Internet that supports both the access to information and a format for dialogue so that counter-hegemonic social action is encouraged. It seems, however, that contemporary GIS technology, with its propensity for Cyborg (Haraway, 1991) and Archimedean

(Gregory, 1994) views, enables a particular form of imperialism to be perpetuated. In particular, the contrivance of scale and the seemingly natural delineation of regions are particularly susceptible to the vagrancy of these new technologies. An example of virtual tourism off the Californian coast suggests precisely this kind of process.

A World Wide Web Home Page for Catalina Island, California enables interested viewers to navigate through detailed and tangential information and images representing the Island's cultural history, natural resources, recreational attractions and a developing GIS. In time, perhaps it will be possible to tour the Island through virtual GIS without leaving home. Will this virtual experience replace first-hand knowledge? Perhaps, instead, the images will entice more tourists to Catalina Island. Could a GIS accessed through the Internet be used to educate and help protect natural resources? Could this be considered a new form of conservation? The problem with questions like these is that they do not deal with the fiction that is, and always has been, Catalina Island. Of particular interest to this chapter is that the tourist images of Catalina Island mask an implicit control of space and region which is described in an interesting dance of contrast and contradictions between aesthetics and politics. What the virtual GIS represents is an extension of that control to an explicit production of scale. Concern is with how GIS is used to define natural resources, conservation and the scale of interaction between people and the Island (Aitken and Westersund, 1996). It seems that GIS technology is enabling the creation of information and landscapes that conform to certain ideas of how nature is constituted and how tourists and residents, insiders and outsiders, wealthy and poor should relate to nature.

The Catalina Island Homepage constructs a very distinct conception of space to which the virtual GIS systems appeal: a three-dimensional field or surface in which subject positions are definitively located. The space is materially undifferentiated and homogeneous insofar as all locations are intrinsically equal; the only criteria of differentiation are mathematical. Space is also, in this conception, entirely separate from the objects, events and relations that occur 'in' space. The subject hovers in 'his' Archimedean position of supposed control. Creating these virtual places as if they were real (or even a mirror on reality) loses the rich variety of social, political, economic and cultural meaning bound up with geographical differentiation. Scale, as an active political process, is either summarily dismissed by virtual GIS or trivialized to the optical metaphor of 'zooming in and out'. As such, virtual GIS technology is offering ways to subvert scales and to dismiss regions. This is disconcerting given the previous argument that we only have a limited understanding of how the processes of scale production impact the material conditions of lived experience. For example, should we trivialize scale to the switch of a key that enables us to see city traffic densities at one moment and then, at the next, the current agenda at the city hall (Gelernter, 1992)? Virtual GIS can trivialize the material and social process that produce differential space because it reifies and uncritically accepts a natural hierarchy of scales.

Many advocates applaud the ability of virtual GIS and silicon graphics to 'overcome' the scale issue. For example, software marketer Alan Dressler's approach (quoted in Pickles, 1995) is to get people to drop the human scale completely: "... if you are going to think galaxies you've got to be galaxy-like." Virtual GIS permits the transcendence of scale and creates as 'natural' a space wherein it is possible to lose the body politic. The creation of this kind of virtual space ignores the material conditions of our lived world and of hierarchical oppositions that politically segment that world. As was made clear in the first half of this chapter, space and scale are social constructions and the very notion of 'fixing' them so that we may travel through or up and down them with ease presupposes a particular way of thinking about the world that is forged out of instrumental and strategic reasoning.

If we want spatial issues to provide a fundamental organizing concern, then we need to overcome social domination exercised through the exploitative and oppressive construction of scale. The beginnings of a virtual GIS for Catalina Island suggest the beginnings of an exploitative and oppressive construction of scale.

13.5 JUMPING SCALE

What needs to be considered more fully to realize Pickles' optimism for a Global Village on Internet is a communicative model that accommodates dissent and difference, and enables communities and local coalitions to 'jump scale' (Smith, 1993). To this end, there may be possibilities in a more communicative form of GIS (Sheppard, 1996).

Jürgen Habermas (1984, 1987, 1989) calls for a 'paradigm shift' from a philosophy of consciousness and self to a philosophy of language and communication. The philosophy of consciousness operates with the methodologies of instrumental and strategic rationality that drive most contemporary GIS applications. Habermas distinguishes instrumental and strategic action from communicative action. The former relates means to ends and techniques to goals without reflection on the rationality or justness of the goals themselves. It is rooted in a self-oriented, subjective goal to dominate and control nature and other people. Alternatively, communicative action is oriented towards understanding, agreement and uncoerced consensus. The validity inherent within the practice of communication is based upon the speaker's claims of truthfulness, correctness (when compared with social norms) and sincerity. Communicative action is a move by two or more parties to reach an understanding concerning a particular context. This focus decentres the individualistic and self-interested philosophies inherent in instrumental and strategic rationality by acknowledging that the real world operates through consensus and negotiation between collective identities (Aitken and Michel 1995, p. 23).

John Pickles (1995) envisages Internet and 'systems of informatics' becoming a foundation for the re-emergence of a civic culture, a Habermasian community of dialogue. Informatics can be conceived as part of the Habermasian project because they are a potential source of new power for marginalized groups to whom traditional modes of communication have been inaccessible. Informatics permit community and dialogue to emerge for those who would otherwise have no voice or space for communicative action.

There is an important point associated with the freedom to communicate that remains hidden within the Habermasian formulation: if freedom of speech is important within GIS and policy-making, then we must move away from consensus building models to models that openly accept dissensus and contestation (Aitken and Michel, 1995, p. 25).

13.6 CONCLUSION: THE POSSIBILITIES FOR DISSENT

In the United States, several members of the GIS community have collaborated with geographers interested in social theory to suggest a constellation of alternative forms of GIS production that focus communicative and dissent rather than consensus building. Work in this area is as yet embryonic, but it encompasses a series of critical questions that relate to the construction of scale, community and the space of GIS. I close this chapter with some of the questions that have been raised concerning the potential of these new GIS.

In recent years a series of political, economic and social changes makes it increasingly important to think about the ways in which the logics, systems and representational content of GIS support particular types of social practice and inhibit others.

These new GIS focus upon the possibility of developing a communicative forum wherein differential power relations are contested, and space and scale are problematized. Most GIS representations to date tend to be contrived from a fairly narrow way of knowing based upon strategic and instrumental rationality, and Cartesian space. Given this base for GIS, those whose values and culture differ from the interests of the dominant sectors in society because of age, physical abilities, ethnicity, race, gender and sexual preference may be disadvantaged when, for example, locational conflict resolution involves a significant GIS component. Concerns raised in the production of new GIS need to relate issues of efficiency versus equity, and access versus privacy, with the new ways of understanding space and scale discussed in this chapter.

Tied in with the larger nature–society debate in geography, and related to the examples described above, the new GIS must question which particular knowledge systems are privileged in existing GIS and how GIS can change to accommodate other conceptions of space, environment and nature. How, for example, are biophysical resources culturally constructed, politically contested and scale-dependent and how do current GIS representations distort these constructions? Can we develop a GIS that incorporates grassroots perspectives, particularly in terms of natural resource access and technological or environmental risk? Can 'voices from everyday experience' be digitally represented when local knowledge is fuzzy, imprecise and ethereal?

Can a socially differentiated world be produced and represented in a new form of GIS?

Can we incorporate GIS as a more democratic and communicative model for planning? Ultimately, GIS will be useful only if it is constituted in an everyday language of practice that accommodates dissent, rather than the specialized speech of strategic rational planning. If space is social constructed and the production of scale is the process through which our social world is spatially bound and temporally contextualized, then new communicative models are needed to enable disempowered groups to jump scale and contest the oppressive production of particular spaces.

ACKNOWLEDGEMENT

Research for this chapter was supported in part by Grant SES–9113062 from the National Science Foundation of the United States.

REFERENCES

Aitken, S.C., 1992, The personal contexts of neighbourhood change. *Journal of Architectural and Planning Research*, 9(4), pp. 339–360.

Aitken, S.C., 1998, *Family Fantasies and Community Space*, (New Jersey: Rutgers University Press).

Aitken, S.C. and Bjorklund, E.M., 1988, Transactional and transformational theories in behavioural geography. *The Professional Geographer*, 40(1), pp. 54–64.

Aitken, S.C. and Fik Tim, J., 1988, The daily journey to work and choice of residence. *The Social Science Journal*, 25(4), pp. 463–475.

Aitken, S.C. and Prosser, R., 1990, Residents' spatial knowledge of neighbourhood continuity and form. *Geographical Analysis*, 22(4), pp. 301–325.

Aitken, S.C. and Rushton, G., 1993, Perceptual and behavioural theories in practice. *Progress in Human Geography*, 17(3), pp. 378–388.

Aitken, S.C. and Michel, S., 1995, Who contrives the 'Real' in GIS? Geographic Information, planning and critical theory. *Cartography and Geographic Information Systems*, 22(1), pp. 17–29.

Aitken, S.C. and Westersund, A., 1996, Just passing through: Virtual tourism, justice and 'Informatics'. Paper prepared for the Specialist Meeting of Initiative 19, Social Theory and GIS, of the NCGIA, Koinonia Retreat Center, South Haven, Minn.

Aitken, S.C., Stutz, F.P., Prosser, R. and Chandler, R., 1993, Neighbourhood integrity and residents' familiarity: Using a Geographic Information System to investigate place identity. *Tijdschrift voor Economische en Sociale Geografie*, 84(1), pp. 2–13.

Askov, D.C., Stutz, F.P., Aitken, S.C. and Stutz, C., 1994, Freeway alignment and community residents' receptivity: A GIS distance buffering application. In *Proceedings of GIS/LIS '94*, pp. 14–23.

Baudrillard, J., 1989, *America*, (London: Verso).

Campari, I., 1993, Morphological time in urban development. In *Time in Geographic Space: Report on the Specialist Meeting of Research Initiative 10*, edited by Egenhofer M. and Golledge, R. (UC Santa Barbara: NCGIA).

Cheylan, J.-P. and Lardon, S., 1993, Towards a conceptual data model for the analysis of spatio-temporal processes: The example of the search for optimal grazing strategies. In *Spatial Information Theory: A Theoretical Basis for GIS, European Conference COSIT '93*, Lecture Notes in Computer Science 716, edited by Frank A.U. and Campari I. (Berlin: Springer-Verlag), pp. 158–176.

Dyck, I., 1990, Space, time and renegotiating motherhood: An exploration of the domestic workplace. *Society and Space*, **8**, pp. 459–483.

Foucault, M., 1977, *Discipline and Punishment*, (Penguin).

Gelernter, D., 1992, *Mirror Worlds of the Day the Software Puts the University in a Shoebox ... How it will Happen and What it Will Mean*, (New York: Oxford University Press).

Giddens, A., 1979, *Central Problems in Social Theory: Action, Structure and Contradiction in Social Analysis*, (London: Macmillan).

Giddens, A., 1984, *The Constitution of Society*, (Cambridge: Polity Press).

Gregory, D., 1985, *Space and Time in Social Life.* Wallace W. Atwood Lecture Series 1, The Graduate School of Geography, Clark University Worcester, MA.

Gregory, D., 1994, *Geographical Imaginations*, (Cambridge, MA: Blackwell).

Habermas, J., 1984, *The Theory of Communicative Action*, Vol. 1, (Boston: Beacon Press).

Habermas, J., 1987a, *The Theory of Communicative Action*, Vol. 2, (Boston: Beacon Press).

Habermas, J., 1987b, *The Philosophical Discourse of Modernity: Twelve Lectures*, (Cambridge, MA: MIT Press).

Habermas, J., 1989, *The Structural Transformation of the Public Sphere*, (Cambridge, MA: MIT Press.

Hägerstrand, T., 1970, What about people in regional science? *Papers: The Regional Science Association*, **24**, pp. 7–21.

Hägerstrand, T., 1975, Space, time and human conditions. In *Dynamic Allocation of Urban Space*, edited by Karlqvist, A., Lundqvist, L. and Snickars, F. (Farnborough: Saxon House), pp. 3–12.

Haraway, D., 1991, *Simians, Cyborgs and Women: The Reinvention of Nature*, (New York: Routledge).

Herod, A., 1996, Labor's Spatial Praxis and the Geography of Contract Bargaining in the United States East Coast Longshore Industry, 1953–1989, *Political Geography.*

Jameson, F., 1984, Postmodernism, or the cultural logic of late capitalism, *New Left Review*, **146**, pp. 53–92.

Jameson, F., 1992, *The Geopolitical Aesthetic: Cinema and Space in the World System*, (Bloomington: Indiana University Press).

Keating, E., 1995, Spatial Conceptualizations of Social Hierarchy in Pohnpei, Micronesia. In *Spatial Information Theory: A Theoretical Basis for GIS, Int. Conference COSIT '95*, Lecture Notes in Computer Science 988, edited by Frank, A.U. and Kuhn, W. (Berlin: Springer-Verlag), pp. 463–474.

Kitchin, R., 1996, Increasing the integrity of cognitive mapping research: Appraising conceptual schema of environment–behavior interaction. *Progress in Human Geography*, 20(1), pp. 56–84.

Langran, G., 1989, A review of temporal database research and its use in GIS applications. *International Journal of Geographical Information Systems*, **3**, pp. 215–232.

Langran, G., 1993, Issues of implementing a spatiotemporal system. *International Journal of Geographical Information Systems*, **7**, pp. 305–314.

Lefebvre, H., 1984, *Everyday Life in the Modern World*, (New Brunswick, NJ: Transaction Books).

Lefebvre, H., 1991, *The Production of Space.* Translated by Donald Nicholson-Smith. (Oxford: Blackwell).

Marston, S.A., 1995, 'Female Citizens': Middle Class Women and the Domestic Management Movement in 19th Century Urban America. Invited lecture at University of Colorado, Boulder.

Massey, D., 1994, *Space, Place and Gender*, (University of Minnesota Press).

Pickles, J., 1995, Representations in and electronic age: Geography, GIS and democracy. In *Ground Truth: The Social Implications of GIS*, edited by Pickles, J. (New York: Guilford Press), pp. 1–30.

Papagno, G., 1992, Seeing time. In *Theories and Methods in Spatio-Temporal Reasoning in Geographic Space*, edited by Frank, A.U., Campari, I. and Formentini, U. (Berlin: Springer-Verlag).

Pred, A., 1973, Urbanization, domestic planning problems and Swiss geographic research. *Progress in Geography*, **5**, pp. 1–76.

Pred, A., 1984, Structuration and the time-geography of becoming places. *Annals of the Association of American Geographers*, **74**, pp. 279–297.

Pred, A., 1986, *Place, Practice and Structure: Social and Spatial Transformations in S. Sweden 1750–1850*, (Cambridge: Polity Press).

Rose, G., 1993, Some notes towards thinking about the spaces of the future. In *Mapping the Futures: Local Cultures, Global Change*, edited by Bird, J., Curtis, B., Putnam, T., Robertson, G. and Tickner, L. (London: Routledge), pp. 70–83.

Sheppard, E., 1996, President's Plenary Session Paper. Meeting of the Association of American Geographers, Charlotte, North Carolina.

Smith, B., 1995, On drawing lines on a map. In *Spatial Information Theory: A Theoretical Basis for GIS, Int. Conference COSIT '95*, Lecture Notes in Computer Science 988, edited by Frank A.U. and Kuhn, W. (Berlin: Springer-Verlag), pp. 475–484.

Smith, N., 1992, Geography, difference and the politics of scale. In *Postmodernism and the Social Sciences*, edited by Doherty, J., Graham, E. and Mo Malek (London: Macmillan), pp. 57–79.

Smith, N., 1993, Homeless/global: scaling places. In *Mapping the Futures: Local Cultures, Global Change*, edited by Bird, J., Curtis, B., Putnam, T., Robertson, G. and Tickner, L. (London: Routledge), pp. 87–119.

Smith, N. and Katz, C., 1993, Grounding metaphor: towards a spatialized politics. In *Place and the Politics of Identity*, edited by Keith, M. and Pile, S. (London: Routledge), pp. 67–83.

Soja, E.W., 1985, The spatiality of social life: towards a transformative retheorization. In *Social Relations and Spatial Structures*, edited by Gregory, D. and Urry, J. (London: Macmillan), pp. 90–127.

Soja, E.W., 1989, *Postmodern Geographies: The Reassertion of Space in Critical Social Theory*, (London: Verso).

Soja, E.W., 1996, *Thirdspace: Journeys in Los Angeles and Other Real-and-Imagined Places*, (Cambridge, MA: Blackwell).

Stokols, D., 1988, Transformational processes in people–environment relations. In *The Social Psychology of Time: New Perspectives*, edited by McGrath, J.E. (Beverly Hills: Sage), pp. 233–254.

Tivers, J., 1985, *Attached Women: The Daily Lives of Women with Young Children*, (London and Sydney: Croom Helm).

Tivers, J., 1988, Women with young children: Constraints on activities in the urban environment. In *Women in Cities: Gender and the Urban Environment*, edited by Little, J., Peake, L. and Richardson, P. (London: Macmillan).

Valentine, G., 1993, Negotiating and managing multiple sexual identities: lesbian time–space geographies. *Transactions of the Institute of British Geographers*, **18**, pp. 237–248.

Wiener, D., Warner, T., Harris, T. and Levin, R., 1995, Apartheid representations in a digital landscape: GIS, remote sensing and local knowledge in Kiepersol, South Africa. *Cartography and Geographic Information Systems*, 22(1), pp. 30–44.

Young, I.M., 1990, The ideal of community and the politics of difference. In *Feminism/ Postmodernism*, edited by Nicholson, L. (New York: Routledge), pp. 300–323.

CHAPTER FOURTEEN

Spatio-Temporal Analysis of Rural Land-Use Dynamics

Review and Integration of three Methods Involving GIS and Matrices

Denis Gautier

14.1 INTRODUCTION

The objective of this chapter is to discuss three methods that have been used jointly to analyse the contemporary spatial dynamics of small rural territories and to explore how they may be integrated. The landscape is emphasised as the primary level of organisation and perception with respect to a small rural community's sphere of activity. This *space* is comprised of a combination of repetitive yet interacting landscape elements (Forman and Godron, 1986). The landscape element, defined morphologically, is the smallest spatial unit presenting uniform structure and function. Such a unit allows visible features in the landscape to be related to the combination of biophysical and human factors that produce it. In this chapter, the SEU is conceived as a landscape element defined by land ownership rather than as an aggregation of people (though the role of people has influenced the dimension of this element). As a rule, the landscape element is the finest unit available for spatio-temporal analysis. Often the size of a single land-use parcel, the interactive combination of these units can hypothetically reconstitute land-use for a given area over time.

Each method is used in an attempt to understand the impact of Chestnut (*Castanea sativa)* forest management in the Cévennes Mountains of France (South border of Massif Central). This region, a succession of deep valleys, has been exploited by small mixed farming operations, gathered in hamlets, with a combination of extensive livestock, chestnut cultivation and complementary activities. Since the beginning of the century, a series of events (Chestnut tree, vine and silkworm diseases as well as World Wars) has resulted in depopulation. Despite the return to the countryside in the 1970s, this depopulation, combined with an intensification of farming, has led to an abandonment of land, a degradation of terraces and chestnut forests. This degradation is all the more important as the Chestnut forests are situated far from hamlets and on southern slopes.

On the one hand, a spatial approach is required to take into consideration the interaction between the Chestnut forest and other landscape elements like hamlets, fields, rangelands and woodlands. On the other hand, a historical perspective is necessary to understand the relation between the action of man and the biophysical environment. Therefore, a true spatial and temporal analysis is required to understand the short-term and mid-term effect of man's action on the organisation of landscape.

The improvement of spatial management increasingly depends on a better understanding of dynamics. Faced with a specific problem such as rural land-use, it is essential to formalise methods and tools available for this type of analysis. The objective of this work is to contribute to this development by presenting and discussing the three methods applied, as well as the manner by which they organise information, and to demonstrate how they can be integrated in order to improve spatio-temporal modelling:

1. the comparison by GIS of two successive states of the same territory, in order to detect processes that generate spatial transitions;
2. the comparison by a GIS of a territorial reference state with the chronology of land-use events that modelled it, in order to understand the weight of these events in terms of spatial dynamics;
3. the overlaying of function models for chestnut forest types on a surface evolution matrix over the course of time, in order to understand spatial dynamics through the evolution of biological and economical parameters of land-use.

14.2 PRESENTATION OF THE THREE METHODS OF ANALYSIS

The methods applied are presented successively without first evaluating or classifying them. They have been developed in different projects as a response to particular questions forwarded to researchers by actors (agriculturists, residents, managers) in the territory studied. The fact that they have been developed in distinct areas is not important. They all take into account spatial dynamics and can be potentially integrated in the same area. As such, it is possible to distinguish their individual qualities and value, which will be discussed as well.

14.2.1 Transition-State Method: Transition Between Different States of the Same Landscape (Langran, 1992; Langran and Chrisman, 1988)

The trend of abandoning Chestnut forest management gives rise to brush, increased fire risk, a regression of rural activities (grazing, harvesting, gathering, hiking) and, most importantly, a loss of cultural identity with respect to the landscape. In order to slow down a corresponding return to unmanaged ecosystems in a particular Chestnut massif, an association of property owners was created to encourage the area's *reconversion* to managed land-use. This project needed to show these landscape trends in order to justify a proposal for initial financing (the sale of chestnut production could eventually provide long-term financing). Thus, this first method of *transition-states* was developed to demonstrate the abandonment of territory (Gautier, 1995).

The applied method considered the transition between two stages of landscape organisation: in 1991 and in 1970, when the abandonment was most significant.

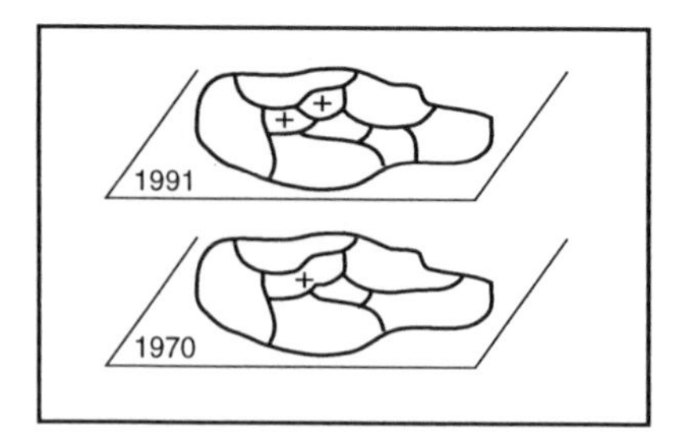

Figure 14.1 Transition states (localized versions of landscape units)

The construction of geographical information layers under GIS, through the interpretation of aerial photos, permitted the constitution of successive versions of elementary spatial units. It also permitted the establishment of a transition matrix between the two landscape states (Godron and Lepart, 1973) by calculating the distribution percent of land-use types in 1991 within each land-use type established in

1970. As such, we may express the area vector in 1991 (V_{1991}, in the last row of the matrix in Table 14.1) as related to the area vector of 1970 (V_{1970}, in the first column of the matrix in Table14.1) by the following multiplication $V_{1991} = M*V_{1970}$ where M = each value in the Transition matrix. Therefore, the Transition matrix in Table 14.1 resumes all the changes that took place on the Massif between 1970 and 1991.

The following graph allows the visualisation of these transitions between the different land-use types. The size of the arrow is proportional to the percent of transition from one landscape element to another. The percentage of area remaining at the same stage of evolution is also expressed in number. (Note: only the transitions higher than 5% are represented.)

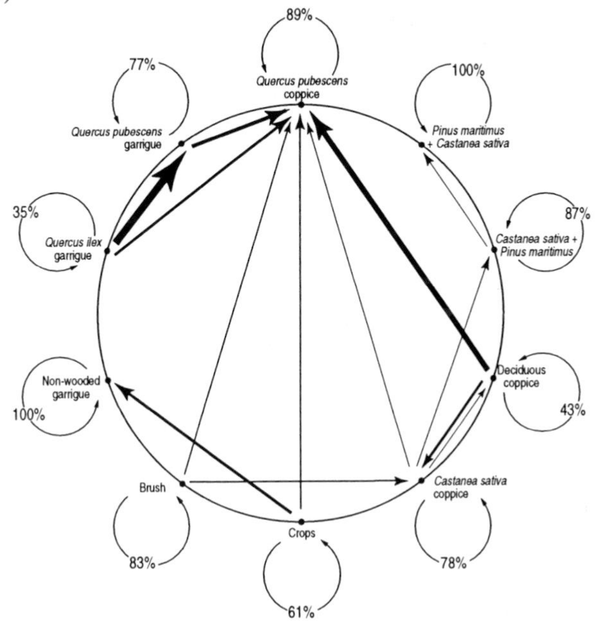

Figure 14.2 Transition between types of land-use between 1970 and 1991

The consideration of transitions between two states of spatial organisation can lead to detect trends in territorial evolution. In the given example, it allowed us to show the dominance of brushing-in and reforestation in the evolution of the Chestnut forest during the last 20 years. If we identify the actions that correspond to each matrix box, we may anticipate the effects of a change in land-use systems. It is this comparison of different spatial states that permits a discussion of landscape dynamics.

14.2.2 The Historical Land-Use Events Method

The territory of a commune, victim of agricultural decline since World War II, continues to be exploited by agriculture (multicropping, grazing) as well as newly introduced land-uses (timber harvesting of Chestnut trees originally intended for chestnut and/or tannin production, clearing for irrigated pastures, gathering, hunting, ecotourism, all corresponding to the contemporary rural reality). Within the framework of a project promoting the sustainable management of this commune's Chestnut forests, it became fundamental to find relations between the territory and the processes that produced it. For example, it is important in terms of land management to understand why a particular

Chestnut forest is denser than a neighbouring one and whether the difference is due to wood processing, ecological potential or other factors. This would allow the prediction of mid-term consequences of any possible management actions. A scientific approach was chosen to define the history of land-use in order to arrive at an explanation of today's managed landscape (Arnaud and Gautier, 1996).

Operationally, this involved detecting the events that have impacted the landscape (nature of the event, date, spatial unit in which it occurred) relating with a landscape reference (the present state). Land-use events have been identified on the cadastral register and by surveys. Known historical land-use events were attributed to each of the landscape elements which are designed to reflect a combination of biophysical and human processes, including slow changes (biophysical processes, ways of land-use inherited from the past) and/or violent changes (spatial events linked to human actions).

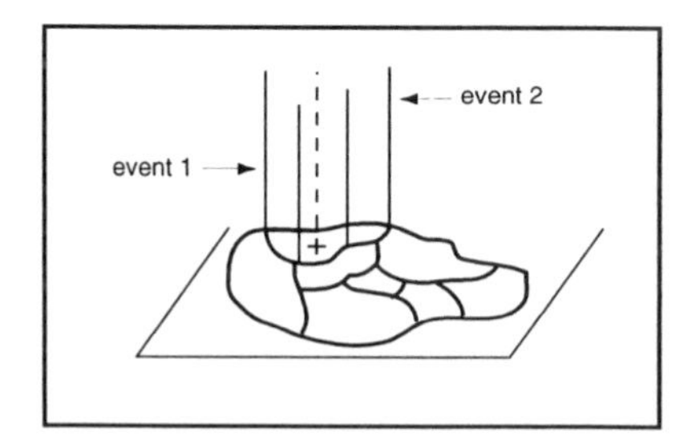

Figure 14.3 Historical land-use events attached to positioned spatial units

The use of a GIS permitted relations to be made between different localised landscape elements with parameters of ecological potential (slope, aspect) on the one hand and with the series of spatial events that contributed to its formation on the other. These historical land-use events are outlined in an attribute table for the GIS layer. For example, a particular Chestnut parcel located on a southern slope may be linked in the table to an abandonment 30 years ago and a fire 10 years ago. The relation between the landscape element and the changes that have affected it permit a discussion of the impact of spatial events on the landscape dynamics. This permits in turn to characterise the effects of man's action on the landscape, either direct (timber cutting, clearing, fire) or indirect (brushing-in after abandoning a type of land-use), temporary or long lasting. It also becomes possible to anticipate the effect of a modification (or mutation) in land-use systems on landscape organisation. It is the historical land-use events, which permit a discussion of landscape dynamics over time.

14.2.3 Function Models Combined with Spatial Change

Consensual local development requires a search for compromise between biophysical potential and territorial management actions in order to arrive at precise objectives. This leads to the analysis of the transformation of the landscape system and its make-up over time in order to elaborate various evolution scenarios. As previously mentioned (in Section 14.2.2 above), this involves linking the condition of various locations of a given territory to its spatio-temporal dynamic. However, in the case of land-use planning, it is not only the events affecting the territory and producing spatial differentiation, which interest the decision-makers. They also seek an understanding of the internal functioning of different land-uses as they are related to processes and methods of exploitation. This requires a higher level of abstraction than the landscape element, in order to construct models of function per land-use type and to relate this to the dynamic of a territory.

Table 14.1 Transition matrix of various land-use types between 1970 and 1991

Initial landscape units (1970)	Surface in ha	Crops	Deciduous coppice	*Castanea sativa* coppice	*Quercus pub.* coppice	*Pinus mar.* + *Castanea sativa*	*Castanea sativa* + *Pinus mar.*	*Quercus pub.* garrigue	*Quercus ilex* garrigue	Non-wooded garrigue	Brush
Crops	**283**	61	3	5	7	0	0	3	0	19	2
Deciduous coppice	**130**	0	43	16	37	0	4	0	0	0	0
Castanea sativa coppice	**403**	0,5	6	78	5	3	6	0,5	0	0,7	0,3
Quercus pub. coppice	**205**	0,5	1	1	89	0	3	3	2	0	0,5
Pinus mar. + *Castanea sativa*	**19**	0	0	0	0	100	0	0	0	0	0
Castanea sativa + *Pinus mar.*	**98**		0	4	0	6	0	3	0	0	0
Quercus pub. garrigue	**46**	0	0	0	23	0	0	77	0	0	0
Quercus ilex garrigue	**12**	0	0	0	18	0	0	47	35	0	0
Non-wooded garrigue	**13**	0	0	0	0	0	0	0	0	100	0
Brush	**57**	0	1	9	7	0	0	0	0	0	83
Final surface areas (1991)		**175**	**92**	**360**	**287**	**37**	**121**	**62**	**7**	**69**	**56**

A method has been proposed to that end, based on matrix calculations, by associating land-use unit function matrices to an evolution scenario matrix for the landscape (Godron, 1979). The evaluation of land-use type function over time is presented in the form of a matrix for which the rows are biological and economic parameters and for which the columns are values taken by each of these parameters over time. These function matrices are established by surveys and by equations simulating the evolution of parameters.

Landscape evolution is then resumed by the entire set of function matrices for all land-use types. On a given territory, these function matrices can be associated to the evolution matrix for surface states, whether past or future. Such evolution matrices specify for each land-use type (row) the surface being managed or to be managed over time (column). This can be obtained either by the use of a GIS with several states of one territory, or from hypotheses on the spatial distribution at several dates. The combination of function matrices and evolution matrices for the different surface states leads to results that indicate the consequences of management approaches at the level of the studied territory.

This tool was employed in order to resume the evolution of Chestnut forests in a commune of the Cévennes over the last 50 years and to discuss with the actors of this territory the consequences of further agricultural abandonment versus the adoption of new exploitation strategies. A typology was made by recent aerial photo interpretation in order to arrive at a detailed land-use map where all types of Chestnut forests appear (Gautier, 1996). This map was integrated in a GIS, which permits us to obtain the surface occupied by each type of land-use. The construction of function models for the designated Chestnut forests required surveying of the land-users. The information necessary for this approach was often fragmented or non-existent since the parameters of biological or economic function for the Chestnut forests are not well known for the Cévennes. It was necessary to search for elements of function outside the territory considered, sometimes even outside the study area, in order to formulate hypotheses. Insofar as time is integrated in an explicit manner in the function model, as much in the data (spread across time) as in the construction of the model (where time is not only a variable), this model acts as a filter of hypotheses, which permits through adjustments to discard some and to fill what is lacking in others.

Table 14.2 Function Matrix for the Chestnut orchard at the stage of abandonment

Year:	1971	1981	1991
Number of stems (n)	100	150	200
Basal area (m^2)	30	33	35
Standing volume (m^3)	100	130	150
Growth (m^3/an)	2	2	2
Dominant height (m)	15	15	15
Work-hours for harvest (h)	20	20	20
Total cost of production (kF)	0,8	0,8	0,8
Production (kg)	600	600	600
Price of the Chestnut (F/kg)	3	3	3
Revenue (kF)	1,2	1,2	1,2
Gross Margin (kF)	0,4	0,4	0,4
Days of Grazing sheep/ha/year	750	750	750

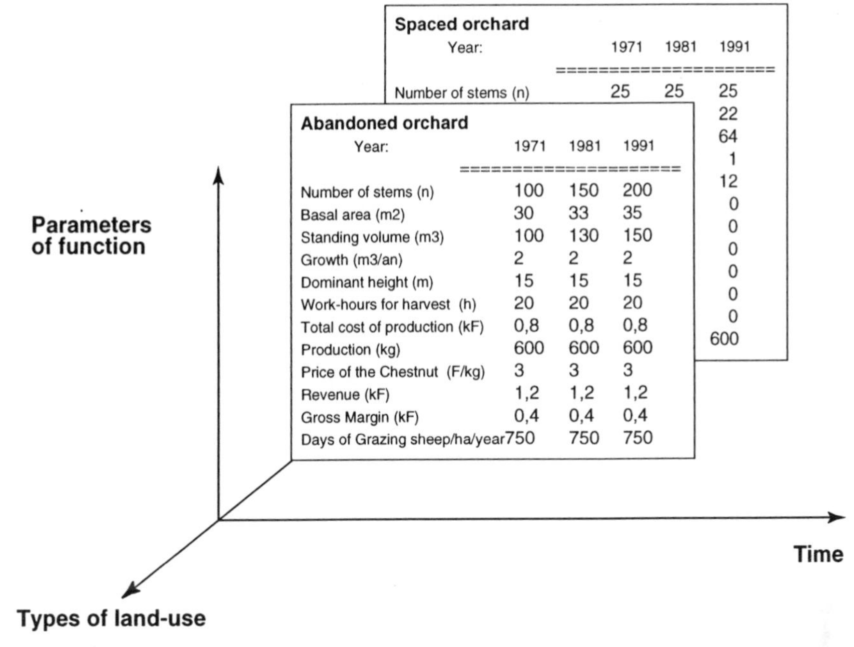

Figure 14.4 Resumé of the evolution of territory

All function matrices of Chestnut forest types are combined with a past evolution matrix, which resumes the surfaces occupied by the different types of Chestnut over the course of time, as related to the data or the hypotheses on their date of initiation or its affiliation.

Table 14.3 Evolution matrix by hectare and by land-use type

Year:	1941	1951	1961	1971	1981	1991
Orchard	455	355	235	165	100	69
Spaced orchard	0	0	49	49	49	49
Recently abandoned orchard	100	100	100	100	100	100
Evolving orchard in coppice	0	100	211	231	243	254
Coppice	0	0	29	29	29	29
Coppice and deciduous	0	0	10	10	10	10
Prairies	100	100	70	80	113	99
Fallow	0	0	0	0	0	14
Scrub (*Quercus ilex*)	147	197	148	148	148	148
Vines	50	0	0	0	0	0
Mix of *Pinus mar.* and *Quercus pub.*	0	0	0	40	60	80
Total surface area	852	852	852	852	852	852

The matrix calculation allows for a logical overlay of all these matrices in order to validate or discard hypotheses regarding landscape dynamics. This is supported by verification with the territory's actors and by testing different management scenarios designed by imagining the state achieved following one or several of the desired actions. (For example, the continuation of land-use abandonment or the transformation of Chestnut fruit production to Chestnut wood production).

Table 14.4 Evolution of parameters over the course of time for all Chestnut forest in the territory

Year:	1941	1951	1961	1971	1981	1991
Number of stems (n)	55500	65500	100725	141025	169275	219075
Standing volume (m^3)	55500	60500	67461	73620	74727	61101
Growth (m^3/an)	1110	1110	1376	1456	1542	1742
Maintenance work-hours (h)	24750	19750	13750	10250	7000	5400
Maintenance product (F)	227500	177500	117500	82500	50000	34000
Harvest work-hours (h)	20200	16200	11400	8600	6000	4720
Total work-hours (h)	40950	31950	21150	14650	9000	6120
Total cost of production (kF)	1900	1500	1020	740	480	352
Production (100 kg)	6060	1860	3420	2500	1000	1416
Gross margin (kF)	950	750	510	370	240	176
Days of grazing sheep/ha/year x 100	1433	1683	2299	2390	2282	2220
Surface occupied:	852	852	852	852	852	852

14.3 CRITICAL ANALYSIS OF THE THREE METHODS OF ANALYSIS

The three methods were put into practice in order to respond to the fo!lowing questions: What are the social and biophysical motors of spatial dynamics? How should they be integrated in a model in order to simulate landscape evolution, according to different scenarios? The three methods employed present particularities and limits, as much from a conceptual point of view as with their application, which are important to analyse in order to find where they can be integrated.

14.3.1 The Transition-State Method

The comparison of several states of the same territory under GIS permits a demonstration of how spatial organisation has evolved over the course of time. For a given elementary spatial unit, we are faced with several versions (in the database). We know if it has changed form, nomenclature (identifier) or spatial context (relation to neighbouring units). At the global landscape level, we may detect changes in spatial organisation as much through the surface transition matrices as through contiguity matrices. This method, utilised alone, does not directly give information on territorial dynamics but rather on its kinematics in the physical sense of the term. By deduction from this succession of states, it is possible to formulate hypotheses on the processes of change that affect the land-use units between two different landscape states and to trace out the dynamic of the territory.

Though the advantage of this method is to facilitate information consolidation, its application leads us to question the significance of the observed transitions. The succession of individual stages of landscape only permits us to propose hypotheses regarding spatial processes, but without resolving the problem of its significance. Actually, the time dimension, which becomes apparent with a succession of territorial states, permits us to trace the stages of landscape evolution and to compare them in order to highlight key features of the spatial dynamics. On the conceptual level, it is important to be sure that we have not missed an important step in the spatial dynamics and that we correctly relate the different landscape states.

In practice, in order that significance may be found in the transition between different landscape states, spatial information must be obtained for definitive dates in the studied territory. This requires the determination of key dates in association with information on their importance. For example, 1970 is an important date in the Cévennes because it is a period of agricultural modernisation (mechanisation, intensification). Thus, the comparison of two landscape states, before and after this date, would help understanding the effects of this modernisation on landscape.

Since it is necessary to compare different states, the layers of geographical information must be comparable, in terms of the accuracy of spatial information (Goodchild and Gopal, 1989) and of the definition of land-use types. The evolution of geographical tools towards a better resolution of detectable information has conversely uncovered difficulties in comparing information of different quality (law of the lowest common denominator) or led to hypotheses being proposed on other states solely on the basis of the most complete state.

When several states of comparable information are collected, they may be linked together on a time scale. The question posed then depends on the quality of the transition estimation. When a change is observed between two dates for an elementary spatial unit, it is important to be aware of error linked to the collection and treatment of information on the GIS. This may be due as much to the interpretation of different transitions (Is the person in charge of the photo-interpretation the same for the two dates? Is the resolution of the spatial information equivalent for the two dates?), as to technical problems (What is the effect of the transfer of photo-interpretation data on a map, and of digitising this map with the detected spatial transitions?). These questions force us to distinguish error and the real evolution of the observed transitions. Therefore, we have to link function to changes in observed structure, which leads to the formulation of function hypotheses or to combining the transition-state method to another.

14.3.2 The Historical Land-Use Events Method

The chronological approach gives the succession of events relative to a reference spatial unit in the landscape. Therefore, for a given landscape element, independently of its identifier and form, we have historical land-use events attached to a point of reference in the GIS. The constitution of chronologies then permits the construction of a dynamic that is not always subject to the existence and the quality of completed spatial information for all the territory at certain dates as previously mentioned. What is important from a conceptual point of view, is to assure the relevance of recorded events in the explanation of spatial dynamics and, accordingly, to describe them well enough. Considering the succession of events that affect a spatial unit permits, by hypothesis, to trace its temporal trajectory (Flewelling, Egenhofer and Frank, 1992) without necessarily elaborating exhaustive information on all the studied territory. We possess the keys that explain in part the dynamic of spatial units constituting the landscape.

On the operational level, the method does not put into practice anything more than a state of land-use and a history of land-use events to respond to territorial planning problems. On the one hand, spatial information is not degraded by the necessity to homogenise between several states and can be of good quality (withal with current information). On the other hand, by knowing the strategies and practices of land-users and the potential vegetation, relations can be established between the morphology and functioning of landscape elements. For example, Chestnut forests in the Cévennes have been historically established for fruit harvesting. Thus, mapping a Chestnut coppice and relating it with an elevator model leads to the assumption that it would be situated in a humid ecological sector where trees have vigorous growth 2–3 cuts after the abandonment of Chestnut forest. If the landscape point of reference is the present state, field checks could be made. The construction of a territorial dynamic is then facilitated by the richness of information that can be collected on the landscape point of reference.

From these landscape states, the construction of histories imposes an organisation of information that the use of a GIS can provide. The spatial events that modified the evolution of a land-use unit can be identified by the diverse sources—oral, written, photographic—that may be encountered. The surveying of the territory's habitants gives precious information on the functioning of production systems in recent history, but poses a problem of reliability regarding the memory of precise dates from those surveyed. The information contained in the cadastral survey permits the completion of these surveys by recording changes in land-use states for dated and mappable land parcels. GIS analysis permits us to establish hypotheses on the relation between an event recorded in the cadastral survey and the dynamic of a landscape element by highlighting commonalties in the coupling of reference landscape states and their corresponding recorded history of land-use events.

It is therefore possible to establish the chronology of events that led to the formation of a land-use unit and to evaluate the impact of a given event on the organisation of the landscape. At the same time, however, we are not assured of having a complete chronology. This history is based on events, meaning mutations that modify the internal system of landscape elements that are not always related in a complete fashion. In particular, slow processes, either biophysical or social in nature, are not taken into account. Otherwise, this method which attaches a chronology of events to a state of reference does not inform us as to changes of form or identifiers that have affected the spatial units, and as a result, as to the evolution of the spatial organisation of territory.

14.3.3 The Matrix Method for Spatial Dynamics

The construction of function models for landscape elements and their association to different evolution scenario matrices allows for the examination of processes that are not detected by preceding methods. The function models are established for each type of landscape element with a limited number of descriptor parameters. For a given landscape element, the evolution of these variables in time is manifested as the spatial land-use system function. This is the result of a combination of ecological and human activity sub-systems and is expressed in the territory by a landscape element. A model of function established for a type of land-use (i.e., the Chestnut orchard) therefore explains the dynamic of the system through biological and economic parameters, and thereby includes the complexity of relations between the biophysical and human processes at play. This model may be adjusted to a particular landscape element (for example, the Chestnut forest of a certain location and with a certain property owner) so that the spatial events occurring in the history of that unit may be then integrated to the function matrix.

The spatial evolution matrix of the past or of the future resumes the evolution of surfaces occupied by the landscape elements over the course of time. It can be established at the territorial level managed by a society, a human group or a farmer, depending on the spatial level of application of the function matrices. Knowledge of initiation dates and evolution rhythms of each land-use type, as well as the transitions between these types, enrich the surface evolution matrix. It makes it easier to link the function models on the evolution matrices and to assure the coherence of the matrix combination. For example, if we know on the one hand that x ha of Chestnut forest were abandoned in 1950, without any human action since that time, and if we know on the other hand that it takes 30 years for an abandoned Chestnut forest to turn into coppice, we can assume that in 1980, the surface occupied by Chestnut coppice will be of x ha more.

However, the landscape dynamic that this model explains is more functional than spatial because the modelling itself needs the generalisation of spatial information that is aggregated by landscape element. The function matrices do not integrate criteria of spatial differentiation. The superposition and the combination of matrices permit a further understanding of the spatial unit dynamic by the evolution of explanatory parameters and by surface changes. It does not translate into spatial relations between landscape units over the course of time, except to detail them by the application of function models to local spatial units.

14.4 DISCUSSION AND RESEARCH PERSPECTIVES

The three methods outlined above respond to specific aspects of land-use dynamics analysis: land-use processes that express themselves in landscape can be modelled by function matrices; events that change those land-use processes can be identified by a chronological approach; and morphological changes of those processes and events can be detected by the transition-state method. For a particular landscape element (before considering the spatial propagation of the identified processes), we need to understand how these three methods can be integrated, in order to establish a link for this element between morphological changes, land-use processes and historical events. This combination can lead to a modelling of landscape dynamics that would be more appropriate for management objectives.

However, in order to realise this combination, we need to specify the limits of the three methods. These limits are partly linked to the structuring of spatial information and the application of corresponding tools that are also interconnected. With regard to spatio-temporal modelling, they bring the defaults of their qualities and it is necessary to try to compensate them by finding where they may compliment each other.

The matrix method of spatial dynamics seeks evolutionary trends through the use of a spatio-temporal model that combines the function matrices with the various surface states. There is an attempt to fully consider the complexity of systems in the spatial use of land. However, the combination of matrices leads to a global result, not localised, in order to at least add spatial differentiation criteria to the function matrices and to structure them in the form of attribute tables per spatial unit.

The use of a GIS in integrating local information, through the methods of transition-states and historical land-use event states, presents the advantage of not distancing from the postulates on the land-use system and of not formulating spatial relations between landscape units. The constitution of information, in local data based on the information layer of a landscape state, requires a minimal construction of the dynamic at play. In the case of transition-states, it is a succession of spatial states that is used to construct the spatial dynamics, emphasising the spatial structure more than the functioning. In the case of the chronological approach, it is the reconstitution of events in time that is utilised, emphasising the dynamics more than the spatial structure. The limit of the GIS tool is that it demands spatial information covering all the space studied, which can become extremely *heavy* to manipulate and require choices to be made in the type of data to be collected.

The following schema is an attempt to illustrate potential relations between the three methods, in classifying the forms of spatio-temporal information structuring put into place by two variables. One is spatial: the information can be localised or not; the other is temporal: the information can be a succession of states or a dynamic.

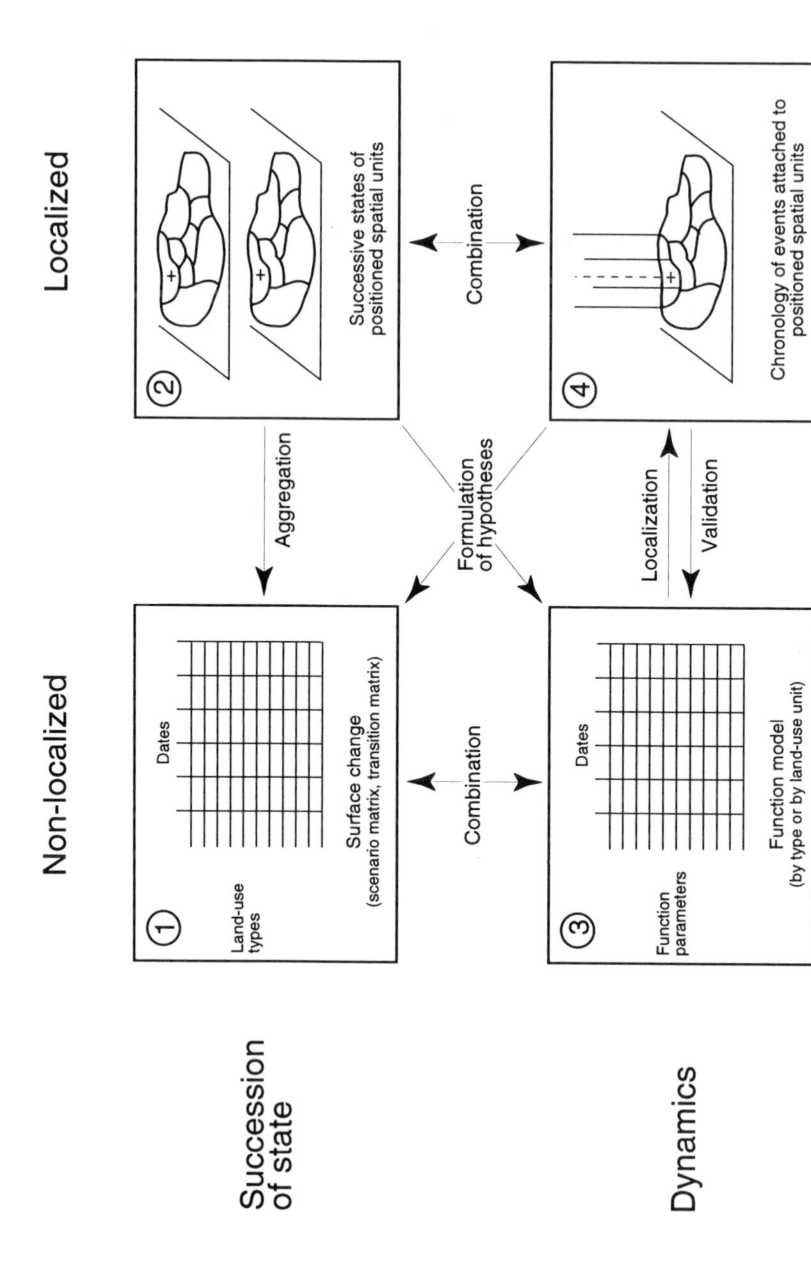

Figure 14.5 Relations between forms of structuring spatio-temporal information

In the examples presented, the spatial dynamics matrix method combines non-located dynamic information (function matrices) (3) with non-located state information (scenario or transition matrix) (1). The transition-state method combines located information of different spatial unit versions (as geographic information layers of the same space for several dates) (2) with information per non-located state (transition matrix) (1). The historical land-use event method only deals with located dynamic information (chronologies of territorial units attached to a GIS layer) (4).

When spatio-temporal information can be collected and structured, other relations may be established between the four poles of this schema. In the field of GIS, the representation of dynamic states (2) can be combined with the representation of events (4). Beginning with the best known territorial state, the construction of the spatial unit dynamic is done by fitting events that have impacted it with its successive spatial states. As soon as this overlay is possible, spatial changes in time can be joined and a spatial and temporal trajectory of spatial units can be constructed (therefore treating life and motion at the same time for each spatial unit).

The GIS land-use layer at several dates (2) permits by aggregation to construct evolution scenario and/or transition matrices (1), which may be combined to a contiguity matrix. The chronologies of land-use events attached to the GIS land-use layer at a given date (4) can be integrated in the function models (3) in order to validate them. Conversely, the function models can be specified for a located land-use unit and become an attribute table of a GIS layer. Knowledge of successive localised spatial units states (2) permits the formulation of hypotheses for the construction of function models (3). In the same way, the knowledge of located chronological spatial units (4) permits the formulation of hypotheses for the construction of transition matrices (1).

Rural management requires the use of all these combinations in order to arrive at an operational model of land-use dynamics. At the landscape element level, corresponding to the base units of the constructed geographic information layers, it is necessary to know the changes of identifiers and form, and to associate them to the functioning of the landscape element, which is product at the same time of slow processes and violent events. The following simple schema demonstrates the manner in which spatio-temporal information can be linked to build a model of land-use dynamics for a particular landscape element. The main perspective of the present methodological proposal is to use the spatio-temporal modelling of the landscape element to formulate rules about landscape dynamics and to implement scenarios of landscape evolution (for example, combining GIS with a spatial multi-agent approach).

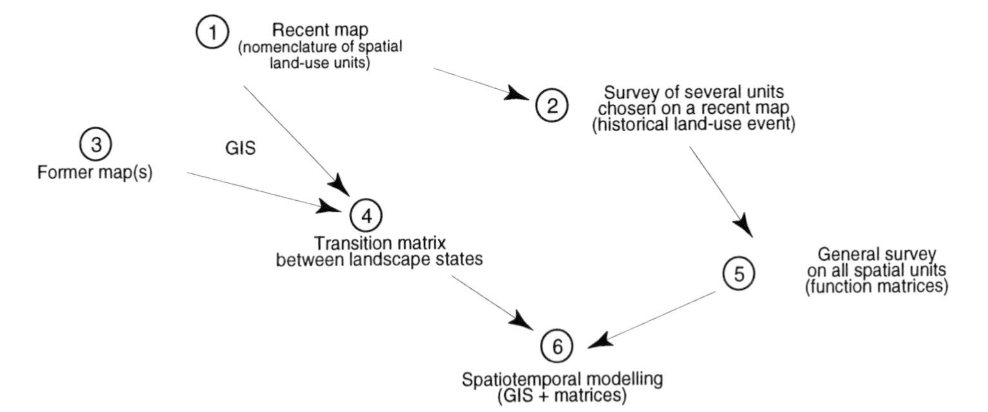

Figure 14.6 Succession of operations to acquire and structure information necessary for spatio-temporal modelling

REFERENCES

Arnaud, M.-T. and Gautier, D., 1996, Impact d'événements spatiaux dans le paysage de Gabriac (Cévennes). *Le paysage, pour quoi faire?* Actes Avignon 3/1996, Laboratoire Structures et Dynamiques Spatiales, Avignon University, pp. 31–39.

Flewelling, D., Egenhofer, M.J. and Frank, A.U., 1992, Constructing geological cross sections with a chronology of geologic events. In *Proceedings of the 5th International Symposium on Spatial Data Handling*, Charleston, SC, (IGU), Vol. 2, pp. 544–553.

Forman, R.T.T. and Godron, M., 1986, *Landscape Ecology*, (New York: John Wiley), p. 619.

Gautier, D., 1995, Dynamique spatiale de mise en valeur d'une châtaigneraie par transition entre deux états d'occupation du sol. *Revue Internationale de Géomatique*, 5(1) 1995, pp. 53–71.

Gautier, D., 1996, *Analyse du rapport entre l'organisation spatiale et la gestion des ressources renouvellables, appliquée à la gestion de la châtaigneraie cévenole* [Analysis of relationships between spatial organization and renewable resource management as applied to the Chestnut forest landscape of the Cévennes]. Ph.D thesis, University of Avignon, Avignon, France, p. 352.

Godron, M., 1979, *...éléments d'écologie des végétaux terrestres*. U.S.T.L, Montpellier, p. 63.

Godron, M. and Lepart, J., 1973, Sur la représentation de la dynamique de la végétation au moyen de matrice de succession. *Bericht zum Internationalen Symposium der Vereinigung zur Vegetationskunde*, (Reinhold Publ.), pp. 269–287.

Goodchild, M. and Gopal, S., Eds, 1989, *Accuracy of Spatial Databases*, (London: Taylor & Francis), p. 290.

Langran, G., 1992, *Time in Geographic Information Systems*, (London: Taylor & Francis).

Langran, G. and Chrisman, N.R., 1988, A framework for temporal geographic information. *Cartographica*, 25(1), pp. 1–14.

CHAPTER FIFTEEN

Representing Geographic Information to Support Queries about Life and Motion of Socio-Economic Units

May Yuan

15.1 INTRODUCTION

This chapter proposes a dynamic GIS representation for the support of spatio-temporal information retrieval and analysis. Hitherto, GIS representations have been mostly developed upon *layer* concepts with map metaphors, which fundamentally depict 2-D static geographic worlds with spatial components as the focal objects, semantic components as attributes of spatial objects, and temporal components as time stamps (Frank and Mark, 1991). Consequently, spatio-temporal information about *life* (birth, death, growth, merge, split, and reincarnation) and *motion* (movement, spread, jump, and flow) (Frank, this volume) is difficult to represent in a GIS. This is because information about *life* and *motion* requires the separation of semantic and temporal components from spatial components to allow description of geographic information from any perspective of semantics, time, or space.

Distinguished from static GIS data layers, the proposed representation comprises dynamically linked data objects in three independently managed data domains, each of which bears data about geographic semantics, space or time. Semantic information conveys geographic concepts (properties, entities, phenomena, events or abstract ideas). Spatial and temporal information marks the realization of the semantic information in space and time. In order to support geographic information query and analysis in a geographic information system (GIS), a data framework is needed to dynamically link the three components in ways that geographic information can be accessed and computed from semantic, spatial and temporal perspectives, independently or correspondingly. Dynamic links among semantic, spatial and temporal objects allow representation of information about *life* and *motion* depending upon if changes occur to semantic or spatial objects over time. In either case, spatio-temporal information represented in the proposed framework can be used to support spatio-temporal information analysis of what, when and where about *life* or *motion*.

Concepts about *life* and *motion* emerge to modelling of geographic information with an emphasis upon an entity, a phenomenon or an event in that determination of an essential semantic component is critical to the identification of a focal data object and description of its behaviour. In most cases, essential semantic components are the nomenclatures or names of geographic entities, such as soil types, administrative units and wildfire events. Once being determined, a semantic object needs to hold persistent object identifiers in a GIS with other semantic, spatial and temporal components as its descriptors to represent geographic meaning carried by the object. Consequently, concepts of *life* can be represented by changing spatial and non-spatial properties of semantic objects over time, whereas concepts about *motion* can be represented by changing locations over time, respectively.

It is easy to discern that layer-based GIS models have limited capabilities to represent information about *life* and *motion*. Most GIS data models apply time-stamping techniques to associate temporal information with spatial objects in a GIS layer. In layer-based GIS data models, geographic information is decomposed into sets of single-theme layers as regular (raster) or irregular (vector) tessellation models (Frank and Mark, 1991). A set of geometry-based spatial objects is used to represent reality. Information other than dimension or geometry (cells, points, lines and areas) is modelled as attributes of these spatial objects. Such location-based layers handicap GIS abilities to represent dynamic phenomena and to support queries about rate, frequency and process. Nevertheless, techniques have been proposed in temporal GIS data modelling include time stamping layers (the snapshot model: Armstrong, 1988), time stamping attributes (the space–time composite model: Langran and Chrisman, 1988), and time-stamping spatio-temporal objects (the spatio-temporal object model: Worboys, 1992). The snapshot approach has disadvantages of data redundancy and operation inefficiency (Peuquet and Duan, 1995). The space–time composite and spatio-temporal object models have problems with object persistency because they need constant updating of object identities upon continual changes in spatial objects and configurations.

All of these data models represent geographic worlds as attributes of geometric primitives, such as points, lines, polygons or cells. With the exception of points and cells, which will be discussed in the next paragraph, these geometrically indexed objects "*force a segmentation of the entities being represented into separate layers whenever they interact in time or space: adopting this representational method forces compromises on most environmental modelling*" (Raper and Livingstone, 1995, pp. 359). As a result, geography is represented as attributes at locations of geometrical objects in GIS worlds. For example, Cleveland County is modelled not as a geographic object but as the attribute value of an attribute 'County Name' with spatial mappings to a polygon set. Consequently, any changes to the spatial extent or shape of Cleveland County will cause further fragmentation of space and result in a new set of spatial objects with new identifiers in the GIS database. Since tracing the uncorrected old and new identifiers of spatial objects is difficult in these data models, query about the changes of Cleveland County boundaries during a period of time is deficient in current GIS.

To enhance the support for spatio-temporal queries and analysis, recent development has attempted to extend the importance of time in GIS data modelling, but the applications of these data models are constrained by their strong ties to chosen data structures and underlined layer-based model frameworks. Smith *et al.* (1993) have proposed a modelling and database system (MDBS) to support high-level modelling of spatio-temporal phenomena by (1) a conceptual domain (C-Domain) for abstract views of entities and transformations and (2) a representation domain (R-Domain) for symbolic representation. MDBS is designed with a strong emphasis on computational support for scientific modelling of GIS data. While MDBS encounters support of hydrographic graphing and modelling, the design of data manipulation and processes is based on raster data and, thus, appears difficult to apply to vector systems and problem domains other than hydrology. Peuquet and Duan (1995) have designed the Event-based Spatio-Temporal Data Model (ESTDM) to record changes of an event at locations through time. However, ESTDM is also a raster system, which needs substantial modification to be applicable to a vector system to maintain entity identities.

Another approach is Raper and Livingstone's OOgeomorph (1995), which models processes by point-based data. OOgeomorph is designed specifically for modelling point-based variations of coastlines, and its applicability to other geographic phenomena of higher geometrical dimensions remains undocumented. All the three approaches cannot fully represent spatio-temporal information about *life* and *motion* due to their emphasis on describing geographic attributes at locations (points, lines, polygons or cells).

To further GIS support for spatio-temporal representation, query and analysis, this chapter applies a three-domain model to represent spatio-temporal phenomena (Yuan, 1996). While change of trade areas among supermarkets is used here as an example to demonstrate the support for spatio-temporal queries of the proposed GIS data model, the same conceptual framework is applicable to other socio-economic and process-based applications, such as changes in county boundaries and wildfire spread. The use of the three-domain model is distinguished by its emphasis of the equality of the three geographic information components in GIS data modelling, whereas the previous studies have started with location-based layer frameworks. In other words, the three-domain model is designed in accordance with the composition of geographic information and spatio-temporal reasoning and analysis. Later in this chapter, the utility of the proposed GIS data model is demonstrated by a hypothetical evolution of supermarket trade areas in a region. The case study simplifies yet represents a wide range of complex spatio-temporal queries about life and motion of socio-economic units necessary to be supported by a temporal GIS. The generality of the proposed representation endows applicability to spatio-temporal data modelling for wildfire history, ghetto development and land use change.

This introductory section has outlined the current approaches to spatio-temporal data modelling in GIS. These proposed methods appear limited to a specific problem domain and ineffective to support a broad range of spatio-temporal query. However, these limitations can be overcome by adopting a three-domain data framework. The following sections discuss (1) use of a three-domain model to represent spatio-temporal information by incorporating geographic entities and constructs into a GIS database, and (2) support of the proposed data model for spatio-temporal query and analysis of information about the evolution of supermarket trade areas in a hypothetical region. A concluding section summarizes the key features of the new approach and outlines further research in using the proposed model to facilitate query and analysis of spatio-temporal information in GIS.

15.2 A THREE DOMAIN MODEL TO REPRESENT SPATIO-TEMPORAL INFORMATION ABOUT 'LIFE' AND 'MOTION'

As discussed in the introductory section, spatio-temporal information can be categorized as *life* and *motion*. Life-based information describes historical changes to a geographic entity throughout the course of its existence, including birth, death, growth, merge, split and reincarnation. In contrast, *motion* emphasizes gradual or eruptive shifts in locations as of movement, spread, jump and flow. Distinction of *life* and *motion* resides in entity identity in that *life* usually involves an increase or decrease in the total number of entities in a database universe, while *motion* does not. The emphasis of entity identity also explains the difficulty of representing *life* and *motion* in the current location-based GIS data layers. The fundamental reason is that both representations of *life* and *motion* require the separation of semantic and temporal components from spatial components to allow semantic or temporal components be the focal objects (with their own identities).

Both life and motion schemata have three constructs of space, time and geographic semantics (definitions of entities and attributes independent of space and time) that can be used in spatio-temporal information modelling. Yuan (1994) proposes a three-domain model based on conceptual schemata for fires forecasting, fire behaviour modelling, fire effect analysis and fire history modelling. The three-domain model structures geographic information in three separate but correlated domains of semantics, time and space (Figure 15.1).

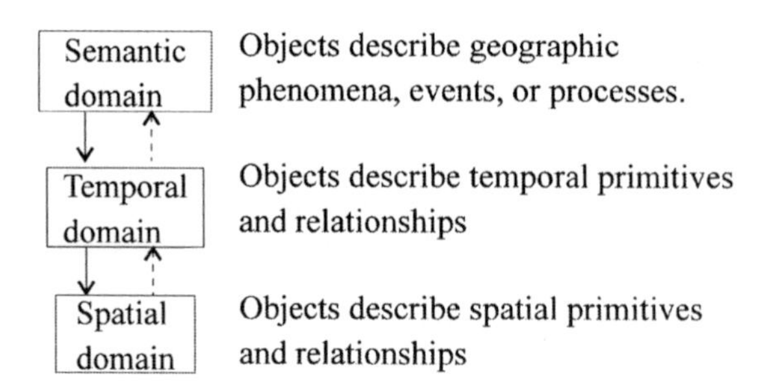
Semantic domain
Objects describe geographic phenomena, events, or processes.
Temporal domain
Objects describe temporal primitives and relationships
Spatial domain
Objects describe spatial primitives and relationships

Figure 15.1 Linking semantic, temporal and spatial objects to model spatio-temporal information in a three-domain model. Solid arrows indicate process or event-oriented data modelling approach. Dashed arrows indicate a location-centred data modelling approach.

Geographic phenomena are represented by linking semantic, temporal and spatial objects to describe the three aspects of geographic properties. Semantic objects, depicting the meanings of geography that the information attempts to convey, represent concrete entities (such as Cleveland County) or abstract concepts (such as land ownership) whose identities are independent of spatial and temporal properties. Temporal objects indicate time of significance to a semantic object or a spatial object. Spatial objects show location, geometry and dimension as points, lines, polygons or cells. The three-domain model can be easily applied to represent spatio-temporal information and to support a wide range of spatio-temporal queries because it anticipates scenarios and questions correlating with *life* and *motion* schemata. A hypothetical example of changes in trade areas among supermarkets is used to demonstrate the model's capabilities in GIS representation and query support. The same framework can be applied to other socio-economic domains such as changes in school districts and voting districts. In the hypothetical example, a simplified socio-economic history presents an evolution of supermarket trade areas from T_1 to T_6 in a hypothetical region. Socio-economic information embedded in the history is summarized as follows (Figure 15.2):

$\mathbf{T_1}$: the initial state with only one Supermarket (A) in the region;
$\mathbf{T_2}$: a new Supermarket (B) entering the region;
$\mathbf{T_3}$: an increase in the trade area of Supermarket A;
$\mathbf{T_4}$: a new Supermarket (C) entering the region;
$\mathbf{T_5}$: both Supermarkets A and B increasing their trade areas, and the opening of second Supermarket A in the southeast region;
$\mathbf{T_6}$: the re-allocation of the original Supermarket A to the northeast, Supermarket B and the second Supermarket A increasing trade areas, and Supermarket C running out of business.

A convenient way to model spatio-temporal information involved in this example is to store six time-stamped layers of T_1 to T_6 as shown in Figure 15.2. However, the approach results in a significant amount of duplicated information because some trade areas remain unchanged throughout a long period of time. For instance, the trade area for Supermarket B remains unchanged till T_4. In addition, it is difficult to support queries like "When did the trade area of Supermarket A occupy more than half of the region?" and "When did we have the most supermarkets in the region from T_1 to T_6?" Use of the space–time composite model or space–time object model can eliminate data duplication, but these models also encounter difficulty in support of the listed spatio-temporal queries. (The snapshot and space–time composite schemata model supermarkets as attributes of locations so that they cannot allow users to trace what happens to individual supermarkets. The spatio-temporal object model can associate supermarkets with ST-objects, but the model cannot handle 'reincarnation'. It needs two independent ST-objects to represent the 'Supermarket A' from T_1 to T_2 and the 'Supermarket A' at T_3.)

The three-domain model, on the other hand, can eliminate data duplication *and* provide sufficient support for spatio-temporal queries. In the example, semantic objects refer to supermarkets because their identities are independent of the spatial configurations and their locations; i.e., the identification of Supermarket A is independent of its location and trade area. Temporal objects indicate time of significance, T_1 to T_6. Spatial objects represent individual trade polygons, which have areal extent, shape, and dimension, and points of locations. The following paragraphs outline the representation structures of the trade area information in a three-domain model. Formal definitions of the domains and objects of the model are elaborated in (Yuan, 1996).

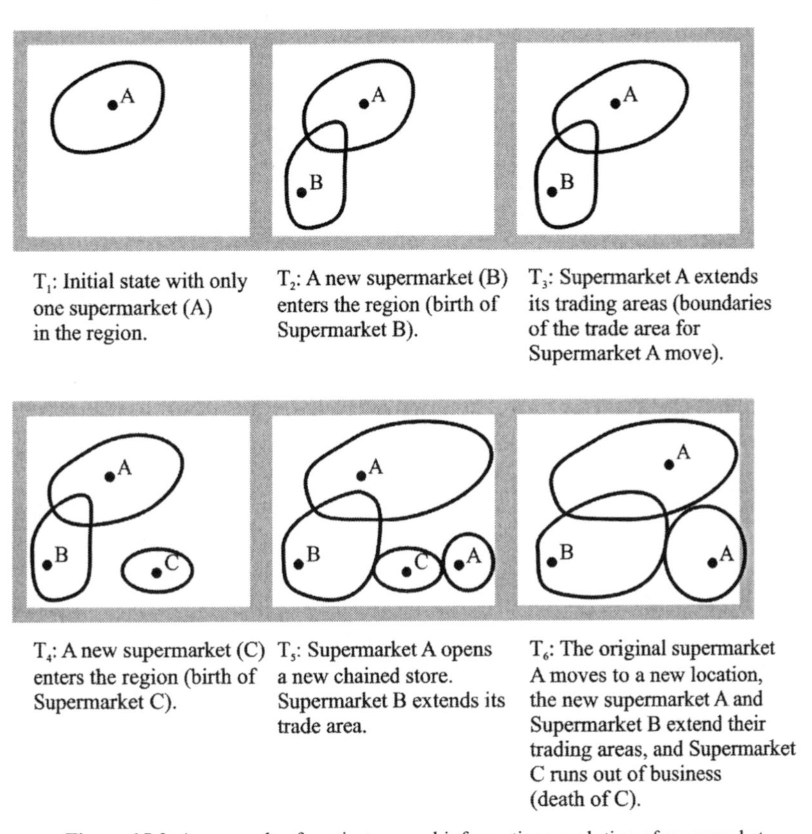

T_1: Initial state with only one supermarket (A) in the region.

T_2: A new supermarket (B) enters the region (birth of Supermarket B).

T_3: Supermarket A extends its trading areas (boundaries of the trade area for Supermarket A move).

T_4: A new supermarket (C) enters the region (birth of Supermarket C).

T_5: Supermarket A opens a new chained store. Supermarket B extends its trade area.

T_6: The original supermarket A moves to a new location, the new supermarket A and Supermarket B extend their trading areas, and Supermarket C runs out of business (death of C).

Figure 15.2 An example of spatio-temporal information: evolution of supermarket trading areas in a hypothetical region

15.2.1 The Semantic Domain

Table 15.1 lists the semantic objects in the semantic domain of supermarkets. In the semantic domain, entities (supermarkets) are represented by semantic objects with unique identifiers throughout the course of the evolution, and they are directly associated with conceptual constructs of supermarkets independent of locations and trade areas. Spatio-temporal information about 'life' encountered in the hypothetical case includes birth of Supermarkets B and C, death of Supermarket C, and re-location and splitting of Supermarket A.

Table 15.1 Semantics Table

Sem. ID	Sem. Name	Owner
1	Super A	James
2	Super B	Brian
3	Super C	Mary

Table 15.2 Time Table

Time ID	Time (dd–mm–yr)
1	01 05 75
2	30 07 83
3	20 10 85
4	31 12 88
5	04 04 89
6	11 03 90

Table 15.3 Polygon Table

Poly ID	Area	Perimeters
3	(value)	(value)
4	(value)	(value)
8	(value)	(value)
9	(value)	(value)
10	(value)	(value)
11	(value)	(value)
12	(value)	(value)
13	(value)	(value)
14	(value)	(value)
15	(value)	(value)
16	(value)	(value)
17	(value)	(value)
18	(value)	(value)
19	(value)	(value)
20	(value)	(value)
21	(value)	(value)
22	(value)	(value)

Table 15.4 Point Table

Point ID	X	Y
1	X_1	Y_1
2	X_2	Y_2
3	X_3	Y_3
4	X_4	Y_4
5	X_5	Y_5

Table 15.6 Location Relation Table (links among temporal, semantic and spatial objects, and '−1' indicating the death of associated semantic objects)

Sem. ID	Time ID	Space ID
1	1	1
2	2	2
3	4	3
1	5	1, 4
3	5	−1
1	6	4, 5

Table 15.5 Trade Area Relation Table (links among temporal, semantic and spatial objects, and '−1' indicating the death of associated semantic objects)

Sem. ID	Time ID	Poly ID
1	1	1
1	2	2, 3
2	2	3, 4
1	3	2, 3, 5
3	4	6
1	5	3, 5, 7, 8, 10
2	5	3, 4, 8, 9
3	5	−1
1	6	3, 8, 10, 11, 12, 13, 14, 19, 20, 21, 22
2	6	3, 4, 8, 12, 13, 14, 15, 16, 17

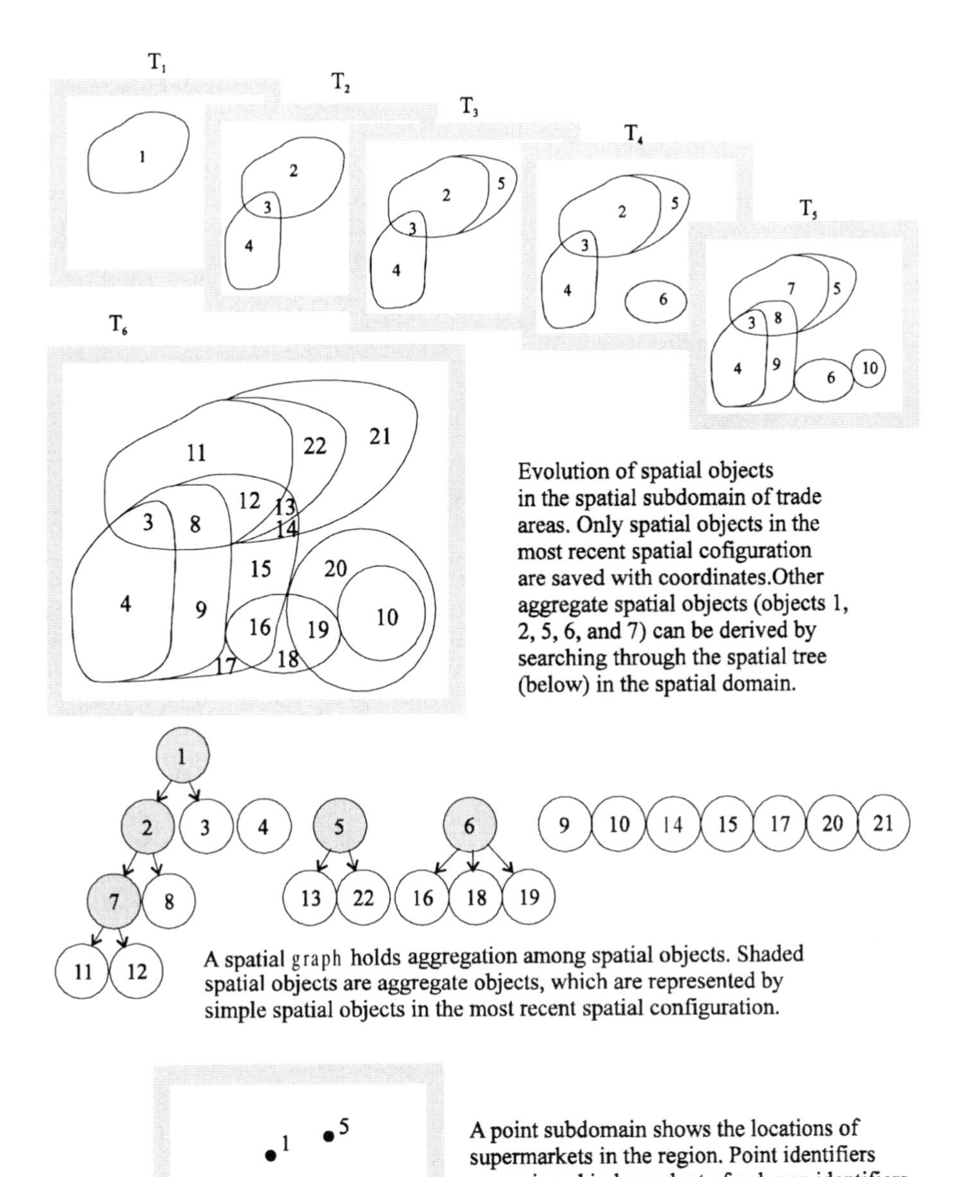

T1
T2
T3
T4
T5
T6
Evolution of spatial objects in the spatial subdomain of trade areas. Only spatial objects in the most recent spatial cofiguration are saved with coordinates.Other aggregate spatial objects (objects 1, 2, 5, 6, and 7) can be derived by searching through the spatial tree (below) in the spatial domain.
A spatial graph holds aggregation among spatial objects. Shaded spatial objects are aggregate objects, which are represented by simple spatial objects in the most recent spatial configuration.
A point subdomain shows the locations of supermarkets in the region. Point identifiers are assigned independent of polygon identifiers. References can be made by separate relation tables, such as trade area relation and location relation.

Figure 15.3 The spatial domain for trade area evolution among supermarkets A, B and C from T_1 to T_6

15.2.2 The Temporal Domain

Temporal objects included in the transition are six time points from T_1 to T_6. It is assumed that there is no knowledge about changes occurring from T_x to T_y unless the semantic or spatial object has a link to another temporal object between T_x and T_y (Table 15.2). Should there be a change, a time point shall be noted between T_x and T_y. However, in other cases, temporal objects can be points (instants as defined in Freksa, 1992) or lines (intervals as defined in Allen, 1983). Temporal structures and relationships have been elaborated in (Snodgrass and Ahn, 1985; Freksa, 1992, and Worboys, 1990).

15.2.3 The Spatial Domain

Figure 15.3 shows spatial objects in the spatial domain, which represents spatio-temporal configurations of trading areas and locations of supermarkets. Spatial objects of the same object type hold unique identifiers throughout the course of their life. For polygon objects, overlaps can result in change of object identifiers, which is resolved by constructing a spatial tree to record parenthood relationships among spatial objects. With the spatial tree, we can trace spatial object 1 from spatial objects 3, 8, 11 and 12. The open nodes in the spatial tree represent spatial objects in the most current spatial configuration of the region as indicated in the polygon sub-domain table (Table 15.3). Only spatial objects in the most current spatial domain are stored with coordinates, geometrical properties, and topological properties. This geometrical and topological information about objects in the most current spatial configuration can be stored by the current GIS layer-table structures or object-oriented frames. Aggregate spatial objects (shaded nodes in the spatial tree) which are excluded from the current spatial domain (and thus do not appear in the polygon table) can be derived by tracing the spatial tree. For example, while spatial object 5 is omitted in the most current spatial configuration, it can be traced through the spatial tree as an aggregate object of spatial objects 13 and 22. Consequently, geometrical and topological properties, area, perimeter and connectivity for the aggregate object 5 can be computed based on its composing objects 13 and 22. On the other hand, the point sub-domain, in which points represent locations, can be fully constructed without a spatial tree since overlapping objects are exclusive among points (Table 15.4).

15.2.4 Domain Links

Through applying a relational database structure, semantic, temporal and spatial objects are organized as shown in Tables 15.1, 15.2, 15.3 and 15.4. Tables 15.3 and 15.4 correspond to polygon and point sub-domains in space, since the sample case of supermarket trading involves both point data for supermarket locations and polygon data for supermarket trade areas. Relation tables (Tables 15.5 and 15.6) illustrate the links between semantic, temporal, and spatial objects, but such a table is inefficient in handling many-to-many relationships among these objects. An alternative for organizing links among the three types of objects would be an object-oriented framework. Tables 15.5 and 15.6 are used here because of simple tabular structures.

The semantic domain (Table 15.1), temporal domain (Table 15.2), spatial domain (Tables 15.3 and 15.4 as well as Figure 15.3) and domain links (Tables 15.5 and 15.6) provide the basic references to infer spatio-temporal information about evolution of supermarket trading. While Tables 15.1–15.4 and Figure 15.3 correspond to the semantic, temporal and spatial domains, Tables 15.5 and 15.6 indicate object links among the three domains. Domain links among the three types of objects can, therefore, represent geographic phenomena from either a location-based or an entity-based perspective (Figure 15.2) to support query and reasoning about 'life' and 'motion'. In the next section, the capabilities of the three-domain model to facilitate a wide variety of spatio-temporal queries will be further illustrated in three sets of sample queries about the hypothetical evolution of supermarket trade areas.

15.3 SUPPORT BY THE THREE-DOMAIN MODEL TO STORE AND QUERY SPATIO-TEMPORAL INFORMATION ABOUT LIFE AND MOTION

A representational schema needs to anticipate all possible queries and analysis to be performed in an information system, but that requirement has long been neglected in the modelling of geographic information. As a result, many GIS cannot provide sufficient capabilities for query and analysis of spatio-temporal data on history or processes. The three-domain model, however, is designed with a strong emphasis on spatio-temporal query and analytical support based on the ways we conceptualize and acquire geographic knowledge. Queries about spatio-temporal information can be categorized as when, what and where (Peuquet, 1994), each of which can be viewed from *life* (change of the total number of entities involved and their properties) and *motion* (change of entity's locations) perspectives. Based on the example of evolution of trading areas among supermarkets as described in Section 15.2 (Figure 15.3), the following paragraphs demonstrate the support of the three domain model for querying spatio-temporal information about when, what and where in the perspectives of life and motion. Computation of these queries is based on semantics, space, time and relation tables (Tables 15.1–15.6) and a spatial tree (Figure 15.3). Procedures are described in a pseudo query language for demonstration convenience. A query language can be designed to allow query computation to be performed internally and automatically in a GIS and, thus, to ease users from having to remember technical operations in a query process. However, a pseudo query language used in the following examples is simply to demonstrate reasoning procedures applied to the proposed representation schemes for query support. No attempts are made to use it as a formal query language in this chapter. Although syntax of the pseudo query language is primitive, it shows the basic algorithms to derive answers for complex spatio-temporal query that is beyond the capability of most spatio-temporal data models proposed in GIS.

15.3.1 Query Spatio-Temporal Information about When

This kind of query is used to obtain information on temporal objects. Answers can be obtained by referring semantic or spatial objects to temporal objects through a proper relation table. Life-oriented questions inquire when birth, death, splitting, merging or reincarnation occur in a certain period of time, while motion-oriented questions ask when a move, jump or spread takes place. The following are sample queries and derivations of the answers:

Query about life: When was Supermarket C opened and when did it run out of business?

Supermarket C is represented by semantic object 3, which is associated with temporal objects 4 and 5 in the location relation table (Table 15.6). The smaller temporal identifier represents an earlier point in time. Therefore, Supermarket C was first open on 31 December, 1988 (temporal object 4) at location 3 (spatial point object 3), but was out of business (indicated by spatial point object with identifier –1) on 4 April, 1989 (temporal object 5). The computation procedures are described as follows:

STEP	QUERY PROCEDURES	RESULTS
1	SELECT sem_id FROM sem_table WHERE sem_name = Super C	*sem_id = 3*
2	SELECT (time_id) AND (point_id) FROM location-area_relation_table WHERE sem_id = 3	*time_id* / *point_id*: 4 / 3; 5 / -1
3	SE!.ECT time FROM time_table WHERE time_id = 4 OR time_id = 5	*time_id* / *time*: *4* / *31-12-88*; *5* / *4-4-89*

Query about motion: When was Supermarket A re-located to the northeast side of the region?

Supermarket A has the semantic identifiers 1 and 4. In the location relation table (Table 15.6) semantic object 1 is associated with spatial point object 1 (location 1) at T_1 and spatial point object 5 at T_6, whereas semantic object 4 is associated with spatial point object 4 (location 4). Therefore, inference can be made that an outlet of Supermarket A (represented by semantic object 1) changes its location by moving from spatial point object 1 to spatial point object 4 at T_6 (11 March, 1990), but another Supermarket A (represented by semantic object 4) stays at location 4.

STEP	QUERY PROCEDURES	RESULTS
1	SELECT sem_id FROM sem_table WHERE sem_name = Super A	*sem_id = 1, 4*
2	SELECT time_id AND point_id FROM location_relation_table WHERE sem_id = 1 OR sem_id = 4	*Sem_id* / *time_id* / *point_id*: *1* / *1* / *1*; *1* / *5* / *5*; *4* / *6* / *4*
3	SELECT time FROM time_table WHERE time_id = 1 OR time_id = 6	*time_id* / *time*: *1* / *1-5-75*; *6* / *11-3-90*

15.3.2 Query Spatio-Temporal Information about Where

This type of query aims at obtaining information about spatial objects for locations and spatial properties of a semantic object at a specific time. *Where* questions can be static (asking whereabouts or states of entities or attributes) or dynamic (asking paths of an entity changing its location through time). Query about static information can be well supported by layer-based data models, but query about dynamic movement cannot be easily derived from current GIS databases. In general, movement or spread is simulated in a raster GIS by techniques like cellular automata (Clarke *et al.*, 1994) instead of being represented and stored in GIS databases. Typical examples of static information include the distributions of soil types and locations of banks, whereas dynamic information describes movement of a flock of sheep, spread of a fire, and expansion or elimination of a region. The following are sample queries of location information about *life* and *motion* and derivations of answers to the queries:

Query about life: Where was Supermarket A when it was first open?

Answer to the query needs to determine when Supermarket A was opened and where it was at that time. Supermarket A is represented by semantic objects 1 and 4 in the supermarket semantics table. The location relation table (Table 15.6) indicates that semantic object 1 is associated with temporal objects 1 and 6, while semantic object 4 is associated with temporal object 5. The smaller temporal identifier represents an earlier point in time. Therefore, Supermarket A was first open on 1 May, 1975 (temporal object 1) when it was located at location 1 (point object 1).

STEP	QUERY PROCEDURES	RESULTS
1	SELECT sem_id FROM sem_table WHERE sem_name = Super A	*sem_id = 1, 4*
2	SELECT MIN(time_id) AND point_id FROM location_relation_table WHERE sem_id = 1 OR sem_id = 4	*sem_id* = *1*; *time_id* = *1*; *point_id* = *1*
3	SELECT (x,y) FROM point_table WHERE point_id = 1	*point_id* = *1*; *x* = X_1; *y* = Y_1

Query about motion: To which extent did Supermarket A expand its trade area from T_1 to T_6?

The search starts with Supermarket A in the semantic table (Table 15.1) and looks for spatial polygon objects that represent trade areas for Supermarket A. As mentioned earlier, both semantic objects 1 and 4 represent Supermarket A in the semantic domain. Only semantic objects with changes in spatial extent (linked to different sets of spatial polygon objects) at a certain point in time will be listed in the trade-area relation table (Table 15.5) to indicate the changes. Otherwise, it implies that no areal change occurs to a semantic object at that time. For example, semantic object 1 is associated with temporal objects 1, 2, 3, 5 and 6, which indicates that changes in the areal extent of semantic object 1 have occurred at these points in time. Since temporal object 4 is exclusive from the list, the areal extent of semantic object 1 at T_4 is the same as the previous record at T_3 (the previous state of semantic object 1 at T_4). Summation of spatial polygon objects associated with semantic objects 1 and 4 at these time points will suggest the largest trade area covered by Supermarket A from T_1 to T_6.

Spatial polygon objects, 2, 3, 5 and 6 are aggregate spatial objects so that they are excluded from the polygon table (Table 15.3). Examination of the spatial tree (Figure 15.3) suggests that spatial polygon object 2 is composed of spatial objects 7 and 8. While spatial object 8 is a simple polygon object, spatial object 7 is further composed of simple spatial objects 11 and 12. In addition, spatial objects 5 and 6 are also aggregate polygon objects, each of which constitutes spatial simple objects 12 and 13, and simple spatial objects 16, 18 and 19, respectively. Although area measurements are unavailable from the polygon table (Table 15.3) for aggregate polygon objects, their values of area coverage can be derived by summation of all their composed simple spatial objects. As a result, the answer for this query is that Supermarket A has its largest coverage of trade area at T_6 when its total trade area is covered by spatial polygon objects 3, 8, 10, 11, 12, 13, 14, 19, 20, 21 and 22.

<table>
<tr><th>STEP</th><th>QUERY PROCEDURES</th><th>RESULTS</th></tr>
<tr><td>1</td><td>SELECT sem_id
FROM sem_table
WHERE sem_name = Super A</td><td>sem_id = 1, 4</td></tr>
<tr><td>2</td><td>SELECT time_id AND poly_id
FROM trade-area_relation_table
WHERE sem_id = 1 OR sem_id = 4</td><td>
<table>
<tr><th>Sem_id</th><th>time_id</th><th>poly_id</th></tr>
<tr><td>1</td><td>1</td><td>1</td></tr>
<tr><td>1</td><td>2</td><td>2,3</td></tr>
<tr><td>1</td><td>3</td><td>2,3,5</td></tr>
<tr><td>1</td><td>5</td><td>3,5,7,8</td></tr>
<tr><td>1</td><td>6</td><td>3,8,11,12,13,
14,21,22</td></tr>
<tr><td>4</td><td>6</td><td>10,19,20</td></tr>
</table>
</td></tr>
<tr><td>3</td><td>SELECT poly_id IN STEP 2 RESULTS
FROM poly_table
SUM area BY time_id</td><td>Summed area</td></tr>
</table>

15.3.3 Query Spatio-Temporal Information about What

This type of query seeks information about changes in which the focal information is semantic objects such as changes of supermarket services for a particular area. We first identify the area of interest, and then examine what has been changed in that area by referring to its corresponding semantic objects at that time. The following are sample queries of information about change and procedures to derive answers to the queries:

Query about life: Which supermarket has ever been the dominant supermarket in the southeast region from T_1 to T_6?

This query can be started with specifying area of interest by interactive user input from a graphic user interface or by translating the verbal description of 'the southeast region' to the coordinates of the minimum bounding box at the southeast quadrangle of the region. In either way, spatial polygon objects in the most recent polygon sub-domain (Table 15.3, Figure 15.3), which reside in the area of interest, can be selected by applying the most commonly used GIS spatial function, overlay. Selection criteria can be set to retrieve either any spatial polygon objects in the area of interest or only those completely within the area of interest. Numerous GIS functions are available or in research to facilitate a wide range of select criteria based on spatial relationships among spatial objects as examples studied by Egenhofer and Mark (1995).

The selected spatial polygon objects are used to retrieve correspondent semantic objects through the links of temporal objects in the trade-area relation table (Table 15.5). In those cases when we are to select all spatial polygon objects completely or partially in the area of interest, spatial polygon objects 10, 14, 15, 16, 18, 19, 20 and 21 are retrieved. These spatial polygon objects are associated with temporal objects 5 and 6 (T_5 and T_6) and semantic objects 1, 2 and 4 in the trade-area relation table (Table 15.5). In addition, the spatial tree (Figure 15.3) suggests that spatial polygon objects, 16, 18 and 19 constitute spatial polygon object 6, which is associated with semantic object 3 at T_4.

<table>
<tr><th>STEP</th><th>QUERY PROCEDURES</th><th>RESULTS</th></tr>
<tr><td>1</td><td>SELECT poly_id
FROM poly_table
WHERE polygons OVERLAP WITH area of interest</td><td>poly_id = 10, 14, 15, 16, 18, 19, 20, 21</td></tr>
<tr><td>2</td><td>SELECT poly_id
FROM spatial_tree
WHICH COMPOSED OF poly_id IN STEP 1 RESULTS</td><td>poly_id = 3</td></tr>
<tr><td>3</td><td>SELECT time_id AND sem_id
FROM trade-area_relation_table
WHERE poly_id IN STEP 1 RESULTS</td><td><table>
<tr><th>sem_id</th><th>time_id</th><th>poly_id</th></tr>
<tr><td>3</td><td>4</td><td>6</td></tr>
<tr><td>4</td><td>5</td><td>10</td></tr>
<tr><td>1</td><td>6</td><td>3,8,11,12,13,14,21,22</td></tr>
<tr><td>2</td><td>6</td><td>3,4,8,12,13,14,15,16,17</td></tr>
<tr><td>4</td><td>6</td><td>10,19,20</td></tr>
</table></td></tr>
<tr><td>4</td><td>SELECT sem_name
FROM sem_table
WHERE sem_id IN STEP 3 RESULTS</td><td><table>
<tr><th>sem_id</th><th>sem_name</th></tr>
<tr><td>1</td><td>Super A</td></tr>
<tr><td>2</td><td>Super B</td></tr>
<tr><td>3</td><td>Super C</td></tr>
<tr><td>4</td><td>Super A</td></tr>
</table></td></tr>
<tr><td>5</td><td>SELECT poly_id IN STEP 4 RESULTS
FROM poly_table
SUM area by time_id AND sem_name</td><td><table>
<tr><th>sem_name</th><th>time_id</th><th>area</th></tr>
<tr><td>Super C</td><td>4</td><td>A_1</td></tr>
<tr><td>Super A</td><td>5</td><td>A_2</td></tr>
<tr><td>Super A</td><td>6</td><td>A_3</td></tr>
<tr><td>Super B</td><td>6</td><td>A_4</td></tr>
</table></td></tr>
</table>

A_1 = area of spatial polygon object 6
A_2 = area of spatial polygon object 10
A_3 = summation of area of spatial polygon objects 3, 8, 10, 11, 12, 13, 14, 19, 20, 21 and 22.
A_4 = summation of area of spatial polygon objects 3, 4, 8, 12, 13, 14, 15, 16 and 17.

Both semantic objects 1 and 4 represent Supermarket A, while semantic object 2 represents Supermarket B and semantic object 3 represents Supermarket C. Therefore, summation of area measurements for spatial polygon objects associated with each semantic object at T_4, T_5 and T_6 will resolve the dominant supermarket in the area of interest at these points in time. That temporal objects 1, 2 and 3 have no associations with selected spatial polygon objects implies no supermarkets exist in the southeast region from T_1 to T_3.

Comparison of A_1 to A_2 can reveal the dominant supermarkets in this region during the period of time. It appears that supermarkets C and A are the dominant supermarkets at T_4 and T_5. In particular, comparison of A_3 and A_4 can reveal if Supermarket A or B covers the largest trading area in the southeast region at T_6.

The example of dominant supermarket coverage shows the support of the three-domain model to spatio-temporal query about attribute distributions over time. In addition, separation of semantic objects and temporal objects from spatial objects in the three-domain model enables representing overlapping areas by the common spatial objects constituting area extent of two or more semantic objects. For example, Supermarket A and Supermarket B have an overlap at spatial object 14 in the southeast region. The ability of a GIS data model to manage geospatial data about overlapping areas is particularly important for the representation of wildfire and other re-occurring geographic events.

Query about motion: What was the supermarket that had gained the largest trade area from T_1 to T_4 after the supermarket was established?

Identification of supermarkets existing from T_1 to T_4 and tracing their trade areas in the period of time is the key to answer this query. The trade-area relation table (Table 15.5) indicates that supermarkets A, B and C existed in the region from T_1 to T_4. Supermarket A extended its trade area from spatial object 1 at T_1, spatial objects 2 and 3 at T_2, and spatial objects 2, 3 and 5 at T_3. Supermarket B covered a trade area of spatial objects 3 and 4 at T_2, and Supermarket C covered a trade area represented by spatial object 6. Based on the spatial objects associated with these supermarkets from T_1 to T_4 and the spatial tree (Figure 15.3), inference can be made that Supermarket A is established at T_1 and gains an area of spatial object 5 at T_3; Supermarket B was opened at T_2 and maintained its trade area through T_4, and Supermarket C was opened at T_4. As a result, Supermarket A is the only supermarket that expanded its trade area from T_1 to T_4. Supermarkets B and C maintained trade areas during the period of interest.

Consequently, deduction can be made that Supermarket A extended its trade area to polygons objects 13 and 22 at T_3. As shown in this example, the three-domain model is able to resolve information regarding 'movement' and 'spread' by referencing links from semantic and temporal to spatial objects. Once the semantic objects of interest are determined, retrieval of associated spatial objects through time can provide answers for complex spatio-temporal queries like advancing wildfire spread, increasing damage areas by flooding events, and movement of sheep flocks, for instance. These types of spatio-temporal query ask for dynamic information, which is difficult if not impossible, to be derived based on the current static structured layer models. The three-domain model, however, represents semantic, temporal and spatial constructs in a separate manner and uses dynamic links to describe geographic phenomena. As a result, the three-domain model allows changes of semantic, temporal and spatial properties to a geographic phenomenon separately (such as change of the name of a supermarket) and correspondingly (such as merge of two supermarkets resulting in change of the name of one or two supermarkets and the resulting trade area).

STEP	QUERY PROCEDURES
1	SELECT sem_id AND poly_id FROM trade-area_relation_table WHERE time_id ≤ 4

RESULTS (Step 1)

sem_id	time_id	poly_id
1	1	1
1	2	2, 3
2	2	3, 4
1	3	2, 3, 5
3	4	6

STEP	QUERY PROCEDURES
2	SELECT sem_name FROM sem_table WHERE sem_id IN STEP 1 RESULTS

RESULTS (Step 2)

sem_id	sem_name
1	Super A
2	Super B
3	Super C

STEP	QUERY PROCEDURES
3	RELATE STEP 1 RESULTS AND STEP 2 RESULTS BY sem_id

RESULTS (Step 3)

sem_id	sem_name	time_id	poly_id
1	Super A	1	1
1	Super A	2	2, 3
2	Super B	2	3, 4
1	Super A	3	2,3,5
3	Super C	4	6

STEP	QUERY PROCEDURES
4	SELECT poly_id FROM spatial tree WHICH COMPOSED OF poly_id IN STEP 3 RESULTS

RESULTS (Step 4)

(aggregate) poly_id	(simple) poly_id
1	2, 3
2	7, 8
5	13, 22
6	16, 18, 19

STEP	QUERY PROCEDURES
5	RELATE STEP 3 RESULTS AND STEP 4 RESULTS BY (simple) poly_id SORT BY sem_id

RESULTS (Step 5)

sem_id	sem_name	time_id	poly_id
1	Super A	1	3,8,11,12
1	Super A	2	3,8,11,12
1	Super A	3	3,8,11,12,13,22
2	Super B	2	3,4
3	Super C	3	16,18,19

As demonstrated in the sample queries above, modelling geographic semantics, time and space separately corresponds to the three categories of spatio-temporal queries. 'Queries about when' inquire temporal information for geographic semantics or locations, so that answers can be derived by checking temporal objects through referencing semantic objects or spatial objects in their relation table. 'Queries about what' look for semantic information at locations, so that answers can be found by deriving semantic objects through referencing spatial and temporal objects in their relation table. 'Queries about where' search for location information about geographic semantics that can be answered by obtaining spatial objects through referencing semantic and temporal objects in their relation table. Each of these query types can be further categorized as searching for life-oriented or motion-oriented information. Life-oriented information involves changes to the total number of entities considered in the database.

On the other hand, motion-oriented information emphasizes the existing geographic objects. The three-domain model can handle both types of information in a comparably simple manner in that addition or deletion of semantic, temporal and spatial objects can be performed independently without interfering the identifier of any object in its own domain or in the other domains. However, this is not the case in the current layer-based data models, in which semantics are attributes of spatial objects and need to be re-organized as changes occurring to spatial objects. In other words, the current layer-based GIS data models have difficulty to maintain persistent object identifiers, which has been identified as the primary problem in building a vector-based temporal GIS (Raper and Livingstone, 1995; Peuquet and Duan, 1995). With the support of dynamic links and spatial trees, the three-domain model is able to overcome this hurdle and provides a generic GIS data framework for representing spatio-temporal information and facilitating complex spatio-temporal query about *life* and *motion.*

15.4 CONCLUSIONS

This chapter presents a new approach in GIS modelling spatio-temporal data. The new approach represents geographic information according to the three primary components of semantics, space and time, which have been identified in geographic information analysis (Berry, 1964; Sinton, 1978). The generic GIS data framework proposed to model spatio-temporal data in a GIS represents the three geographic information components in three independently managed data domains: the semantic domain, spatial domain and temporal domain. Distinguished from other spatio-temporal GIS data models, the three-domain model handles semantic and temporal information separately instead of being attributes of spatial objects as in layer-based time-stamped approaches to GIS data modelling.

The representation and information production capabilities of the three-domain model have been demonstrated through a hypothetical evolution of changes of super-market trading areas and a set of complex spatio-temporal queries on the example. The three-domain model is able to facilitate spatio-temporal queries about when, what and where, of which target information is time, geographic semantics or locations. In addition, the information being asked can be either life-oriented or motion-oriented. 'Queries about what' seek information regarding which geographic semantics are present at locations at different points in time. 'Queries about where' search for spatial information of geographic entities (semantics) at different points in time or over time. 'Queries about when', on the other hand, target to temporal properties or significant time regarding changes to geographic phenomena. They can also be mapped to a movement schema if our questions direct to when a certain geographic semantic has moved from one location to another.

A GIS should be based on a complete and rigorous schema for geographic data modelling (Goodchild, 1992) to overcome the difficulty in handling geographic complexity, scale differences, generalization, and accuracy (Burrough and Frank, 1995). The three-domain model is shown in this chapter, its ability to support multi-value (overlaps of supermarket trade areas) and multi-dimensional (point and polygon data for supermarkets) geographic information, which may be unsupported by simple layer-based GIS data models. Support for representation at different scales can also be achieved by links defined by scale functions to associate a semantic object with proper spatial objects. In summary, the three-domain model can provide a more flexible and powerful alternative schema for geographic data modelling to the layer-based data models. Research is underway on the application of the three-domain framework to environmental data. For example, a tornado or a wildfire can be modelled as semantic objects where their paths or areas of influence represent their locations. Sample spatio-temporal queries include "When did a tornado form?" "Where was its path?" and "Where and when had it dissipated then reformed during the course of its existence?" The dynamics and flexibility of the three-domain model enable a GIS to handle and analyze spatio-temporal data from diverse perspectives because the model handles all three geographic information components independently. The independence of semantic, temporal and spatial data modelling is the key to maintain persistent object identifiers so that the *life* and *motion* of an object can be traced and monitored in a GIS database. The three-domain model enables the support for spatio-temporal information query and analysis but needs a proper query language for query communication. A pseudo query language is used in this chapter for convenience. A research is planned to design a proper query language for accessing spatio-temporal information in the three-domain model.

REFERENCES

Allen, J.F., 1983, Maintaining knowledge about temporal intervals. *Communications of the ACM*, **26** (11), pp. 832–843.

Armstrong, M.P., 1988, Temporality in spatial databases. In *Proceedings of GIS/LIS '88*, **2**, pp. 880–889.

Berry, B.J.L., 1964, Approaches to regional analysis: A synthesis. *Annals of the Association of American Geographers*, **54** (1), pp. 2–11.

Burrough, P.A. and Frank, A.U., 1995, Concepts and paradigms in spatial information: are current geographical information systems truly generic? *International Journal of Geographical Information Systems*, **9** (2), pp. 101–116.

Clarke, K.C., Brass, A.J. and Riggan, P.J., 1994, A cellular automaton model of wildfire propagation and extinction. *Photogrammetric Engineering & Remote Sensing*, **60** (November 1994), pp. 1355–1367.

Egenhofer, M.J. and Mark, D.M., 1995, Modelling conceptual neighbourhoods of topological line-region relations. *International Journal of Geographical Information Systems*, **9** (5), pp. 555–566.

Frank, A.U. and Mark, D.M., 1991, Language issues for GIS. In *Geographic Information Systems: Principles and Applications*, Vol. 1, edited by Maguire, D.J., Goodchild, M.F. and Rhind, D.W. (Essex: Longman Scientific & Technical), pp. 147–163.

Freksa, C., 1992, Temporal reasoning based on semi-intervals. *Artificial Intelligence*, **54**, pp. 199–227.

Goodchild, M.F., 1992, Geographical information science. *International Journal of Geographical Information Systems*, **6** (1), pp. 31–45.

Langran, G. and Chrisman, N.R., 1988, A framework for temporal geographic information. *Cartographica*, **25** (3), pp. 1–14.

Peuquet, D.J., 1994, It's about time: a conceptual framework for the representation of temporal dynamics in geographic information systems. *Annals of the Association of American Geographers*, **84** (3), pp. 441–462.

Peuquet, D.J. and Duan, N., 1995, An event-based spatiotemporal data model (ESTDM) for temporal analysis of geographical data. *International Journal of Geographical Information Systems*, **9** (1), pp. 7–24.

Raper, J. and Livingstone, D., 1995, Development of a geomorphologic spatial model using object-oriented design. *International Journal of Geographical Information Systems*, **9** (4), pp. 359–384.

Sinton, D., 1978, The inherent structure of information as a constraint to analysis: mapped thematic data as a case study. In *Harvard Papers on GIS*, Vol. 6, edited by Dutton, G. (Reading, MA: Addison-Wesley).

Smith, T.R., Su, J., Agrawal, D. and El Abbadi, A., 1993, Database and modeling systems for the earth sciences. *IEEE*, **6** (Special Issue on Scientific Databases).

Snodgrass, R. and Ahn, I., 1985, A taxonomy of time in databases. In *Proceedings of ACM SIGMOD International Conference on Management of Data*, pp. 236–264.

Worboys, M.F., 1990, Reasoning about GIS Using Temporal and Dynamic Logics. NCGIA Technical Report 90–4.

Worboys, M.F., 1992, A model for spatio-temporal information. In *Proceedings of the 5th International Symposium on Spatial Data Handling*, Charleston, SC, **2**, pp. 602–611.

Yuan, M., 1994, Wildfire conceptual modeling for building GIS space–time models. In *Proceedings of GIS/LIS '94*, pp. 860–869.

Yuan, M., 1996, Modeling semantic, temporal, and spatial information in geographic information systems. In *Geographic Information Research: Bridging the Atlantic*, edited by Craglia, M. and Couclelis, H. (Taylor & Francis), pp. 334–347.

CHAPTER SIXTEEN

Elementary Socio-Economic Units and City Planning: Limits and Future Developments in GIS

Mauro Salvemini

16.1 THE SPATIAL-TEMPORAL CHANGES IN THE PLANNING PROCESS

The Socio-Economic Units (SEUs) are deeply related to the real situation of the ground and to the land use classification and definition. According to Clawson and Stewart the classification should be based on what the observer sees on the ground so that his personal interpretation plays a marginal role (Clawson and Stewart, 1965; Stewart, 1959). As far as the information data is concerned, the area on the ground should be "the smallest recognizable and geographically identifiable parcel or tract of land". Of course, these are ideal characteristics. In general the classification and the geographic encoding are related to the decisions taken by survey and planning authorities, and therefore they are strongly related to the data collected in the past and to the way they were collected. As planning process needs time to develop and in its successive steps is producing new data, the compatibility with the old ones has to be taken seriously into consideration. The compatibility is based on the spatial definition of elementary units, on the type of data collected, and on the time they were collected. Time compatibility is easily provided by the frequency of data collecting. Type of data and spatial definition should be sub-modules of older data so as to allow historical database comparative analysis.

If we accept the cybernetic definition of planning of McLoughlin (1969) "as the control of change in a system, the system being composed by those human activities and communications which have a location or a spatial element" the planning sub-areas, such as SEU, may thus be considered as a definite space in terms of activities and then of land use. Only, we want to discuss the sub-areas which have to be adapted to the Geographic Information System (GIS) for planning purposes. If we define the term 'parcel' as the smallest unit of land identifiable with the techniques used in a particular study, Figure 16.1 (McLoughlin, 1969) is representing the composition of the parcels obtained from different analysis performed on the same piece of land.

The managing of small parts of land and of the data connected with them is not a problem as far as large database technology is concerned. In this way it would be possible to satisfy the needs of some planners who would like to deal with the single dwelling unit for planning purposes on a city level planning scale.

By exploiting new technology-like data mining, knowledge discovery, search engines and other techniques of data manipulation, the complexity of querying very large databases can be solved. A very detailed description of elementary parts or dwellings of a city is not adequate to outline the pattern of the city itself. The shape of a city cannot be modelled in theory because of its complexity. Therefore a very small unit, such as the parcel or the dwelling, can be useful for cadastral purposes, taxation managing or network service supplies; however, it is not adequate for comprehensive planning at city level.

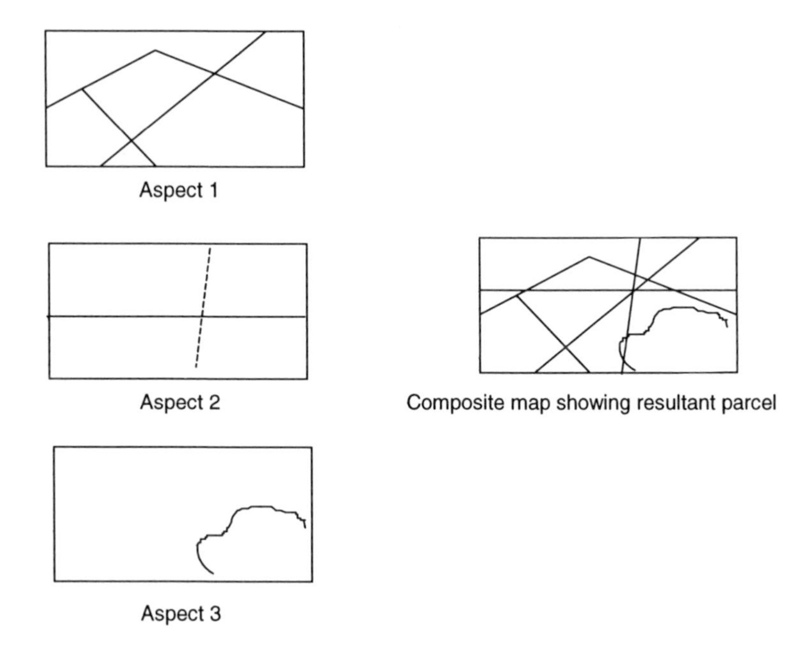

Figure 16.1

The question is how the sum of many elementary units, such as parcels and dwellings, can be generalized and managed if we consider the city changes in terms of both space and time. The definition of the planning process can help us in defining the level of the generalization and management details. After the implementation of a city plan we have to consider further more that the SEU definition changes in terms of space and time. Besides, the planning process must include both guidance and control. These are strictly related to the possibility of performing several analyses, the same type of those performed in the implementation phase; otherwise the control and the comparison with previous analysis will not be valid.

The planning process and the GIS should monitor the spatial and temporal changes of the city. These take place when changes of activities in the elementary units (SEU) or modifications of physical aspects (mainly due to building constructions) occur. As consequence, both of these changes have taken place according to the modifications induced by the realization of the planning process.

The monitoring of the changes undergone by SEU depends on the planning process and is strictly related to the GIS architecture developed for planning activities.

16.2 THE COMPLEXITY OF HUMAN SETTLEMENTS AND THE PROBLEM OF REPRESENTATION

The high complexity of human settlements and the inadequacy of modelling a tree structure are widely demonstrated. Such representation can lead to a planning process that produces cities not functional for human behaviour (Alexander, 1969).

The representation of descriptive elementary units for the planning, the control and the guidance of a city by a GIS should take into account the complexity of the described object: the city. To represent a city is much more difficult than representing any specific knowledge. Every place described by a geographic database is given its character by a certain set of events that keep on taking place. These patterns are always interlocked with other geometric patterns in the space. Each town is ultimately made out of these patterns, and they are the molecules from which a town is made. The patterns take form in the elementary units that have to be spatially represented according to the existence of data related to the mentioned patterns.

Alexander (1969) pointed out that the more patterns are in a city, the more it stands out, and the entropy increases in the city itself, and becomes part of nature. At that point the quality is reached, the city is integrated in the environment, and the buildings are integrated in the cities.

In view of the complexity of patterns, how is it possible to represent the SEU and which patterns have to be selected for the representation? It is obvious that the representation of the SEU depends on the patterns, and that an SEU represented in a specific way is giving the representation of several patterns which are not all the constituent patterns of the considered part of the city.

The representation of changes of SEUs depends on the patterns monitored in the city and is strictly related to the patterns formalized in the GIS architecture developed for analysis and planning activities.

16.3 THE REPRESENTATION OF GEOGRAPHIC INFORMATION

The representation of data and geographic information depends of the amount of data which have to be represented. Assuming that the elementary unit of analysis is **P** function of **(x,y,z)** of physical unit of the territory, and that **t** is the time of observation, the number of data describing a city or a region is easily approaching infinite, either considering very small dimensional areas or very close time of monitoring.

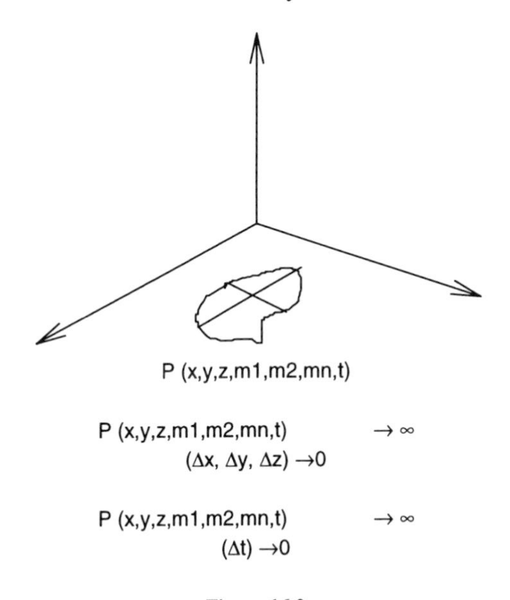

Figure 16.2

On the other hand, if we consider a very large amount of physical or socio-economic territorial aspects to be monitored *(m)*, the analysis, in presence of a relatively large physical dimension, and a discontinuous time of observation, is always producing a definite amount of data which can be handled through the normal data-base technology.

$$P\ (x,y,z,m1,m2,mn,t) \rightarrow \text{constant}$$
$$mi \rightarrow \text{constant}$$

From the above considerations we can gather that the representation is influenced by the amount of data, and that it will exert its influence on the planning activities and the consistency of GIS operations.

In terms of spatiality, the data representation of SEU is affected by a peculiar characteristic not easy to manage: the uncertainty of spreading evenly over an area a datum obtained as result of the sum of data distributed in the same area, representing the features of the human settlement. The data representation of the sub-SEU, such as the cadastral parcels, are, on the other end, affected by the problem that the geocoded area information have generally a point address which is transferred into the above-mentioned area representation for being spread on an area. As far as time is concerned, it may happen that many of the information relative to an SEU get lost when one of the following conditions occurs:

- the time of monitoring is more than one;
- the SEU are changing their dimension during the monitoring; or
- the elementary parcels changed their patterns between two times *t* and *t*.

In case we are able to manage the information regarding the above three points, the content of information itself will greatly increase, and the information system will have the characteristics of a spatial temporal information system.

With respect to some researchers (Armstrong *et al.*, 1991), the assumption outlined above is very relevant in terms of limitations given to the planning process and to the GIS. Some authors are considering limitations to the GIS, deriving them from the fact that the compilation of different layers of geocoded information, using the typical function of topology, is not able to capture explicitly relations such as city patterns which are characteristic of the geocoded data.

Keller (1994) discusses Exploratory Spatial Data Analysis (ESDA) and the software performing it. He is defending the hypothesis that this type of analysis and the GIS can offer innovative exploratory capabilities for the scientific investigation of large geocoded database. The two techniques (ESDA and GIS) are expected to undergo many improvements in the near future, especially in relation to the comprehensive planning process.

Harvey (1994) discusses the conceptual design for visualization aid in spatial decision support systems, focusing on the representation as being the most relevant component. Under these assumptions it seems very important to define the limits of GIS architecture related to the geographical addressed data, as the geographical tools for the planning process.

16.4 THE TOPOLOGICAL, GESTALTIST AND PHENOMENOLOGICAL DEFINITION OF THE GEOGRAPHIC INFORMATION

Most of the commercial GIS offer rough facilities to perform searching and topological functions, and offer a simple user interface mainly oriented to produce simple maps from database. At the moment, some possibilities for knowledge discovery in large database are offered by the modern DB systems, and some GIS systems are offering macro-programming language to develop application-oriented functions to perform the real data mining, and the discovering of relations in the geocoded data (Salvemini, 1990).

In running a GIS, the data of SEU should have three essential characteristics, to be a valuable tool to usefully manage the geocoded information: They are:

- topological definition:
- gestalt definition;
- phenomenological definition.

The topological definition is the easiest one to achieve, due to the fact that all GIS entities have to be topologically defined to run applications. In terms of validity, the topological definition has to guarantee the understanding of the analysis of the spatial relations among data.

The gestalt definition has to ensure that the modifications made on one of the elementary components will modify the related components as well. This condition is taking into account that the robustness of the GIS and its data is not indifferent to the links and connections among the parts of the system, which have to give different results in view of the modification of the system parts. The phenomenological definition has to ensure that the system has to comprehend all spatial definitions structured in a finite way among the system components.

This condition considers the fact that the validity is guaranteed when the space and the geocoded data are structured in a finite way under the control of the end user.

Under the above-mentioned conditions, a GIS and the connected DB of SEU become a planning tool, and in this way it may be usefully employed in planning. The representation has to follow the task of the GIS use and has to be coordinated with the other tools of the planning process. The representation of the data guarantees the validity of the entire GIS used for planning.

16.5 THE STATIC AND THE DYNAMIC REPRESENTATION

Representation of SEU data can be dynamic or static and in both cases this definition can be related to space and time.

Unfortunately, only a few times the planning is using the dynamic representation, not only because of the shortage of time-distributed data to be represented, but also because in the planning science the simulation and the analysis have been restricted to well defined, motionless places (Salvemini, 1995).

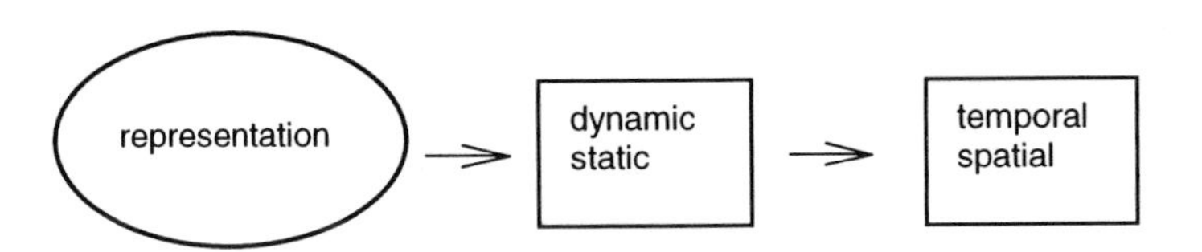

The dynamic representation of data necessary to the planning process would greatly help to understand physical and socio-economic phenomena, which remain hidden because of the way the analyses are carried out.

16.6 CURRENT TRENDS IN DATABASES INFLUENCING THE USE OF SEU IN GIS

Current trend in database and intelligent data analysis technologies may offer a valuable support to the next generation of GIS. On one side, the new database technology improvements may assist the proper data representation and querying activities. On the other side, data mining techniques may help the user to discover hidden and useful patterns, and the relationships present among the huge amount of collected data. In the rest of this section we will discuss some current trends on database research and application areas which may, when necessary, be helpful to assist the development of next GIS.

Active database may help to model the dynamic of the stored data. It was mentioned above how the dynamic nature of the stored GIS data may greatly improve the overall planning process. In an active database some behaviour rules (also known as *triggers*) may be defined in order to be executed, as soon as certain conditions among the data are satisfied. Such triggers have an 'if... then' form. The 'if' side collects the conditions under which the trigger has to be fired, the 'then' part states some actions to be performed over the data (Giuffrida and Zaniolo, 1994). The natural application of *active rules* is to keep the stored data set valid and up to date. All the semantic relationships will certainly hold at any time during lifetime of the database.

Temporal database is also a current and a very active research area. It is mainly concerned with the development of new data models and query languages including the *time* dimension as core part of the data itself. Classical database may be extended to include the temporal information without any need of time specific features. However, in many cases the work required for a proper modelling of the time dimension is so difficult as to discourage anyone. In a temporal database a *timestamp* (or, most often, a set of timestamps) is associated with each tuple in the database at the time the tuple itself is created/modified/deleted. Such timestamp information is used to retrieve data. For instance, a temporal database query language allows the user to easily formalize things like: "retrieve all the tuples occurred between March 1992 and December 1992" or "retrieve all tuples where the attribute X has doubled its value within two working days". Contrary to a classical database, a temporal implementation keeps track of a deleted tuple, that is, a deletion does not *physically* remove the tuple itself but it marks the tuple as deleted at a certain time (logical deletion). Ideally, a temporal database increases monotonically. This makes the temporal database the ideal system for storing the historical data that may be required by good planning activity.

Due to the amount of data contained in a database, new technologies have been developed to assist the user in analyzing such data. The user is properly supported in querying the database according to his current goal. While the database grows in size, the user may lose accordingly the ability of successfully extracting all the desired information from the grown data sets. Knowledge Discovery and Data Mining (Piatetski-Shapiro and Frawley, 1991) technologies are directed to discover *hidden* and *useful* information among large data set. For instance, a futuristic GIS system may be educated to find out automatically things like: "the crossing between X street and Y avenue gets very crowded between 5 p.m. and 7 p.m. on Monday and Wednesday".

Such discoveries are strongly related to the *contents* of the database, and not to its *structure* (this latter is much easier to understand by humans). Since the contents of the database change continuously, interesting relationships become dynamic. They may be true at a certain time, and become false at a later time or vice versa. Furthermore, to make hidden relationships sound, a large collection of data needs to be analyzed. As data grows in size, the number of possible meaningful relationships grows accordingly. The expertise of a user skilled in certain tasks may be successfully integrated in an intelligent data analysis system. Such artificial intelligence techniques are mostly appropriate to select the most interesting hidden relationships among all the possible ones. Rule-based systems (Giuffrida *et al.*, 1992; Giuffrida, 1994), and neural network approaches are currently preferred.

16.7 AN ELEMENTARY EXERCISE OF SPATIAL-TEMPORAL MANAGING DATA FOR PLANNING

The project is referring to a city, Anzio, placed in the metropolitan area of Rome. In this project two analysis and planning tools have been used: the SEUs and the master plan. In this study the SEUs are the census tracts used in 1991 Italian Population Census, and the plan is the master plan of the city drafted in 1978.

At the moment, many Italian cities have an old master plan, having a shortage of funds to re-design the plan. In past decades, in many cities the self-help housing, and the 'without permit' building activities have occurred. Therefore, many cities decided to determine the exact perimeter of the land misused by construction activities.

The simple definition of the built-up areas does not make the master plan more effective, as it would be necessary to have other data associated with them.

Under this assumption the data coming from the population census seems useful as it records the construction activities in each census tract.

Three problems were arising from the use of the census data:

1. the area dimensions of the tract are often too large, specially in rural areas;
2. the information useful for planning, and contained in the surveyed data, are generally related only to the number of housing, and to the number of the total rooms in the census tract; and
3. the boundaries of the census tracts generally do not coincide with the homogeneous zones of the master plan.

In spite of these limits, the planning project of Anzio has shown that some planning parameters may be directly obtained from the census data, and these can be very useful to local planning, such as:

- density of population
- rate of dwelling occupation per family or per inhabitant.

Indirectly, two more important planning parameters are defining the average volume of the residential room:

- rate of residential volume per inhabitants;
- density of construction per unit of total area of census tract.

The result of this study is that in the absence of better and up-to-date urban planning data, we can use the census tracts.

> *Since the spatial definition of the master plan zones is based on the above set of parameters, the census tract, considered as SEU, can be useful if used in monitoring the temporal changes of the construction activity, and the rate of residential volume per inhabitant.*
>
> *The utilization of census tracts has one limit. In terms of spatial changes, assuming that the master plan has not been changed, the SEU (census tracts) may be usefully used only in case they are sub-modules or sum of homogeneous zones of the master plan.*

Figure 16.3 is representing two townships (Anzio and Nettuno) in the metropolitan area of Rome, divided in census tracts according to the 1991 Population Census definition. Even though the two cities are very similar in terms of population (about 32.000 inhabitants), it is possible to detect the considerable difference in the dimensions of the census tracts, which also is obviously affecting their number.

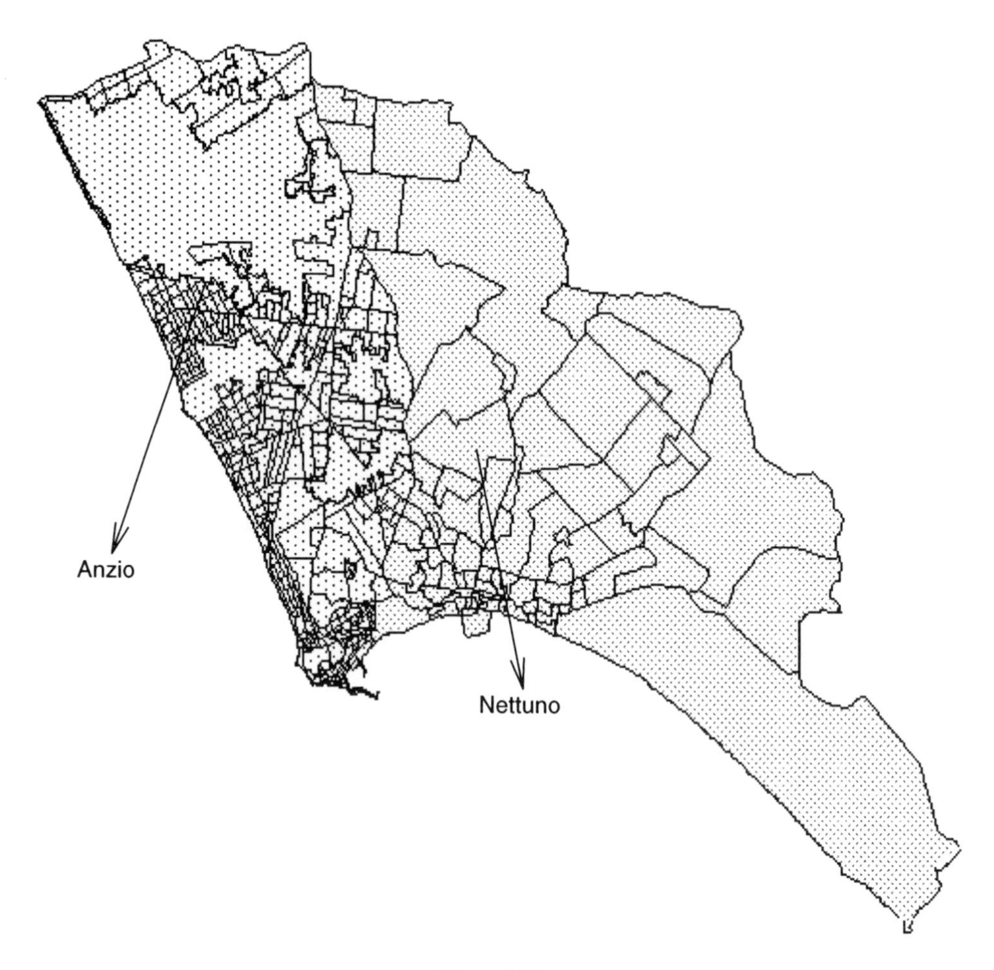

Figure 16.3

Figure 16.4 is representing in gray tone the self-built and the illegal built up areas superimposed on the borders of the 1991 census tracts in the city of Anzio. The project demonstrates also that the overlapping of the Census Tract borders in the zones of the master plan may be very useful in the planning process.

Figure 16.4

REFERENCES

Alexander, C., 1969, *The Timeless Way of Building*, (New York: Oxford University Press).

Armstrong, M.P., Densham, P.J. and Panagiotis, L., 1991, Cartographic visualization and user interfaces in spatial decision support systems. In *Proceedings of GIS/LIS '91*, Atlanta, Georgia.

Clawson, M. and Stewart, C.L., 1965, *Land Use Information*, (Baltimore, MD).

Giuffrida, G., 1994, Expert system shell to reason on large amount of data. In *Proceedings of the Third CLIPS Conference*, NASA's Johnson Space Center, Houston, Texas, September 1994.

Giuffrida, G. and Zaniolo, C., 1994, EPL: An event pattern language. In *Proceedings of the Third CLIPS Conference*, NASA's Johnson Space Center, Houston, Texas, September 1994.

Giuffrida, G., Salvemini, M. and Stothouber, L.P., 1992, G-CLIPS: A rule based language for expert GIS applications. In *Proceedings of GIS/LIS '92*, San Jose, CA, November 1992.

Harvey, F., 1994, Supporting planning tasks in regional planning: A conceptual design for visualization aids based on a task hierarchy. In *Proceedings of GIS '94*, Vancouver, Canada.

Keller, P.C., 1994, Exploratory spatial data analysis (ESDA)—the next revolution in GIS. In *Proceedings of GIS '94*, Vancouver, Canada.

McLoughlin, J.B., 1969, *Urban and Regional Planning. A System Approach*, (Bristol, UK).

Piatetsky-Shapiro, G. and Frawley, W.J., 1991, *Knowledge Discovery in Databases*, (Cambridge, Mass: The MIT Press).

Salvemini, M., 1990, *Ambiente, Territorio e Informatica*, (Milano: Pirola Editore).

Salvemini, M., 1995, Consistency of data representation in GIS for comprehensive planning. In *Proceedings of 1st Conference on Spatial Multimedia and Virtual Reality*, Lisbon, Portugal.

Stewart, C.L., 1959. The size and spacing of cities. In *Readings in Urban Geography*, edited by Mayer, H.M. and Kohn, C.F. (Chicago: University of Chicago Press).

CHAPTER SEVENTEEN

Why Time Matters in Cadastral Systems

Khaled Al-Taha

17.1 INTRODUCTION

The concept of 'time' is studied in various fields, such as philosophy, psychology, linguistics, artificial intelligence (AI) and database management systems (DBMS). The research framework is called *temporal reasoning*. In recent years, the necessity to include temporal information into databases has provided different models for temporal databases. Researchers have been examining aspects of temporal reasoning, temporal propositions and temporal query languages. However, researchers from various fields concerned with this topic have found it difficult to become familiar with the work of others. Bolour *et al.* (1982) presented a survey on *time* from the different fields as a first step towards an increased communication and exchange of ideas about information processing and *time*.

Today, temporal reasoning attracts many scientists dealing with databases and information systems. In computer science, it is a crucial topic in information systems, program verification, artificial intelligence, and other areas involving process modelling (Allen, 1983). Recently, temporal reasoning has been getting increased attention in geographic information systems where it has a wide range of applications: forest resource management, urban and regional management, and the management of electronic navigation charts and transportation (Langran, 1989). Ever since 1978, when Basoglu and Morrison built an U.S. Historical County Boundary for the last 200 years, researchers have been addressing the need for temporal geographic information systems (GIS). Langran has explored a wide range of applications for temporal GIS, such as forest resource management, and urban and regional management (Langran, 1990). Armstrong (1988) addressed *time* in spatial databases and discussed the types of spatial change. Worboys (1990) showed the role of modal logic in GIS applications. Hunter and Williamson (1990) have designed and built a *historical cadastral database.*

"An estate in land is a *time* in land." This statement by Paul Creteau (1977) confirms our belief that cadastral systems provide a rich example for temporal reasoning in a GIS. Records about rights in land are chronologically ordered and indexed. While some information systems may destroy old records in the system periodically, cadastral systems keep their records no matter how old these records are. It is not unusual to keep cadastral records for hundreds of years. The relatively manageable number of transactions that occur in these systems may have contributed to maintaining this tradition. However, the vital importance of these records for assuring rights in real estate dictates their being kept over a long time.

This chapter demonstrates with examples how cadastral systems depend on temporal issues and how powerful logical systems can be in formalising and reasoning about interests in real property. In addition, we also summarise object identity and its evolution over time representing and extending temporal constructs we feel necessary to represent objects in cadastral systems within a temporal database.

17.2 FEATURE IDENTITY AND TEMPORAL EVOLUTION

Features in geographic information systems (GIS) have an identity and other spatial and non-spatial properties. Other than the identity, feature properties might vary over time. For example, a river may alter its course, a town may decrease its population or adopt a new name. The changes to a feature are not restricted to modifications of the value of these three components. The type of a feature may change through time; for instance, an old house may become a national monument, a river may cease to be a political boundary, or a protectorate may become a free country. Yet, the old house, river and the country are in some sense still the same features. Thus, each feature requires the existence of a property, called its *identity*, with three characteristics: uniqueness, immutability and non-reusability.

Temporal databases have recently gained a lot of interest among GIS researchers (Al-Taha *et al.*, 1993; Chrisman, 1993; Frank *et al.*, 1992; Hazelton, 1993; Kelmelis, 1993; Langran, 1993; Worboys, 1992). Researchers from the database community have developed different types of temporal databases (Barrera and Al-Taha, 1990). Temporal databases vary in their representation of time perspectives and in their treatment of erroneous data. Examples of different time perspectives are transaction and valid time (Snodgrass and Ahn, 1985). Temporal databases developed in the past decade are subdivided into four main designs: snapshot, transaction-time, valid-time and bitemporal. The major difference stems from the time perspectives that exist in these databases: valid time or real-world time describes when an action happened in the real world, and transaction or database time describes when the information was entered into the database (Figure 17.1). A snapshot database does not deal explicitly with time. A transaction-time database deals only with the time of acquisition of a fact. A valid-time (historical) database is only concerned with when a fact became valid and a bitemporal database deals with both valid and transaction times. Another difference between these databases is seen in the way errors are corrected; e.g., in a valid-time database, an error is corrected without maintaining records about when and what has been modified.

		Transaction Time
	Snapshot Database	Transaction-Time Database
Valid Time	Valid-Time Database	Bitemporal Database

Figure 17.1 Time perspective and the various temporal databases

An integrated GIS should implement several functions in order to manage the identity of temporal features:

1. Support mechanism for assigning identities to temporal objects. Only an integrated GIS can guarantee identity uniqueness through time. It is convenient that the identity of a temporal feature be an abstract type (having the methods cited in Section 17.3).
2. Maintain mappings between the Temporal Feature's identity and the keys or identifiers used by the member databases.
3. Implement mechanisms to alert the integrated GIS of the birth and death of features in the member databases, since these transitions may translate into transitions in the integrated systems.
4. Support constructs that model the birth and death of temporal features; their split from other features and merging into others.

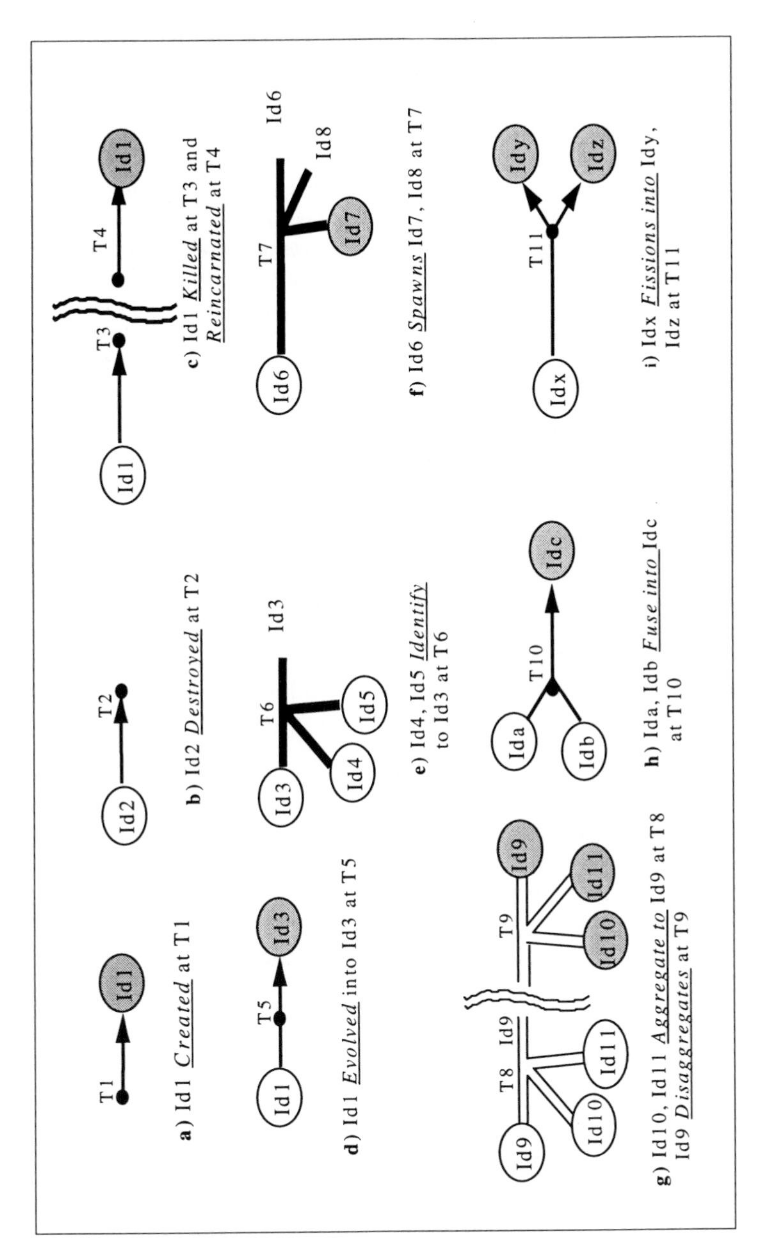

Figure 17.2 Temporal constructs of identities

There is a very reduced bibliography on temporal constructs of features. Clifford and Croker (1988) propose two operations called **CREATE** and **DESTROY** for creating a new identity and for disposing, forever, of the essence of an object. They also propose two operations, called **KILL** and **REINCARNATE,** for the temporary disappearance (and reappearance) of the identity of an object that has been created and not yet destroyed. Finally, they propose an operation called **IDENTIFY** that takes an ordered set of 'n' identities and merges the last 'n–1' of them into the first identity. The last 'n–1' identities disappear permanently from the system.

Chu *et al.* (1992) present an evolutionary model for objects identified from X-ray images. They present three more constructs called **EVOLUTION**, **FUSION** and **FISSION**. A feature that *evolves* into a new one has its identity permanently removed from the system, and the new identity resulting from this process preserves a temporal relationship with its ancestor. A feature that *fissions* causes the creation of several new identities and the permanent disappearance of the identity of the fissioned feature, and the new identities preserve a relationship with their generator. Finally, *Fusion* can be modelled by an *identification,* followed by an *evolution*.

Three additional operations are suggested here that should be included: **SPAWN**, to model a feature that generates new objects while keeping its identity, and a pair of operations: **AGGREGATE, DISAGGREGATE,** to represent the temporary transition of a feature from the status of being a primary object (with identity) to that of a secondary object (embedded into a primary object) and vice versa.

All the constructs described in Figure 17.2 generate relationships among identities. These relationships, coupled with a temporal query language, should empower the user with the capability of tracking the evolution of a geographic feature across time. With the following example, we will demonstrate how some of these temporal constructs are used for spatial features.

Example: A developer purchased four acres of land from a farmland. The farm is located on Parcels 7 through 10 on Block 35 as shown in Figure 17.4. A one-acre patch taken from Parcel 7 (Parcel 34) was joined with three acres taken from Parcel 8 (Parcel 35) to form parcel number 36, as shown in Figure 17.4. The developer made a subdivision that resulted in six new parcels, numbered 37 to 42, and a service road.

There are several ways to carry out a subdivision and union of parcels to technically record changes in parcel management depending on cadastral regulations in a particular state. For example, a possible way is to produce new parcels from an original parcel while retaining the original parcel number. This case is demonstrated in Figure 17.3. In another case, it may be necessary to destroy the original parcel number while only retaining the new parcel ID's. This scenario is shown in Figure 17.4.

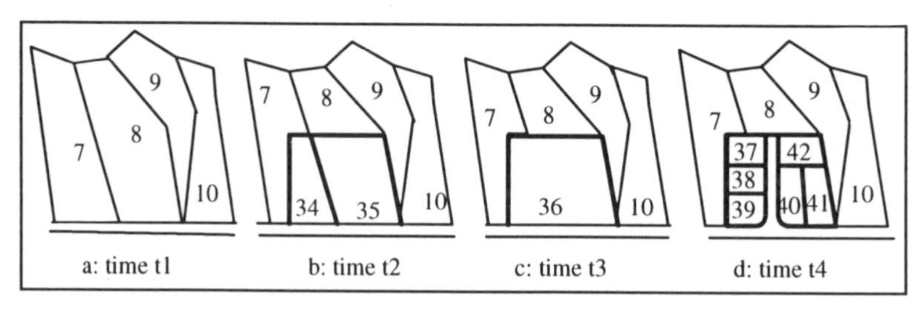

Figure 17.3 A possible ID evolution for Parcels 7 and 8 and their descendants

In the above example shown in Figure 17.3, Parcels 34 and 35 were extracted from Parcels 7 and 8 respectively without the original parcels (7 and 8) being destroyed. To represent the parcel identities as they evolved through time as seen in Figure 17.3, the following temporal constructs are used (see Figure 17.4):

- Create Parcels 7–10 at t1.
- Parcel 7 spawns Parcel 34 and Parcel 8 spawns Parcel 35 at t2.
- Parcels 34 and 35 fuse into Parcel 36 at t3.
- Parcel 36 fissions into Parcels 37–42 and a road at t4.

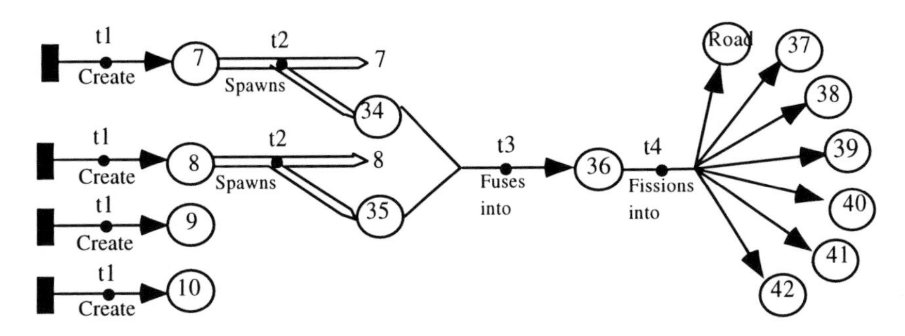

Figure 17.4 First scenario of temporal ID evolution retaining ID's of subdivided parcels

In another scenario of ID evolution shown in Figure 17.5, Parcels 7 and 8 were destroyed at t2 as soon as they have been subdivided. Only ID's of the new parcels are retained (33–36). The ID evolution of parcel subdivision and union in this case can be represented using the following temporal constructs (see Figure 17.6):

- Create Parcels 7–10 at t1;
- At t2, Parcel 7 fissions into Parcels 33–34, and Parcel 8 spawns Parcels 35–36;
- Parcels 34 and 35 fuse into Parcel 37 at t3;
- Parcel 37 fissions into Parcels 38–43 and a road at t4.

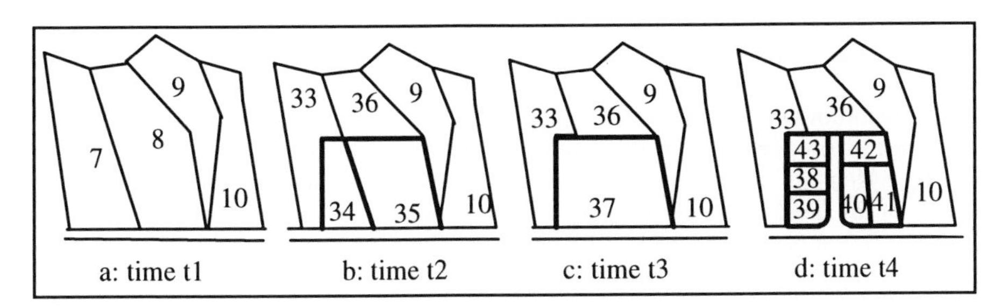

Figure 17.5 Another possible case of parcel subdivision and union

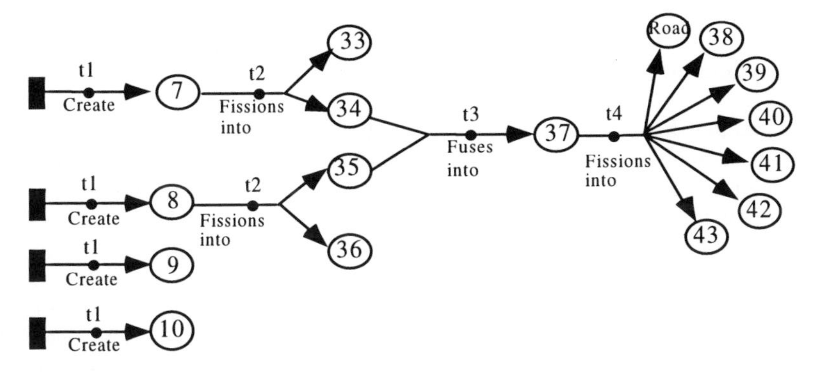

Figure 17.6 Second scenario of temporal ID evolution destroying ID's of subdivided parcels

17.3 TEMPORAL ISSUES IN CADASTRAL SYSTEMS

Cadastre is a technical term for a collection of records showing the extent, value and ownership (or other basis of occupancy) of land. The word, according to Binns (1953), is derived from the Latin word *capitastrum,* which was a register of *capita* or units made for Roman land tax (*capitatio terrena*). Dale (1976) has defined cadastre as follows: "a cadastre is a general, systematic and up-to-date register containing information about land parcels including details of their area, value and ownership". We understand cadastral systems to be administrative systems for the management (the making, keeping or updating) of land records in a county, a state or a country. Modern perspectives of cadastral systems call for a multi-purpose cadastre for a unified source for cadastral, topographic, infrastructural, and other large-scale and detail-oriented information. A cadastral system that provides various types of information to serve a variety of users is called a *multi-purpose-cadastre* (National Research Council, 1980).

Cadastral systems deal with making and maintaining records for real estate. They provide users from a variety of disciplines with land ownership data, use, value and topographic details. Throughout the world, cadastral systems differ from one country to another (Binns, 1953). However, there are two distinct approaches for such systems: the deed recording and the title registration approach. In deed recording systems (such as the systems used in the USA), the title for land (the opinion of ownership of land) is deduced from recorded deeds (Hintz and Onsrud, 1990; Onsrud, 1985). Interpretation of these records results in an attorney's title opinion (or a court decision), which presents, with a great deal of qualifying language, the status of rights in the land. On the other hand, in a title registration system as used in many European countries (such as Germany, Austria, and Switzerland), the ownership and other forms of tenure are determined by a governmental act and often guaranteed by the state.

17.3.1 Deed Recording System

The deed recording system described in this section is the one used in the USA. Although a large part of property law in the United States traces its origin to England, the U.S. recording system is unique, not following the pattern of any particular country. The principle of the deed recording system is the registration of evidence-of-title (as distinct to the registration of the title itself). Registration of this evidence in public records serves as notice to the world that the registrant claims an interest in real property. The recording of a deed does *not* attest to the validity of the deed itself or the claim as described in it (U.S. Dept. of Agriculture and Commerce, 1974).

17.3.2 Recording Statutes

Recording statutes differ in their reliance on temporal data and the events of making and recording an instrument. States have their own recording statute to regulate priorities of rights among purchasers of the same land and to protect an innocent purchaser *(bona fide purchaser*). An innocent purchaser is "one who, by an honest contract or agreement, purchases property or acquires an interest therein, without knowledge, or means of knowledge sufficient to charge him in law with knowledge, of any infirmity in the title of the seller" (Black, 1979). Most statutes require the purchaser to examine the history of the parcel to insure that his grantor still holds the title to the parcel and to be aware of any previous recorded transfer to another party. To protect his rights, most recording statutes require a purchaser to record his instrument prior to others who might claim similar rights in the same property (Unger, 1974). Therefore, the government's time stamps on a real estate instrument can be a decisive element concerning whether the instrument is effective or made void by an earlier recorded instrument.

The titling process and the determination of ownership are the most important processes in cadastral systems because ownership is the major right one can have on real properties. Temporal information is highly important in determining a *good title* in the deed recording system used in the United States (U.S. Dept. of Agriculture and Commerce, 1974).

Time-dependent information and how it affects a decision made in the titling process is better explained by an example. Depending on the real estate law in a particular state, the recording of deeds falls into one of three different recording statutes: the race statute, the notice statute and the race-notice statute.

The Race Statute

In this recording system, the first legal-person to record his deed has priority whether he has a notice of a previous unrecorded transfer or not. This type of statute is unpopular because it does not protect an innocent purchaser who failed to record a deed before another purchaser did (U.S. Dept. of Agriculture and Commerce, 1974). For example, A sold his parcel to B who failed to record his deed. A died. Before B recorded his deed, A's heirs sold the parcel to C who immediately recorded his deed. Under the race statute, B lost ownership of the parcel because he failed to record his deed before C did.

The Notice Statute

In this system, a second purchaser for a property gets a good title if he does not have notice of a prior transfer. For example, A sold his parcel to B who failed to record his deed. A died. Before B recorded his deed, A's heirs sold the parcel to C, an innocent purchaser, who did not record his deed either. Under the notice statute, B lost ownership to the parcel because he failed to record his deed before the title passed to an innocent purchaser (Creteau, 1977).

The Race-Notice Statute

In this hybrid system, a good title is restricted for a later purchaser on taking title without notice, and on recording that title first. Under a race-notice statute, the race to record a deed is valid only between innocent purchasers and an earlier purchaser does not lose his title instantly at the time a deed is given to a second purchaser because he still may record (U.S. Dept. of Agriculture and Commerce, 1974). For example, A sold his parcel to B who failed to record his deed. A died. Before B records his deed, A's heirs sold the parcel to C, an innocent purchaser. Whoever records his deed first (whether B or C) will acquire a good title to the parcel (Creteau, 1977).

17.4 A TEMPORAL LOGIC FOR DEED-RECORDING SYSTEMS

Temporal reasoning in any field requires a certain type of temporal logic. Temporal logic provides its users with tools for achieving a better understanding of the nature of time itself. The primary aim of temporal logic is to clarify the content, to elaborate the consequences, and to elucidate the interrelationships among the axioms of time in general (Rescher and Urquhart, 1971).

17.4.1 Predicate Calculus (First-Order Logic)

The temporal logic used in this section is based on predicate calculus (also called first-order logic). Predicate calculus is the most important and commonly used logical system (Davis, 1990; Frank *et al.,* 1991; Smullyan, 1968). It is based on the propositional calculus where one deals with *statements* (or sentences) only. A *statement* is a declarative sentence that is either *true* or *false,* but not both. A sample set of axioms that are later used in this chapter are:

T1 $P \supset Q \equiv \sim Q \supset \sim P$ transposition
MP $(P \wedge (P \supset Q)) \supset Q$ modus ponens
Where: $\wedge$ stands for *and,* $\sim$ stands for *not,* $\supset$ stands for *implies*

In first-order logic, however, one uses *formulas* with free variables that are neither true nor false, unless their variables are bound (being assigned to constants). A formula with bound variables is equivalent to a statement; otherwise it is called a predicate. An *n-place predicate* is a formula with exactly *n* distinct free variables (Margaris, 1967). In this section, we will use capital letters (such as X) for constants and small letters (such as x) for variables.

17.5 A TEMPORAL REASONING MODEL FOR DEED-RECORDING CADASTRAL SYSTEMS

In this section we will introduce a formal model for temporal reasoning in cadastral systems. This model will be based on the first-order logic. First-order logic is complete (Davis, 1990). We will use logical symbols, axioms and theorems presented in the predicate calculus, temporal, tense and modal logic. In the following two sub-sections, we present a first-order logic for the titling process in a cadastral system and demonstrate its use with an example.

17.5.1 A First-Order Logic for the Titling Process

The titling process differs in technical details from one country (or state) to another. We first introduce a generic logic that will provide us with the basic processes applicable for most cadastral systems. Then we will extend the logic for the case of a deed recording system.

The logical system presented here is applicable to a limited number of operations using a basic set of rules. Later, based on our first logical scheme, we develop a more detailed axiom scheme powerful enough to support the major titling-process operations in a real deed-recording cadastral system.

Symbols Used

- *C* is a constant symbol if it has the form A, B,..., Z, T1, T2, John, or F-Bank, etc.;
- *v* is a variable symbol if it has the form a, b,..., z, t1, t2, legal-person, parcel, etc.;
- τ is a term if it is a constant, a variable symbol, or a formula;
- *f* is a formula if it has the form owner (τ1, τ2,..., τn). A formula must have at least one variable or constant symbol.

Axioms for defining sorts of objects in cadastral systems:

TP1. $[\exists x\ (\text{social-security}(y) = x)] \supset \text{Legal-person}(y)$
or for short TP1. $SS \supset LP$
Where SS: stands for '$\exists x$ social-security(y) = x',
LY: stands for 'Legal-person(y)'

Which reads: if y has a social security number, then y is a legal-person.

TP2. $[\exists x\ \text{parcel-number}(p) = x] \supset \text{Parcel}(p)$
or for short TP2. $PN \supset PL$
Where PN: stands for '$\exists x$ parcel-number(y) = x', PL: stands for 'Parcel(p)'

Which reads: if p has a parcel number, then p is a parcel.

TP3. $[\exists x\ \text{registry-name}(y) = x] \supset \text{Registry}(y)$
or for short TP3. $RN \supset RY$
Where RN: stands for '$\exists x$ registry-name(y) = x',
RY: stands for 'Registry(y)'

Which reads: if y has a registry name, then y is a registry.

TP4. $[\exists x\ \text{court-name}(y) = x] \supset \text{Court}(y)$
or for short TP4. $CN \supset CT$
Where CN: stands for '$\exists x$ court-name(y) = x', CT: stands for 'Court(y)'

Which reads: if y has a court name, then y is a court.

TP5. $[\exists x\ \text{state-name}(y) = x] \supset \text{State}(y)$
or for short TP5. $GT \supset ST$
Where GT: stands for '$\exists x$ state-name(y) = x',ST: stands for 'State(y)'

Which reads: if y has a state name, then y is a state.

Axioms TP1 through TP5 define the basic classes or objects in cadastral systems; namely, a registry, the state, the court, parcels, and legal-persons.

TP6. $[(\text{sort-of}(x) = \text{State}(x) \vee \text{sort-of}(x) = \text{Court}(x) \vee \text{sort-of}(x) = \text{Legal-person}(x))\ \wedge$ $\text{sort-off}(y) = \text{Legal-Person}(y) \wedge \text{sort-of}(p) = \text{Parcel}(p) \wedge \text{sort-of}(d) = \text{Time}] \supset [\text{sort-of}(f(x, y, p, d) = \text{Instrument}]$
or for short TP6. $[(ST \vee CT \vee LX) \wedge LY \wedge PL \wedge DT] \supset IT$
Where ST: stands for 'State', LX: stands for 'Legal-person(x)',
DT: stands for 'Date(t)', IT: stands for 'Instrument(x, y, p, d)'

Which reads: if x is a court or a state or a legal-person and y is a legal-person and p is a parcel and d is a time point, then a contract between legal-persons on a parcel at time d is an instrument. Axiom TP6 defines the generic class *instrument* that can be a grant, a making of a deed, or mortgage. Assumptions implied in this axiom are: an instrument must be dated and must deal with parcels.

TP7. $[\text{holds}(x,p,t1) \wedge \text{sort-of}(x,y,p,t2) = IT \wedge t2>t1] \supset [\text{claims}(y,p,t2)]$
or for short TP7. $(HXt1 \wedge IT \wedge LD) \supset CYt2$
Where HXt1: stands for 'holds(x,p,t1)', LD: stands for 't2>t1'
CYt2: stands for 'claims(y,p,t2)'

This generic axiom defines the transfer of rights on a parcel from one legal-person to another. The rights on a parcel being transferred are unspecified at this level. We will show specific rights later, such as ownership and mortgage rights.

TP8. $[holds(x,p,t1) \wedge sort\text{-}of(f(x,y,p,t2) = IT \wedge t2>t1] \supset \sim[ho(x,p,t2)]$
or for short TP8. $(HXt1 \wedge IT \wedge LD) \supset \sim HXt2$
Where $HXt2$: stands for 'holds(x,p,t2)'

Axiom TP8 states that if a legal-person has passed the rights on a property to another legal-person at a later time t2, then the first legal-person no longer holds these rights at time t2.

TP9. $[claims(y,p,t2) \wedge \sim\exists x(holds(x,p,t2)] \supset [holds(y,p,t2)]$
or for short TP9. $(CYt2 \wedge \sim\exists x\ Hxt2) \supset HYt2$
Where $\exists x\ Hxt2$: stands for '$\exists x(holds(x,p,t2))$',
$HYt2$: stands for 'holds(y,p,t2)'

Axiom TP9 states that if a legal-person has a claim of a right on a parcel at time t2 and there is no other legal-person who holds this right on the same parcel, then the first legal-person holds this right at time t2.

The set of axioms TP7 through TP9 allow for a distinction between the claiming of a right and the holding of it. In other words, they imply that a legal-person will hold a certain right exclusively on a parcel when the legal-person is the only one who has a claim on that parcel. While more than one legal-person can have a claim on a parcel, only one of them can hold that right exclusively. For example, if two legal-persons claim exclusive ownership of the same parcel at the same time, one of the two claims will make void the other. The advantage of the two-stage process of transferring rights is to distinguish between legal-persons having a claim to a property and those who actually hold its title. These cases are resolved by court decisions or by mutual agreement. For example, a person may own a property and another may at the same time acquire an adverse position of the same property. Another example, in the case of a notice statute, later purchasers will acquire a good title to the property if they made a deed without notice. In this case, more than one purchaser may claim a good title to the same property, but only one will hold that right.

17.5.2 Sample Deed Recording System

The sample logic presented in the previous section lacks the expressiveness to represent real operations and assertions in a cadastral system. In this section, we extend this logic to deal with these issues by introducing the following concepts:

- The concept of *persistency*. Persistency is the property that allows a proposition to hold its validity over time. If P holds now and P is persistent, then P holds at all times in the future, or as long as P satisfies a certain condition.
- The concept of *time perspective* allows us to differentiate between valid time, transaction time and other times. In cadastral systems, more than one time perspective exists, namely, the creation time (when an action took place), and the recording time (when it was reported to the system).
- The concept of *sub-classes*. In addition, we need to extend our logic to allow for the definition of sub-classes. For example, we need to differentiate between the various types of legal-persons (such as grantor, grantee) and types of instruments (such as deed, mortgage).
- The 'as of' concept of *dynamic knowledge*. Since we need to reason about facts at an assumed time different than the time the query is placed in the system, the logical system should allow us to infer dynamically (when the query is assumed to be asked). For example, if we ask a query "Who owned the island

in 1985," the answer will depend on when the query time was specified to the system. Since our knowledge might change at a later time, we want this change to be reflected in the answer at a later time as well. In the previous example, assuming that we knew that Jackson owned the island in 1980 and that we first learned about him selling it to Smith in 1986 although the sale actually took place in 1985, the answer to the previous query would be different for the following two cases:

Who owned the island in 1985, as of 1985? Jackson.
Who owned the island in 1985, as of 1986? Smith.

These concepts being added to the previous logical system, TP will result in the following logical system for deed-recording (DR). For simplicity, this chapter will only deal with the race-notice statute as a sample deed recording system.

Assumption

Definitions, logical and non-logical symbols previously defined in the TP system are also available to the DR system.

Inference Rules

MP	$\vdash [(f \supset g) \wedge f] \supset g$	modus ponens
RP	P$\vdash \forall$t' (t'>t) [ft ^ (~$\exists$t'~ft')] $\supset$ ft'	persistency
RN	S$\vdash \forall$j [~nc(f,j) ^ rf(f,j)] $\equiv$ good-title(f)	race-notice statute
N	$\vdash \forall$j [~rf(j,f)] $\supset$ ~nc(f,j)	purchase without notice.

Where P: stands for 'cadastral data is persistent'
S: stands for 'race-notice statute applies'

RP reads: If P is an assumption of persistent information in cadastral systems, then for all the cases that t' is later than t if the formula f holds true at time t and it is not the case that there exists a time t' where f does not hold true at t', then f holds true at t'. In other words, when f holds true at time t, it holds also true at any later time t' as long as its truth value did not change at any time point later than t and is less than or equal to t'.

RN reads: If S stands for 'the race-notice statute applies', then iff for all instruments j an instrument f is made without a notice about j and has been recorded first, then f leads to a good title.

N reads: If for all instruments j an instrument f is made without j being recorded before it, then f is made without a notice.

This inference rule illustrates our principal assumption that recording a deed is a notice to the world that it was made.

Axioms

DR1. [holds(x,p,t1,s) ^ sort-of(f(x,y,p,t2,s) = IT ^ t2>t1] $\supset$ [claims(y,p,t2,s)]
or for short DR1. (HXt1 ^ IT ^ LD) $\supset$ CYt2
Where HXt1: stands for 'holds(x,p,t1,s)', LD: stands for 't2>t1'
CYt2: stands for 'claims(y,p,t2,s)'

Which reads: if x holds a right of type s on a property p at time $t1$, and x has transferred this right to y by the instrument f at a time $t2$ later than $t1$, then y has a claim on property p at $t2$.

DR2. [sort-of(x) = ST ^ sort-of(f(x,y,p,t2,s) = IT] ⊃ [holds(y,p,t2,s)]
or for short DR2. (ST ^ ITyt2) ⊃ HYt2
Where ITyt2: stands for 'sort-of(f(x,y,p,t2,s) = IT'

Which reads: if x is a state and x has transferred a right s on a property p to a legal person y at a time $t2$ by the instrument f, then y holds the right s on the property p at $t2$.

DR3. [∃x {book-num(b) = x ^ Registry(y) ^ part-of(b,y)}] ⊃ Book(b)
or for short DR3. (BN ^ RY ^ RB) ⊃ BK
Where BN: stands for 'book-num(b) = x', RB: stands for 'part-of(b,y)'
BK: stands for 'Book(b)'

Which reads: if b has a book number and b is part of registry y, then b is a Book.

DR4. [∃b {Book(b) ^ sort-of(i) = IT ^ in(i,b,d) ^ sort-of(d) = Time] ⊃ recorded(i,d)
or for short DR4. (BK ^ IT ^ IB ^ DT) ⊃ RD
Where IB: stands for 'in(i,b,d)', DT: stands for 'sort-of(d) = Time'
RD: stands for 'recorded(i,d)'

Which reads: if there exists a book b and i is an instrument placed in b at time d, then i is a recorded instrument at time d.

DR5 recorded(i,t1) ^ ~∃j{recorded(j,t) ^ t< t1 ^ grantor(i) = grantor (j)} ⊃ recorded-first(i,j)
or for short DR5. (RDit1 ^ ~∃jRDjt) ⊃ Rfij
Where RDit1: stands for '~∃j{recorded(j,t) ^ t <t1 ^ grantor(i) = grantor (j)}'
RFij: stands for 'recorded-first(i,j)'

Which reads: if the instrument i was recorded at t1 and it is not the case that there exists another instrument j recorded at time t before t1 and the grantor in i is the same as the grantor in j, then i is recorded before j.

DR6 recorded-first(i,j) ⊃ ~recorded-first(j,i)
or for short DR6. RFij ⊃ ~Rfji
Where RFij: stands for 'recorded-first(i,j)'
~RFji: stands for '~recorded-first(j,i)'

Which reads: if the instrument i is recorded before the instrument j, then it is not the case that j is recorded before i.

17.5.3 Temporal Reasoning Examples

In this section, we present a summary of a logical system DR for temporal reasoning in the deed recording system and a number of examples about how this system can be used to reason about rights on real estate property. The following examples describe cases of temporal reasoning in the titling process. Consider the following story:

> The Jackson Family owned Three-Trees Island on 1 January, 1980 by grant from the State. They sold it to the Smith Family on 6 March, 1985. The Smith family recorded their deed in the registry on 15 May, 1985. Another purchaser, Jim, an old friend of the Jackson Family, who lives on the island while the Jacksons are out of town, had a previous deed from the Jackson Family on 5 April, 1982 but never paid the full price of the property. Jim, hearing that the Smith Family was selling the island to someone else, recorded his deed in the registry on 15 March, 1985. Jim

does not intend to steal the property, but he wishes to complete the payments and get hold of the place in which he lives. He believes he has priority on the property since he spent time in its renovation.

The problem explained in the above story contains details about rights dispute. However, real-life disputes could involve mortgages, second mortgages, wills and the like. We will use this story only to demonstrate the power of our formal language DR.

Example 1: Good Title in a Race-Notice Statute

Assuming a race-notice statute interpretation, and that the registry data is persistent, prove that Smith acquired a good title to the Three-Trees Island on 3 June, 1985.

Proof

proof steps:		justification:
s1.	LPid	assumption: names of families are given.
s2.	Plid	assumption: name of island is given.
s3.	$\sim XY$	assumption: Jackson is different than Jim.
s4.	S	assumption: race-notice statute RN.
s5.	P	assumption: persistency PR.
s6.	HXt1	DR2: Jackson, grant in 1980.
s7.	RDit4	DR3, DR4: Jim recorded on 15 March, 1985.
s8.	$\sim\exists j RDjt3$	no recording prior to 15 March, 1985.
s8a.	$\sim\exists j RDjt2$	no recording prior to 5 April, 1982.
s9.	RFij	DR5, s7, s8, *MP:* modus ponens, Jim recorded first.
s9a.	$\sim nc$	s8a, *N, MP:* modus ponens, Jim purchased without notice.
s10.	$S \supset \{(RF \wedge \sim nc) \supset GTt4\}$	Th1, *RS:* deduction theorem, R for Δ, $(RF \wedge \sim nc)$ for P, and GT for Q.
s11.	$(RF \wedge \sim nc) \supset GTt4$	s4, s10, *MP:* modus ponens.
s12.	GTt4	s9, s9a, s10, *MP:* RFij for RF, Jim has a good title on 15 March, 1985.

This proves that "It is the case that Jim acquired a good title for the island on 15 March, 1985." We now continue to prove that "Jim has a good title on 3 June, 1985 (the date of query)."

s13.	$P \supset (GTt4 \supset GTt6)$	Th1, *RP:* deduction theorem, P for Δ and $GTt1 \supset GTt2$ for $P \supset Q$.
s14.	$GTt4 \supset GTt6$	s13, s5, *MP:* modus ponens.
s15.	GTt6	s12, s14, *MP:* RFij for RF, Jim, good title 3 June, 1985.

This completes the proof of the assertion that "It is the case that Jim has a good title for the island on 3 June, 1985."

Example 2: Losing Title in a Race-Notice Statute

Assuming a race-notice statute interpretation, and that the registry data is persistent, prove that Smith Family have a claim but have not acquired a good title to Three-Trees Island on 3 June, 1985.

Proof

Since most proof steps in the previous example are valid for this example and will be used here as well, we will continue with step s16.

proof steps:	justification:
s16. ~RFji	s9, DR6, *MP:* Smith did not record first.
s17. CYt3	s6, DR1, *MP*: Smith, claim on 3 June, 1985.
s18. P ⊃ (CYt3 ⊃ CYt6)	Th1, *RP:* deduction theorem, P for Δ and
s19. CYt3 ⊃ CYt6	s18, s5, *MP:* modus ponens.
s20. CYt6	s17, s19, *MP:* RFij for RF, Smith has a claim on 3 June, 1985.

This proves that "It is the case that Smith has a claim for the island on 3 June, 1985 (the date of query)." We now continue to prove that "it is not the case that Smith has a good title for the island on 3 June, 1985."

s21. ((RF ^ ~nc) ⊃ GT) ^ (GT⊃ (RF ^ ~nc))	*S,* P16: equivalence.
s22. GT⊃ (RF ^ ~nc)	s21, P4: simplification.
s23. ~(RF ^ ~nc) ⊃ ~GTt3	s22, P6: transposition.
s24. (~RF ∨ nc) ⊃ ~GTt3	s23, P12, De Morgan's law.
s24a. (~RF ∨ nc)	s16, P5, *MP*: addition, ~RF for P and nc for Q.
s25. ~GTt3	s24a, s24, *MP:* modus ponens.

This proves that "It is not the case that the Smith Family acquired a good title for the island on 3 June, 1985." We now continue to prove that "it is not the case that the Smith Family has a good title on 3 June, 1985 (the date of query)."

s26. P ⊃ (~GTt3 ⊃ ~GTt6)	Th1, *RP:* deduction theorem, P for Δ and ~GTt3 ⊃ ~GTt6 for P ⊃ Q.
s27. ~GTt3 ⊃ ~GTt6	s26, s5, *MP:* modus ponens.
s28. ~GTt6	s25, s27, *MP:* Smith has non-good title 3 June, 1985.

This completes the proof to the assertion that "It is not the case that the Smith Family has a good title for the island on 3 June, 1985."

17.6 SUMMARY

The focus of this work was to incorporate reasoning models developed in AI and DBMS, and to examine their applicability for GIS and particularly for cadastral systems. Temporal reasoning in cadastral systems deals with dynamic information being updated by legal, topological, metric and other changes. By studying change causes and change effects in cadastral systems, it was possible to determine their temporal aspects.

In this chapter, we presented a sample first-order temporal logic for temporal reasoning in the deed-recording cadastral system. A more comprehensive logical system formalising deed recording statutes is found in (Al-Taha, 1992). The logic extends propositional and first-order logic by a persistency rule, which dictates that the truth value for a statement on time t remains unchanged at time t' later than t if no other statement exists between t and t' that causes the truth-value to be reversed. In other words, if P is true (or false) now, it will remain true (or false) for all times in the future until its truth-value is known to be reversed. Although persistency may not apply to all applications of temporal reasoning, it applies to cadastral records because they remain valid unless made void by other recent records. Some cadastral records prescribe their valid duration, such as an estate for years; however, if their valid duration is unspecified, such as a life estate, then it persists in being valid until later a new record exists that voids the validity of the previous record. The major characteristics of the temporal logic described in this chapter are:

- Ability to distinguish between two time perspectives: a creation and a recording time. The two time perspectives are needed for the reasoning about a good title;
- Ability to distinguish between having a claim to an estate and holding it. Because many legal-persons may have a similar claim on an estate on land at the same time, the logic will allow only one legal-person to hold that estate (or right); and
- Ability to capture basic elements of the titling statutes: recording a deed first, acquiring a deed without a notice, and obtaining or losing a title to a real property.

The basic principle of determining a title in a deed recording system is in building and analysing a sequence of claims, deeds and other instruments related to the real estate property being considered. The process begins with building a chronological list of events that occurred and that can influence whether a deed is a good deed or not. This list is called 'the chain of title' or 'abstract'. A backward and forward searching process on the grantor and the grantee indexes along the time axis is the basic operation needed to determine the chain of title. Without the temporal information being indicated on any instrument, it would be impossible either to create title changes or subsequently to determine the proper rights on any property.

REFERENCES

Allen, J., 1983, Maintaining knowledge about temporal intervals. *Communications of the ACM*, **26** (11), pp. 832–843.

Allen, J., 1984, Towards a general theory of action and time. *Artificial Intelligence*, **23** (2), pp. 123–154.

Al-Taha, K., 1992, *Temporal reasoning in cadastral systems*. Ph.D. Thesis, Department of Surveying Engineering, University of Maine.

Al-Taha, K., Snodgrass, R. and So, M., 1993, Bibliography on spatiotemporal databases. *International Journal of Geographical Information Systems*, **8** (1), pp. 95–103.

Armstrong, M., 1988, Temporality in spatial databases. In *Proceedings of GIS/LIS '88*, San Antonio, (ACSM) Vol. 2, pp. 880–889.

Barrera, R. and Al-Taha, K., 1990, *Models in Temporal Knowledge Representation and Temporal DBMS*, NCGIA Technical Report 90–8. Department of Surveying Engineering and NCGIA, University of Maine.

Binns, E., 1953, *Cadastral Surveys and Records of Rights in Land*. FAO Agricultural Studies No. 18. Food and Agricultural Organization of the United Nations.

Black, H.C., 1979, *Black's Law Dictionary* (Fifth Edition). (St. Paul, Minn: West Publishing).

Bolour, A., Anderson, T., Dekeyser, L. and Wong, H., 1982, The role of time in information processing: a survey. *SIGMOD Record*, **12** (3), pp. 27–50.

Chrisman, N.R., 1993, Beyond the snapshot: Changing the approach to change, error and process. In *Proceedings of NCGIA I–10 Specialist Meeting*, Lake Arrowhead, CA, May 10, 1993, edited by Egenhofer, M. (NCGIA, University of Maine, Orono), p. 7.

Chu, W.W., Ieong, I.T., Taira, R.K. and Breant, C.M., 1992, A temporal evolutionary object-oriented data model and its query language for medical image management. In *Proceedings of 18th International Conference on Very Large Databases,* Vancouver, Canada, August 1992, edited by Yuan, L.Y. (Morgan Kaufmann), pp. 53–64.

Clifford, J. and Croker, A., 1988, Objects in Time. *Database Engineering*, **11** (4), pp. 189–196.

Creteau, P.G., 1977, *Principles of Real Estate Law*, (Portland, ME: Castle Publishing).

Dale, P.F., 1976, *Cadastral Surveys within the Commonwealth*. Overseas research publication No. 23. (Her Majesty's Stationary Office).

Davis, E., 1990, *Representations of Commonsense Knowledge*, (San Mateo, CA: Morgan Kaufmann).

Frank, A.U., Egenhofer, M. and Hudson, D., 1991, *The Design of Information Systems*, Lecture Notes SVE 451, University of Maine.

Frank, A.U., Campari, I. and Formentini, U., Eds, 1992, *Theories and Methods of Spatio-Temporal Reasoning in Geographic Space*, (New York: Springer-Verlag).

Hazelton, N.W.J., 1993, Some Operational Requirements for a Multi-Temporal 4-D GIS. In *Proceedings of NCGIA I–10 Specialist Meeting*, Lake Arrowhead, CA, 10 May, 1993, edited by Egenhofer, M. (NCGIA, University of Maine, Orono), pp. 11.

Hintz, R. and Onsrud, H., 1990, *A Methodology for Upgrading Real Property Boundary Information in a GIS Using a Temporally Efficient Automated Survey Measurement Management System*. Department of Surveying Engineering, University of Maine.

Hunter, G. and Williamson, I., 1990, The development of a historical digital cadastral database. *International Journal of Geographical Information Systems*, **4** (2), pp. 169–179.

Kelmelis, J.A., 1993, Process dynamics, temporal extent, and causal propagation as the basis for linking space and time. In *Proceedings of the NCGIA I–10 Specialist Meeting*, Lake Arrowhead, CA, May 10, 1993, edited by Egenhofer, M. (NCGIA, University of Maine), pp. 14.

Langran, G., 1989, *Time in Geographic Information Systems*. Ph.D. Thesis, University of Washington.

Langran, G., 1990, Tracing temporal information in an automated nautical charting system. *Cartography and Geographic Information Systems*, **1** (4), pp. 291–299.

Langran, G., 1993, One GIS, many realities. In *Proceedings of GIS '93*, Vancouver, Canada, (Ministry of Supply and Services), pp. 757–762.

Margaris, A., 1967, *First Order Mathematical Logic*, (Waltham, MA: Blaisdell Publishing Company).

National Research Council, 1980, *Need for a Multipurpose Cadastre*, (Washington, DC: National Academy Press).

Onsrud, H., 1985, Choosing and evaluating property line survey methods. *Surveying and Mapping*, **45** (2), pp. 139–144.

Rescher, N. and Urquhart, A., 1971, *Temporal Logic*, (New York: Springer-Verlag).

Smullyan, R., 1968, *First-Order Logic*, (New York: Springer-Verlag).

Snodgrass, R. and Ahn, I., 1985, A taxonomy of time in databases. In *Proceedings of SIGMOD Conference*, Austin, Texas, May 1985, edited by Navathe, S. (ACM), pp. 236–246.

Unger, M.A., 1974, *Real Estate: Principles and Practices*, (South-Western Publishing).

U.S. Department of Agriculture and Commerce, 1974, *Land Title Recording in the United States*, (Special Studies No. 67).

Worboys, M.F., 1990, The role of modal logics in the description of a geographical information system. In *Cognitive and Linguistic Aspects of Geographic Space*, Series D, Vol. 63, edited by Mark, D. M. and Frank, A.U. (Dordrecht: Kluwer), pp. 403–413.

PART FIVE

Definition of Socio-Economic Units

INTRODUCTION

This part addresses the particulars of spatial socio-economic units. It concentrates on spatial socio-economic units that are widely used, such as census tracts and administrative units, and discusses methods to define them in general, and to create and change them. The last chapter presents a practical approach on how data related to one unit in one form and other data related to the same unit, but in a transformed position and form, can be made comparable.

Eric Stubkjær continues many of the issues Jonathan Raper introduced in the first part and concentrates on spatial units formed and used for administration and planning. He first differentiates 'jurisdiction', place', 'region' and 'district' and characterises the authority creating them and the purpose they are formed for, but also relates them to speech communities that are aware of them. Spatial units have boundaries, either 'fiat' or 'bona-fide' boundaries in the terminology of Smith (Chapter 6). Units with fiat-boundaries depend on an act of creation and on continued memory of them (and thus documents with respect to the creation of parcels must be kept permanently, as Al-Taha pointed out in Chapter 17). The four different spatial units fall to different degrees in the category of dominance.

A case study of a Danish administrative instrument, the Development Enquiry, helps to give an example for these relations. This enquiry was repeated several times in the 1970s and early 1980s to collect data about the capacity of building zones for additional construction. The case is of special interest, as new spatial socio-economic units were created by law and then used for a few years. This points to yet another type of movement, namely the creation, change and ultimate revocation of the rule, which creates the spatial socio-economic unit (which is related to Libourel's change in the database schema in Chapter 11). Modelling the dynamic behaviour of spatial socio-economic units sometimes uncovers the lack of rigorous definitions and procedures when they are created, or apparent changes in the SSEUs are really rather changes in the definitions used. For this Development Enquiry, precise rules described how the units should be delimited by the town, and these definitions changed over the years, inducing changes in these units and the values reported for them.

The following chapter discusses the design of optimal zoning systems. Openshaw and Alvanides argue that each zoning structure employed for spatial mapping affects the outcome of the analysis. The transition from a map that represents each individual to a generalised map that aggregates individuals to groups is problematic. The aggregation of fine enumeration district data to larger reporting zones, as is routinely done by national census, introduces noise in the data (the well-known Modifiable Areal Unit Problem (MAUP). The data changes from data about individuals to data that describe the zoning units. Hence, the description available in aggregated data is different if different zoning units are selected. Most obvious is the averaging effect, where the maximum and minimum values are reduced when larger units are selected. Openshaw proposes to use different methods of aggregation to form larger zones: methods which are more 'uniform' (on some scale, for example, equal size zones, or equal population number zones); or zones which take into account some properties influencing the phenomena considered (for example, when mapping unemployment, then form zones such that accessibility is used to form zones). With modern GIS software it is possible to experiment with different zones and to select a zoning design which facilitates the analysis and helps to see meaningful patterns. It is not imperative that the standard zoning structure (for example, political subdivisions, which reflect historic processes) is used, just because it is provided. It is evident that there is not a single most appropriate subdivision in zones, which objectively maps reality. With modern software systems, users can select their own zoning rules—one of the effects of the cartographic revolution (Morrison, 1994)—and they should then deconstruct their motivation for selecting them themselves. It remains a research problem to construct zones remaining constant for long periods of time to serve as framework in which changing survey results can be presented; this is partially addressed in the next chapter.

Coombes and Openshaw report on a practical project concerning the definition of a specific set of SSEUs in Britain, the so-called Travel-to-Work Areas (TTWAs). The TTWAs are groups of basic spatial socio-economic units within which the flows of workers from home to work are predominantly internal; in other words the commuting flows are either within the base spatial socio-economic unit or within the TTWA aggregation of spatial socio-economic units. The multiple and conflicting demands of this project were: a large enough number of units to depict the phenomenon properly, but also a demand for few units to assure stability over time. In this application Coombes and Openshaw show a method for identifying concentrations of intercommuting (generally urban areas) and then iteratively allocating all base spatial socio-economic units (including rural spatial socio-economic units) to the prototype concentrations. This contribution shows that spatial socio-economic units can sometimes have a core and periphery structure where 'edge' components of the SSEU change 'allegiance' to the core over time. The results of a temporal analysis of British spatial socio-economic units reveal changes in the size of spatial socio-economic units associated with longer journeys to work over the last two decades. This is an illustration of how traffic growth has led to a restructuring of space physically (through less dense suburban growth) and then administratively (by a change in the form and TTWA 'allegiance' of spatial socio-economic units).

It is—as Stubkjær has described—sometimes necessary to create new spatial socio-economic units. Reis describes some prototype geometric processes under development in Portugal to create spatial socio-economic units, in this case for the postal service. These units follow the postal delivery routes; i.e., they follow the road network as this is one of the basic geometries that is universally available. Reis addresses the theoretical problem of zone creation from a detailed practical perspective looking at the forms of geographic information typically available and how they can be used to drive an automated regionalisation of a city. He shows how the incremental development of the urban fabric over centuries has led to anomalous address numbering and the consequent difficulties of building zones from a reference system built up over time. He also addresses the problem of space exhaustion: should spatial socio-economic units exhaust space—if they do, then when new development takes place the units must be changed—if not, the zones must be added to when new development takes place. Both scenarios affect the analysis of statistical data referenced to the created zones.

The last chapter by Jostein Ryssevik discusses a pragmatic solution to the general problem of time varying enumeration districts. The boundaries of the districts which are used for collecting and reporting socio-economic data change during time—smaller changes due to adaptation to changing urbanisation and the like, and large changes due to major reorganisation of national administration. Many countries have seen efforts to form larger communes, to simplify administration, which had profound effects on the reporting of statistical data. Ryssevik reports on the solution the Norwegian Social

Science Data Services have found. Demographic data from 1769 onwards have been integrated in a single database, even though clearly the spatial units this data refers to have changed several times since. To make this data comparable, tools are provided for the scientists who use this data collection, which simplifies standardisation. The preferred method is to consider how much of basic resources have changed by the redistricting and to use these changes as standardising coefficients for other variables. In practical terms, population transferred between zones is easy to acquire and as population in general influences most other variables, it is an acceptable estimate of redistribution. For some variables this is not appropriate and 'missing value' must be reported. The system includes some automatic guidelines, a sort of expert system, to achieve the best possible adaptation to changing boundaries and to avoid nonsense. A report to assess the viability of the applied adjustments is automatically produced. The system produces for the scientist comparable time series of data for any spatial subdivision (the original historical one or the current one). The assumptions necessary for such an adaptation can be checked by the scientist. A number of practical tests demonstrate that the adjusted data are quite reliable. The service facilitates social science research and is widely used by scientists in Norway.

REFERENCE

Morrison, J., 1994, The Paradigm shift in cartography: The use of electronic technology, digital spatial data and future needs. In *Proceedings of Sixth International Symposium on Spatial Data Handling, SDH '94*, Edinburgh, UK, (AGI).

CHAPTER EIGHTEEN

Spatial, Socio-Economic Units and Societal Needs—Danish Experiences in a Theoretical Context

Erik Stubkjær

18.1 INTRODUCTION

18.1.1 The Need for Operational Definitions of Spatial, Socio-Economic Units

Socio-economic units are the result of processes constructed and controlled by humans. The definition of these units is by no means a trivial task. To illustrate the complexity, mention is made of the fact that the World Wide Web technology makes it possible to rent a room in a 'Web-Hotel' where one can present information on, e.g., company activities. The user company may be Danish, the company in charge of the Web hotel may be registered in the USA, and the computer where the home page, etc., is stored, may be located in Germany. The web hotel room is a spatial, socio-economic unit, but 'where' is it located, what are the boundaries of the unit, and in what legal system shall the definition of the unit be stated?

Definition issues are inherent in international deliberations. The United Nations Convention on the Law of the Sea is presently being supplemented with an 'Agreement on Straddling Fish Stocks and Highly Migratory Fish Stocks'. The resolutions on such 'Highly Migratory Fish Stocks' need an operational definition to be implemented. To establish such a definition seems to be a difficult task, taking into consideration that the fish stock is moving relative to ocean streams.

Maybe a less complex task is to account for the flowing water alone. This is in fact needed: The UN Conference on Water and the Environment, held 1992 in Dublin, stated that "Fresh water is a finite and vulnerable resource…" (The Dublin Water Principles, Principle No. 1), and a recent Report of the secretary-general states that "…there is already growing perception of water as an economic good and as a tradable commodity…" (UN/dpcsd/dsd, 1997, section 15). Bottled water is a common-place commodity, but here reference is made to more subtle socio-economic units, including "…major river basins, and groundwater aquifers (which) cross national boundaries." (section 17). A statement calls for, among others, "water information systems, legal and institutional arrangements and water demand management" (UN SG/SM/6185 ENV/DEV/404 21 March 1997), that is: a GIS for water management with appropriate definitions of spatial, socio-economic units.

The remainder of the chapter treats the more moderate task of discussing the definition of spatial, socio-economic units that are related to the surface of the Earth, e.g., real estates, census tracts, and parishes. Such socio-economic, spatial units have been recorded at least since the age of the Northern Italian principalities and city-states. To illustrate some aspects of the chapter, Table 18.1 below renders the most important spatial SEUs in Denmark.

The dioceses, parishes, and court districts belong to the oldest layer of administrative subdivisions in Denmark. The unit 'township' was defined in the context of the establishment of the Danish fiscal cadastre. A 'township' was originally a corporation of farmers, rather than a contiguous area. However, a place name of the unit had to be available, too, to constitute a cadastral township. Therefore, you may find a cadastral township which included two geographically distinct corporations (Frandsen, 1976). The Municipalities were established through a municipal reform in the 1960s by amalgamation of parish communes. A variety of spatial units, including planning districts, were defined in these and subsequent years, mainly to serve the planning needs of the municipalities.

Table 18.1 Administrative subdivisions and other spatial units in Denmark

No. of units in Denmark	Jurisdictions	Other SEU
1–29	Dioceses, Counties	
30–300	Court districts, Municipalities (from 1970)	
10^3 x 1.0..2.0	Parishes (from ~1100)	
10^4 x 1.0..2.0	Townships (from 1803)	
10^5 x 1.0..2.0		Settlements, Named roads, Planning districts
10^6 x 1.0..2.0	Real Estates, Dwellings	

The relation among the units can be hierarchical: The Danish cadastral parcel reference number (Dale and McLaughlin, 1988: 39f) consisted of the elements: Parcel number and superscript letter (7^{a}), Township, Parish, Court District, and County, until it was simplified as a consequence of the municipal reform. Generally, however, you cannot be sure that smaller administrative units fit into the framework of the larger units because different needs determine the geographical structure of the units.

18.1.2 Emerging Concern for the Definition Issue

The practice of defining spatial, socio-economic units has been embedded in the socio-economic regimes, or the legislation, of the different nations, with the outcome that transfer of data and experiences across regimes was rudimentary. A universal and formal description of the definition practice is still to be developed. The following mentions some of the initiatives and research that have been made in this direction.

The growing need for international comparisons after World War II raised the issue of common definition for statistical units. For example, the Nordic Conference of Statistical Agencies ('nordiske statistiske chefmøde') in 1960 adopted a common definition of the spatial unit of a 'locality' or urban district (Danish: 'bymæssig bebyggelse') to be used for the 1960 population census. This was made in accordance with deliberations of the UN Statistical Commission (Danmarks Statistik, 1968). This definition is independent of references to the administrative structures mentioned above, as it refers to the number of persons (200) and distances between buildings (200 m). By 1970 the definition was further supported by the UN Economic Commission for Europe (Danmarks Statistik, 1975).

The need for common definitions, etc., is especially urgent for countries that establish comprehensive trade agreements, or agree on common policies, like the EU countries. The development of joint statistical definitions in the field of agriculture, forestry and environment, may be traced through the EU Commission's General Report. In 1994, programmes were adopted for environmental statistics and for agricultural statistics, respectively (European Commission, 1994, clauses 88 and 90).

The introduction of the computer during the last decades made it more generally acknowledged that general and formal descriptions of spatial SEU were needed and also possible. Data base theory, and research on spatial topological relationships (Egenhofer and Herring, 1991), among others, has supported this expectation, and recently the issue of formal modelling has been addressed (Frank, 1996). Also, the more administrative approach for defining standards in the field of GIS contributes in this effort (Mark, 1993; Skogan, 1995; David *et al.*, 1996). However, some of the mentioned research has focused on information systems, that is, on rather formalised representations of phenomena, rather than on the phenomena themselves. Geographical information systems must account for changes in the geographical reality. However, to represent such changes in the formalised context of an information system, you need an understanding of reality and perception that is deeper than available in present research, cf. the introductory chapter of the present volume.

18.1.3 Outline of Chapter Content

The present chapter aims at addressing the definition of socio-economic, spatial phenomena. The phenomena can be addressed only through some language that is used by those who are concerned with the phenomena. Burrough and Frank mention five GIS user groups and conclude, among others, that "methods of handling spatial information must be linked to the paradigms of the users' disciplines..." (Burrough and Frank, 1995, p. 114). Likewise, Ferrari concludes that "...the description, and consequently the identification and delimitation, of geographic objects is more strictly related to the corresponding concept or cognitive model than to perceptive reality" (Ferrari, 1996, p. 107). David Mark addresses the definition problem explicitly, and, pointing to the relation between conceptions and natural language, he quotes Benjamin Lee Whorf: "We dissect nature along lines laid down by our native languages". These languages refer to different 'speech communities' (Mark, 1993, p. 272).

The first main part of the chapter takes its point of departure in natural language, more specifically in the linguistic notion of 'sublanguage' (Kittredge, 1983; Stubkjær, 1994). It is posited that each of four sublanguages has corresponding socio-economic, spatial units. The formal notions of mathematics go beyond natural language. However, for the purpose of the chapter it is not necessary to mark a boundary, and hence the notions of mathematics are counted as a fifth sublanguage.

In a following section the ontology of these units is discussed on the basis of the dichotomy of Barry Smith between 'bona fide' and 'fiat' boundaries (Smith, 1995). The concept of real estate is devoted special treatment, and it is posited that ownership is a three-category entity of land, person, and society, respectively.

The second main part of the chapter presents Danish evidence on the use of one of the sublanguages: the administrative sublanguage. From the 1960s to the early 1980s Denmark had a construction boom. In order to manage this boom (in terms of urban planning) an administrative procedure, a development inquiry, was put into effect. The spatial unit of this development inquiry is presented and discussed. In a further section, the life cycle of the inquiry is discussed. As the SEUs themselves have a life span, from creation to deletion (cf. Opening Chapter), so the definition of the SEU develops, as the above-mentioned account of Danish administrative practice shows. A conclusion closes the chapter.

18.2 FOUR CATEGORIES OF SOCIO-ECONOMIC UNITS

How can you discern sublanguages within a natural language when languages are so difficult to describe? A method is to refer to 'prototypes' rather than to formal definitions. In search for 'Geographic Entity Types' David Mark refers to categories modelled at mathematical sets. He quotes Cassirer for an early critique of this model, and points to 'prototypes' of cognitive categories as a fertile basis for work in cognitive models of geographic space (Mark, 1993, pp. 271–272).

In this line it is posited that sublanguages concerning geographical matters can be discerned on the basis of the spatial unit(s) applied. Furthermore, that there exist at least four classes of socio-economic, spatial units: The jurisdiction, the place, the region, and the district. These classes are defined with reference to their social implications (Raper, this volume).

18.2.1 The Jurisdiction

The **jurisdiction** is the term for a spatial unit that denotes a domination. The prince has command over his principality. The bishop rules in his diocese, and the owner of a real estate disposes of all rights of his or her property. The way the ruler dominates, the 'content' of domination differs among the examples. Common is, however, that real world domination—or, in other words, social control is in effect. The borders of the dominated area may fluctuate or be firmly established, depending on the kind of society and the technology available.

The best known unit within this category, that is the prototype, is likely to be the country or *nation*. Within the national boundaries the complex, modern society has established many jurisdictions, almost one hierarchy of units for every ministry of government. Some countries look for that many jurisdictions share boundaries, e.g., boundaries of local governments (municipalities) or parishes.

18.2.2 The Place

The **place** is the term for a spatial unit that is denoted by a place name. The urban square or marketplace, or the *town* are likely to be the prototypes. However, the term applies to the spatial units to which we refer in ordinary conversation: towns and streets of all kind and sizes, parks and other objects described, e.g., by Kevin Lynch (1960): The Image of the City, or more recently by Tuan and Johnson (cf. Raper, this volume).

The diverse use of place names appears from the Swedish handbook on the decimal classification system for the research libraries (Tekniska Litteraturselskabet, 1977). The handbook puts place names, parcel reference numbers, street names, and names of nations within the same clause: 801.311, that is, within Language Science (800). Referring to this, Stubkjær (1992) proposed a distinction between 'natural place names' and 'technical place names' like cadastral designations, address codes, etc. The former develop from the talk of the town (as permitted by the ruler), while the latter are established through an administrative procedure. The conception of place names which is proposed here puts emphasis on who generates the place name rather then on the use of the name. This is consistent with the overall position that an understanding of socio-economic units should be achieved through investigation of language use, but refines the position to include a concern for change of language elements, e.g., the introduction of new vocabulary.

The jurisdictions mentioned above need a place name, unless they carry the name of the ruler. The distinction between the two categories, jurisdiction and place, is blurred when place names are used by the ruler to impose his world view on everyday affairs: The naming of Ho Chi Minh City (former Saigon), or some Stalin Avenue may serve as example of how the relationship between controller and controlled is influenced through means of territoriality (Sack, 1986; Malmberg, 1980; Stubkjær, 1992).

Finally, it is noted that the relation between the place (the real-world phenomena) and the place name has a more technical function for way finding, route descriptions, etc. During the last decennia rationalization of mail services and the use of computer technology have motivated a coding of place names in terms of post codes, street coding, etc. (Raper *et al.*, 1992).

18.2.3 The Region

The **region** is considered the prototype within geographic and other spatial research. Terms like 'zone' or 'field' are frequently applied instead of 'region'. According to Bunge geographic regions may be uniform, experimental, nodal, or applied (Bunge, 1962: 14ff.). The conception of 'region' has developed substantially since to include, among others, a reflection of perception, representation, language, code, and implementation of code (Burrough and Frank, 1995; cf. Raper, this volume). The term 'region' is used here to denote the analytic, spatial unit that the researcher uses as a base for empirical or theoretical statements on geographical phenomena. The 'region' may be delimited by physical features, or by the product of power relations; it may be conceived as an object or a field. It is only decisive that the unit is defined and used by the scientific community for the purpose of scientific inquiry.

By definition 'region' is outside the realm of domination. Etymologically, the term *regio* refers to direction, visible boundary, or tract. It may, however, be related to *regius*, royal (Worboys, 1995). 'Region' has no relation to place names.

18.2.4 The District

The **district** is the term proposed for an area unit that is defined and used by an administration to perform or improve its functions. Census tract and planning zone or planning district count among the prototypes. Used by a governmental body, defining 'districts' for land use zoning has a flavour of governmental dominance, but a 'district' differs from a 'jurisdiction' in that a 'district' does not denote the spatial demarcation of an authority. Furthermore, a transportation company, or a sales department, divides its territory in 'districts' to rationalise their tasks.

'District' has common traits with 'region'. This is not surprising as administrations apply scientific knowledge to fulfil their task. Whether the application of 'Broadbent's rule' and 'zone design' methodologies (Raper, this volume) will result in 'regions' or 'districts' is eventually determined by the purpose: Searching for new knowledge, or solving an administrative task, respectively.

18.2.5 Summary and Discussion

Four classes of socio-economic, spatial units have been presented: The jurisdiction, the place, the region and the district. The main discerning criteria was the 'actor' or body who applied the unit: The ruler, the public, the scientist, and the officer or company staff, respectively. This may be considered an interpretation of Habermas' position, which states that human interests structure knowledge and that different epistemologies are appropriate for each (Raper, this volume). For example, public administration, at least in its classical, bureaucratic form, takes an instrumental or 'technical' interest in the world and, consequently, positivist assumptions are maintained. The real estate, which from the owner's point of view is a 'jurisdiction', becomes, through the eyes of many European cadastral officers, a 'district' with clear cut boundaries for which the officer, and not the owner, takes responsibility.

A fifth class, labelled by the term **area**, may be counted, although this is a mathematical and not a socio-economic unit. 'Area' is similar to 'region' as both are used by scientists, but 'region' refers to the surface of the Earth while 'area' may refer to any delimited surface. Therefore, 'area' is probably used as the most general term for spatial units. Different disciplines of mathematics each coin their terms. Analytic geometry conceives 'area' as the well-known (metric) measure of extent of a plane surface. The discipline of graph theory is not concerned with metric properties (part of it may be called 'rubber sheet geometry'). The different conception of a delimited surface is denoted by the term 'face' (e.g., Wilson, 1985; and for applications in GIS: Bartelme, 1989; Laurini and Thompson, 1992).

18.3 THE ONTOLOGY OF SOCIO-ECONOMIC, SPATIAL UNITS (SEU)

Socio-economic units are not easy to comprehend; they do not contrast to their surroundings as a moving object like a person or a car. Therefore, one has to ask what kind of existence the SEUs have, and what calls them into being, in other words, to investigate their ontology.

18.3.1 Fiat Objects

In a recent paper Barry Smith presents a typology of spatial boundaries, which is based on an opposition between bona fide or physical boundaries on the one hand, and fiat or human- demarcation-induced boundaries on the other hand (Smith, 1995).

The **physical boundaries** are boundaries in the things themselves. They exist independently of all human cognitive acts. They are "a matter of qualitative differentiation or discontinuities in the underlying reality" (Smith, 1995, p. 476).

The **'fiat' boundaries** owe their existence to acts of human decision or fiat, to laws and political decrees, or to related human cognitive phenomena, so "(f)iat boundaries are boundaries which exist only in virtue of different sorts of demarcations effected cognitively by human beings" (p. 477).

The mentioned opposition does not rule out combinations of the two types of boundaries. Rather, the bona fide–fiat dichotomy serve as a means for discussion of different types of boundaries, and for discussing bounded objects. The 'fiat objects' come into being "... through human cognitive operations of certain special sorts, in such a way that both boundaries and objects *exist* only in virtue of these operations" (p. 477 his italics). This short account of the paper by Barry Smith suffices to note that the mentioned four classes of SEU (labelled by jurisdiction, place, region, and district, respectively) are all basically fiat objects. The objects owe their existence to human acts, namely acts of the ruler, the public, the geographer and the staff member, respectively. Thus the notion of fiat objects seems to correspond to the worldview of this chapter, based as it is on a linguistic approach.

18.3.2 The Representation of fiat Objects and the Reality

The ontological status of fiat objects seems to depend on the remains of the initial human act: what is kept in memory or documented on paper or similar records. The representation of fiat objects is thus the only tangible access to fiat objects. This tempts the legislator to rely on the representation: the deed, the recording in the 'Grundbuch', rather than on the transitory human 'fiat'.

By declaring a fiat object something new has come into being. The representation of fiat objects may contain entities (e.g., names) that do not have a counterpart in the bona fide world (cf. Eschenbach, this volume). However, the bona fide–fiat dichotomy articulates a modest claim on human creative power. While some claim "(t)he Social Construction of Reality" (Berger and Luckmann, 1967) we can twist the words, and restrict us to talk of the "human construction of a social reality" (Searle, 1995), accepting the bona fide world as a given reality. Furthermore, the dichotomy poses a question to the Marxist position that "the superstructure is largely irrelevant and analysis of it simply perpetuates the social relations which it reflects" and the similar position by Focault that "truth is defined by the systems of power which create and sustain it". (Raper, referring to Harvey, 1989 and Focault, 1980; cf. Stuart Aitken, this volume). The quoted positions are surely true when real estate and other 'jurisdictions' are meant, but the positions do not hold for 'places' and bona fide objects. Reference to such objects is essential, e.g., in criminal inquiries, so even in the post-modern era it appears valid to talk about truth when the issue is whether or not a certain event happened at a certain place at a certain time.

18.3.3 Possession and Ownership

A real estate has an area, it covers a section on the surface of the Earth, but the essence of the real estate is the relation between the owner and the land: the domination, the cultivation, the base for living. Furthermore, without a society a person would hold land in possession, rather than own it. The possession is visible to the eye, but mere inspection by eye does not reveal whether you own a thing or have it in your possession. The thing may be in your possession because you borrowed it. It is the society, which acknowledges and protects the owner's right in the thing, provided that the owner follows the prescribed procedures for the acquisition of ownership. So ownership, in this view, is a three-category entity of land, person and society.

Similarly, a nation would not be a nation without a land, a people (frequently of common language, history and legislation), and the recognition by other nations. Furthermore, personal integrity is a three-category relation between an imaginary sphere which surrounds the human, the human, and (the conventions of) the society.

The distinction between possession and ownership is common in legal contexts. It supports, however, the relevance of the bona fide–fiat dichotomy, and introduces light and shade into the meaning of the class of 'jurisdiction'.

18.3.4 Fiat Object and 'Shadows'

It is noted that SEUs often appear with an ontology similar to 'shadows': the SEUs are non-physical properties of an area which can be moved without any movement of material (Frank, this volume: referring to Casati and Varzi, 1994). "Shadows are cast on the object and can move without affecting the object. This is ontologically similar to legal assignment of areas to urban zones, to protected areas, etc., ...Poland (or any other nation) is a shadow, and as such can move..." (Frank, this volume).

A crucial question is what the 'object' of the quotation refers to. A likely interpretation is: some section of the surface of the Earth, an area. However, the shadow metaphor seems to understate the fact that SEU objects are determined by a societal, as well as a physical or 'bona fide' reality. An implication of this stand is that you cannot 'move' these objects without recognising the broken bonds, e.g., between the Polish people and their land. Similarly, the "legal assignment of areas" to urban zones does regard non-physical properties of an area. More interesting may be, however, whether a planning officer who draws urban zones on a cadastral map, is satisfied with the colourful outcome of such drawing exercise (a shadow plan if you like), or if he contributes his part to the implementation of democratic decisions, that is to change the rights and behaviour of owners.

In the preceding main section five categories of spatial, socio-economic units have been presented and discussed, mutually and relative to the fields of language, philosophy, and spatial planning. The following main section extracts Danish experience to elucidate the ontology of SEUs and the fact that not only SEUs have a lifetime: The man-made definitions of the SEU change with the need of society as well.

18.4 THE DANISH DEVELOPMENT INQUIRIES 1967–1982

The following review of an administrative instrument, the Danish Development Inquiry (Danish: 'Areal- og byggemodningsundersøgelsen') serves the purpose of testing the above speculative reasoning against empirical evidence. This specific evidence was selected because it demonstrates the life cycle of an administrative instrument. Furthermore, the inquiry was rather crucial in relating planning intentions to the physical reality. Finally, the spatial unit of the development inquiry was transient, and not formally defined which makes an interesting point of departure for a discussion of the ontology of the unit.

18.4.1 Urban Planning, the Measure of 'Capaciousness' and a Digression on Planning Instruments

In the 1960s Danish planning authorities faced the problem of urban sprawl. Zoning of land was carried into effect to reduce public (municipal) costs of roads and sewers to a scattered habitation, and to protect agricultural land and recreational resources. Three zones were instituted: Urban zone, rural zone and summer-cottage zone. Construction and subdivision in the rural zone was forbidden, except for agricultural and similar purposes. The zoning triggered compensation in some, but few instances, and has fiscal implications as well.

The crucial question was how wide the urban zone should be extended. If drawn too wide, the effect of the zoning would be lost. If drawn too narrow, the prices on land would escalate, and the legislation could likely be ignored. Monitoring of the capacity of the urban zone to accommodate the need for dwellings was, therefore, a key issue.

The 'capaciousness' (Danish: 'rummelighed') of the urban zone was perceived as a ratio between the land of the urban zone, which was more or less prepared for construction, on the one hand, and the yearly consumption of land for urban development as recorded in statistics on the other hand. According to legislation (the Danish 'Lov om by- og landzoner') the planning authorities should at any time make available the land which was needed for 5 years consumption. This was achieved, among others, by including rural land into the urban zone in the context of revision of plans every 3 to 5 years, and by the municipalities' construction of gathering roads and main sewer systems simultaneous with the planning process.

An objective of the Danish municipal reform of the 1960s was precisely to collect into one functional unit a town with its surrounding area, and to establish municipal units which could afford to employ engineers and city treasurers. The instruments of Danish planning include the practice that municipalities buy land, develop it to 'parcel level', and sell it to families at cost price to influence the price level of the larger amount of parcels which are provided by the private sector. Also, building societies (Danish: 'sociale boligselskaber') play an important role: They are financially supported by government, and prepared to build dwellings where the private construction sector has no incentive to go, e.g., the less attractive areas of Greater Copenhagen. This means altogether that the lines and colours of the planner on the map could be followed up by municipal investments in infrastructure, by municipal operation on the land market, and by dwellings offered by building societies.

An inquiry into urban development was made with rather regular intervals to monitor the capaciousness of the urban zone. In the early 1960s the planning staff of the Greater Copenhagen Planning Council (Danish: 'Egnsplanrådet') made an assessment of the capaciousness of the existing urban plans (Egnsplanrådet, 1968: 7). A report on developed areas was prepared and issued by the staff in 1966, and later supplemented (Meddelelser, 1968: 5). In 1967, a committee was installed to prepare a more complete inquiry on the capaciousness. The first inquiry to be based on responses from the municipalities was prepared as of 1 June 1967 for the Greater Copenhagen Area. A second inquiry was made regarding this area as of 1 January 1970 with a slightly more complete guidance. In 1974, 1978, and 1982 the methodology was applied by central government, and addressed every county and through the counties every municipality in Denmark. The national development inquiry came to an end when in the 1980s the pressure on land ceased. The inquiry was, however, continued by the Greater Copenhagen Statistical Office (Hovedstadsregionens statistikkontor, HSK) within its area (HSK, 1966).

The life cycle of the administrative instrument followed a quite ordinary path: Inaugurated within an administrative body (Egnsplanrådet, it was first extended as a specific procedure which involved the municipalities within the jurisdiction of this body. Then it was extended to the whole nation as a procedure that involved central government, counties, and municipalities. Finally, the procedure was largely abolished.

The life cycle of the development inquiry followed in a remarkably close way the socio-economic reality in Denmark: During the 1960s the amount of completed constructions increased from 4.8 mio sq. meter floor area in 1960 to 10.0 mio in 1970. During the 1970s the level of activity remained between 9.4 mio and 12.2 mio. The construction activity dropped from the 1980-level of 10.3 mio to 6.0 mio sq. meter floor area in 1983 (Danmarks Statistik) (see Figure 18.1). This close relation can be considered as an example of administrative responsiveness or efficiency.

This survey of an administrative procedure, its inauguration, extension, and abolishment, provides evidence to support the claim that not only SEUs have a life cycle, the definitions of the SEU have a life cycle as well. But how was the topical SEU defined, what was its ontology?

18.4.2 The Procedure of the Development Inquiry

The procedure of the Development Inquiry must be presented, before the ontology of the spatial unit of the inquiry can be discussed. This is due to the fact that the spatial unit was never explicitly defined, and, furthermore, was of a transient nature.

In 1967, and again in 1970, the Greater Copenhagen Planning Council issued a circular to the municipalities of its jurisdiction, asking for information on development of land, and on constructions for dwellings. The data were to be delivered in a table, and on a map in scale 1:10.000, and supplemented with further information in an accompanying letter.

The map should show the zone boundary between urban and rural land, and a borderline enclosing built-up areas (existing urban area, including plots of which more than 50% of the area was already built, or which had a size of less than 2 ha). On an illustrative map (see Figure 18.2), which followed the 1970-circular, the area between the zone boundary and the built-up border was tiled into plots of sizes ranging from 3 ha to 34 ha, most areas between 8 and 15 ha. Using colour code and other signs, the map should portray three periods of development (preceding 3 years, next 3 years, and further 3 years), four classes of land use (detached residential (Danish: 'haveboliger'), multi-storage residential ('etageboliger'), industrial ('industri og håndværk'), and public purpose ('off. formål og lign.'), and three classes of ownership (owned by individuals ('privat eje'), by building societies ('sociale boligselskabers eje'), and by governmental bodies ('offentlig eje'), respectively), (Egnsplanrådet, 1971, Bilag D)

The stock of 'developed area' (to be defined below) as of the appointed date should be calculated, and estimates of urban development should be provided for each of the two subsequent periods. The amount of present and expected developed area should be given in the table, Figure 18.3, in hectares (ha) without decimals.

The information asked for included, furthermore, accounts of population and existing dwellings, as well as the number of dwellings in started constructions.

You may say that the Council asked the municipalities to establish a kind of cadastre which instead of parcels depicted much larger entities which might or might not relate to property boundaries: The plots of developed area, and area to be developed. The inquiry illustrates very well the universal role of the cadastral method of combining table and map in an administrative context to depict some socially relevant aspect of reality. Possibly you cannot establish a 'district' without establishing a cadastre (in the present, restricted sense) as well.

18.4.3 How the Spatial Unit of the Development Inquiry Came into Being

The spatial unit of the Development Inquiry was described with reference to 'developed areas'. A developed area was defined as "areas which can be connected to an existing main sewer pipe and which are included in an approved land use plan (Danish: 'by- eller dispositionplan')..." (Egnsplanrådet, 1968: 3). It was further stated that the investigation should not be concerned with whether the land was actually for sale.

The 1970-inquiry applied a slightly more precise definition: "areas which can be connected to existing main sewer pipe, gathering roads, and other necessary technical infrastructure" (Egnsplanrådet, 1971, page 11, and Annex D, pt 1.3).

To comply with the inquiry a municipal officer had to prepare a map of scale 1:10.000. This presumed some mapping effort, as municipal maps generally are in scale 1:1000 or 1:2000, to serve the needs of the municipal engineer. The national coverage of cadastral maps are mostly in scale 1:4000, occasionally in scale 1:2000, and the topographical maps were of scale 1.20.000, now 1:25.000. The presence of an independent land surveying profession may have facilitated the provision of the new map type.

On the 1:10.000 scale map the border of built-up area and the boundary between the urban and the rural zone then had to be drawn. Next the existing main sewer pipes had to be taken into consideration, and the land which was owned by the municipality and other public bodies, and by building societies (Danish: 'sociale boligselskaber'), respectively. Finally, the units could be delimited, the size of the areas calculated, and the sum of the areas specified for the different periods, land use classes, and owner classes, respectively.

The provision of a suitable map appears to be decisive with regard to the task of defining instances of spatial SEUs. This is in accordance with the conception of 'fiat objects' of (Smith, 1995).

18.4.4 Reflections on the Ontology of the Spatial Unit of the Development Inquiry

Compared to other 'districts', e.g., census tracts or planning zones, the spatial unit of the development inquiry lacks certain features: It is delimited by the judgement of a municipal officer largely without formal definition or guidance. It has no identifier or 'name', no legal implications, and its purpose is completed when the area is calculated. (Of course, this is stated without intending a criticism of the methodology of the inquiry). The existence of the unit thus is restricted to the world or domain of certain planning offices. How many persons were involved? And what reasoning determined the extent of the spatial unit?

An answer to the question **how many persons were aware of the spatial unit** appears from the reports on the inquiries where account is provided for the responses to the inquiry: In 1967–68 it was necessary to remind about 50% of the 90 municipalities in writing or by telephone calls. Of the received answers about 70% were characterized as complete, about 12% were incomplete yet sufficient, while 12% were 'very insufficient', and 6% of municipalities simply did not answer. The lacking information was estimated by the staff of the Egnsplanråd, in two cases after visits at the planning department of the municipalities.

The next inquiry in 1970–71 was almost complete: 92 of the then 94 municipalities completed their task. This may, among others, be due to more complete guidelines issued with the inquiry. The staff prepared the information for the remaining two municipalities. The 'speech community' (Whorf; Mark above) to which the spatial unit was a reality may count some hundreds of persons, compared to the more than 1 million inhabitants in the area of Greater Copenhagen.

The other question regarded **the ontology of the unit of the development inquiry**. The existence of the unit is not restricted to the lines you can read on the set of maps in scale 1:10.000. Rather, the inquiry forced the planning officers to look at the area of their jurisdiction with the question in mind: Where is development to occur, relative to the terrain, to present and future shopping centres, schools, and other facilities, to the different owner classes, and to planning objectives.

It is interesting that the spatial units of the development inquiry were of a transient nature: When the municipalities had completed their response to the inquiry, and the data were used for the report on capaciousness, nobody seemed to care for the material any more. The subsequent inquiries do not mention the material prepared for previous inquiries. Reasons are likely to include that the material was at least partly outdated, and that before the computer age there was no economic incentive to reuse it.

No doubt, however: The unit is a 'district' in the sense set forth in Section 18.2.4 above, but, depending on the education of the planning officer, he may have thought in terms of 'regions' and 'location-allocation' problems during the creation process. The existence or lack of appropriate place names may have influenced the way he drew the boundary line, and even if he was instructed to ignore the question whether the land was actually for sale, the implementability of the plan would benefit by recognising the presence and interests of dominating landlords.

Furthermore, the question on where development actually should take place became closer to implementation when the 'where' was made manifest by drawing lines on a map (cf. Smith, 1995). The physical or 'bona fide' objects in the field, existing roads, streams, sewers, etc., were neither the only, nor the major determinants of the units. Rather, the units are another example of 'fiat objects' (Smith, 1995) as *it was the planning officer's judgement of the complex of information which called the unit into being.*

The reasoning above is supported by remarks in the report on the 1967-inquiry: The report summarises the outcome of the inquiry as follows:

> "In the period 1.7.1967 to 1.7.1970 the amount of developed area is expected to be 6286 ha, while the figure for the period 1.7.1970 to 1.7.1973 was 3862 ha. For the latter period the municipalities have quite naturally not been able to state all of the areas where development is to be expected during this period. This is due to the fact that they were not able to answer the demand of the inquiry to designate specifically on a map the areas where development is intended." (Egnsplanrådet, 1968: 4)

The demarcation of the spatial units on the maps was a kind of test of the existence and implementability of the plans. The sense of realism of the planning officer for good reasons sets limits for this demarcation.

18.4.5 How Were Plans and Reality Related?

The development inquiry was an inquiry into intentions and expectations. These products of the mind can easily have a rather superficial connection with reality. To illustrate this issue the author of the present chapter was informed in 1973 that during the previous years the figures of municipal reports for statistics on new dwellings were likely to be a function of the figures reported in the previous year and the municipal engineers' assessment of the construction activity during the present year. More accurate accounts appeared when the 1970 Danish Municipal Reform resulted in better staffing, and when the Law on Registration of Buildings and Dwellings of 1976 took effect. The hard problem existed, however, before this administrative consolidation.

The demarcation of the spatial units on the maps of the development inquiry served the deeper purpose of relating the plans and expectations to—if not the reality itself—then at least to a rather faithful representation of this reality: the paper map in scale 1:10.000. The repetition of the inquiries, and the method of relating the development figures to specific spatial units on maps, made it possible to test the realism of the plans and estimates. In fact this testing was done, but at a more general level: For example, the municipalities estimated around 25.000 completed dwellings for 1970, while the actual number was close to 19.000 (Egnsplanrådet, 1971: 29). Also, the report reveals that some planning intentions were not completely fulfilled: For example, about 7% of developed area as of 1970 was situated in the rural zone where in principle no construction was to take place (same: 13, derived from Table 2).

The fact that the reports on the 1967 and the 1970 inquiry state deviations between plans and factual figures confirms the validity of the whole effort. The stated deviations prove that the planning and the estimation of future development were not paper exercises, but were anchored in reality, even if the number of members of the planning or speech community was relatively low.

The decisive test is, however: Did all the planning effort have effect with respect to dike the urban sprawl? Did the urban development comply with the fiat zone boundary? Figure 18.4, an aerial photo of the outskirts of Copenhagen provides an answer: The present boundary of urban constructions is fairly clear cut: The fiat boundary became visible in the field as well. The white line separates the urban area (Danish: Byområdet) from the rural zone according to the recent (1997) plan of the region. Within the rural zone, construction, etc., is restricted, cf. Section 18.4.1 of the text. The village of 'Ledøje'with its 18th century radial field structure has thus largely been kept. The structures seen in the right, centre part of the photo, outside the urban area, are neither residential houses nor summer cottages, but allotment gardens (cf. Byplan Guide, 1973, p. 75).—The photo shows that 'fiat' boundaries, cf. Section 18.3.1 of the text, have grown into 'bona fide' boundaries at several locations.

18.5 CONCLUSION

The chapter has drawn on recent research from the fields of GIS, philosophy, linguistics and geography (territoriality). It has demonstrated how these diverse contributions can be integrated into a seemingly consistent comprehension of the definition of spatial, socio-economic units. This was achieved by understanding socio-economic units through an investigation of language use, more specifically the sublanguage of four speech communities and their corresponding spatial units: The Jurisdiction, the Place, the Region, and the District. The proposed conception of Jurisdiction includes ownership. Ownership was perceived as a three-category entity of land, person, and society.

Ontologically, the different units were conceived as 'fiat objects' (Smith, 1995). The ontological status of the fiat objects depends on the representation and coding (in the sense of Burrough and Frank, 1995) of these objects, as the representation and coding is the only tangible access to the objects.

The chapter, furthermore, noted that definitions of socio-economic units have a life cycle that relates to the administrative rules that call the SEU, in fact the District, into being. The life cycle of the rules and definitions puts the life cycle of the single SEU into perspective.

The theoretical discussion has been illustrated and supported by several references to evidence, ranging from UN deliberations to Danish practice. Danish evidence concerning a specific socio-economic, spatial unit, a District, was presented. The presented example illustrates very well the universal role of the cadastral method of combining table and map in an administrative context to depict some socially relevant spatial unit. The ontology of the development unit has been thoroughly discussed. The unit is interesting because of its transitory nature and the informal definition applied. It is posited that the demand to specify the units on the map contributed substantially to the realism of the planning activities.

REFERENCES

Bartelme, N., 1989, *GIS-Technologie, Geoinformationssysteme, Landinformationssysteme und ihre Grundlagen*, (Berlin: Springer-Verlag).

Berger, P.L. and Luckmann, T., 1967, *The Social Construction of Reality—A Treatise in the Sociology of Knowledge*, (New York: Anchor).

Bunge, W., 1962, *Theoretical geography*. Lund, Glerup. Lund Studies in Geography, Series C, No. 1.

Burrough, P.A. and Frank, A.U., 1995, Concepts and paradigms in spatial information: are current geographical information systems truly generic? *International Journal of Geographical Information Systems*, **9** (2), pp. 101–116.

Byplan Guide/Townplanning Guide Denmark, 1973, (Copenhagen: Danish Town Planning Laboratory). ISBN 87–87487–00–4.

Dale, P.F. and McLaughlin, J.D., 1988, *Land Information Management*, (Oxford: Clarendon Press).

Danmarks Statistik, 1968, Folkemængden 27. September 1965 og Danmarks administrative indelling. Statistiske Meddelelser 1968, no. 3, (København: Danmarks Statistik).

Danmarks Statistik, 1975, Danmarks Administrative Indelling. Folke-og Boligtællingen 9. November 1970. Statistisk Tabelværk 1975: II., (København: Danmarks Statistik), 341 p.

David, B., van den Herrewegen, M. and Salgé, F., 1996, Conceptual models for geometry and quality of geographic information. In *Geographic Objects with Indeterminate Boundaries*, GISDATA Series No. 2, edited by Burrough, P.A. and Frank, A.U. (London: Taylor & Francis), pp. 193–206.

Egenhofer, M. and Herring, J., 1991, High-level spatial data structures for GIS. In *Geographical Information Systems: Principles and Applications*, Vol. 1, edited by Maguire, D.J., Rhind, D. and Goodchild, M. (London: Longman), pp. 227–237.

Egnsplanrådet, 1968, Notat om Byggemodning i Hovedstadsregionen 1967–1973 (EPA Jnr D–3–2; NØ/BS 1.6.68). 8 p, 3 bilag.

Egnsplanrådet, 1971, Areal- og byggemodningsundersøgelsen 1970. Egnsplanrådets Planlægningsafdeling, (København). 36 p, 6 bilag.

European Commission, 1994, General report on the activities of the European Union, (Luxembourg).

Ferrari, G., 1996, Boundaries, concepts, language. In *Geographic Objects with Indeterminate Boundaries*, GISDATA Series No. 2, edited by Burrough, P.A. and Frank, A.U. (London: Taylor & Francis), pp. 99–108.

Frandsen, K.-E., 1976, Atlas over Danmarks administrative inddeling efter 1660. Vojens, Danmark. (Vol. II, p. 11–12 mentions Nybølle, Hillerslev parish.).

Frank, A.U., 1996, An object-oriented, formal approach to the design of cadastral systems. In *Proceedings of the 7th International Symposium on Spatial Data Handling, SDH '96, Advances in GIS Research II*, Delft, the Netherlands, 12–16 August, 1996, edited by Kraak, M.-J. and Molenaar, M., (IGU), Vol. 1, pp. 5A.19–5A.35.

Hovedstadsregionens Statistikkontor Beretning, 1996, HSK-tryk, København. ISBN 87–89644–25–5.

Kittredge, R., 1983, Sublanguage. American Journal of Computational Linguistics, **8** (2) (April–June 1982), pp. 79–82.

Laurini, and Thompson, 1992, *Fundamentals of Spatial Information Systems*. Apic Series. (San Diego: Academic Press).

Lynch, K., 1960, *The Image of the City*, (Cambridge: MIT Press).

Malmberg, T., 1980, *Human Territoriality—Survey of Behavioral Territories in Man with Preliminary Analysis and Discussion of Meaning*, (The Hague: Mouton).

Mark, D.M., 1993, Toward a theoretical framework for geographic entity types. In *Spatial Information Theory—A Theoretical Basis for GIS*, Lecture Notes in Computer Science 716, edited by Frank, A.U and Campari, I. (Berlin: Springer-Verlag), pp. 270–283.

Meddelelser, 1968, 1000 m2 by for hver ny bolig. Meddelelser fra Egnsplanrådet No. 3 (1 June 1968). (Jfr. Meddelelser No. 19, p. 2; No. 8).

Raper, J.F., Rhind, W.D. and Sheperd, J., 1992, *Postcodes—The New Geography*, (London: Longman).

Sack, R.D., 1986, *Human Territoriality—Its Theory and History*, Cambridge Studies in Historical Geography, No. 7. (Cambridge: Cambridge University Press).

Searle, J.R., 1995, *The Construction of Social Reality*, (New York: The Free Press).

Skogan, D., 1995, Basic principles of geographic information sharing and standardisation. In *Proceedings of the 5th Scandinavian Research Conference on Geographical Information Systems (ScanGIS '95)*, (Trondheim, University of Trondheim, Dept. of Surveying and Mapping), ISBN 82–993522–0–7, pp. 277–286.

Smith, B., 1995, On drawing lines on a map. In *Spatial Information Theory—A Theoretical Basis for GIS*, Lecture Notes in Computer Science 988, (Berlin: Springer-Verlag), pp. 475–484.

Stubkjær, E., 1992, The development of national, multi-purpose spatial information systems—Danish experiences in a theoretical context. *Computers, Environment and Urban Systems*, **16** (3) (May/June 1992), pp. 209–217.

Stubkjær, E., 1994, Employing the linguistic paradigm for spatial information. In *Proceedings of the 6th International Symposium on Spatial Data Handling*, Edinburgh, UK, 5–9 September, Vol. 1, pp. 572–587.

Tekniska Litteraturselskabet: Universella Decimalklassifikationen (UDK), 1977, (Stockholm).

UN/dpcsd/dsd, 1997, Comprehensive assessment of the Freshwater Resources of the World. Report to Commission on Sustainable Development, Fifth session, 5–25 April, New York. (http://www. un.org/dpcsd/dsd/freshwat.htm).

Wilson, R.J., 1985, *Introduction to Graph Theory*, (London: Longman), ISBN 0–582–44685–6 uib, viii, p. 166.

Worboys, 1995, Personal communication.

CHAPTER NINETEEN

Designing Zoning Systems for the Representation of Socio-Economic Data

Stan Openshaw and Seraphim Alvanides[1]

19.1 INTRODUCTION

From a socio-economic perspective GIS has a major deficiency. It is very poor at handling and displaying data about people rather than the cartographic entities that appear on maps. Additionally, geographers have been slow to appreciate the importance of spatial representation in their attempts to describe and visualise map patterns in what is already badly handled and damaged socio-economic data. Certainly, the map is a wonderful basic data visualisation device. Computer mapping packages and more recently GIS have greatly increased its popularity, but the socio-economic maps we have today are in essence little different from those that could have been drawn by hand, 20 or even 50 years ago. Map design has changed only a little. The map parameters that can be tuned to modify the visualisation are essentially the traditional ones: class intervals, choice of symbolism, nature of representation (viz. choropleth, point, surface) and scale of the display. All of this is now highly automated and computer mapping in a world of GIS is extremely easy. The problem is that there have been very few attempts to exploit the capabilities of the technology as a modern visualisation tool designed for the representation of socio-economic data rather than of the physical environment. The development of a GIS based human cartography has been neglected. One of the principal exceptions is the work by Dorling (1995) and the increasing use of computer map animation as a display and spatial analysis device designed to capture the dynamics of human behaviours (see, for instance, Dorling and Openshaw, 1992; Openshaw *et al.*, 1994).

19.1.1 Some Fundamental Spatial Generalisation Problems

The most fundamental problem with all cartographic based displays of spatial information is the strong (if not complete) dependency of the results on the nature of the spatial data and the inherent coding, spatial distortion, and generalisation filters that are unavoidably involved. These problems are greatest when socio-economic data are being mapped because the underlying objects of interest (people, society) are invisible in GIS terms but the spatial representation that is used to make them visible are essentially static physical map features that are really describing and representing something quite different. Furthermore, there is a major generalisation issue. Dorling (1995) explains part of the problem as follows: "Imagine that in your town or village a symbol is painted on every roof top showing the ages, occupations, wealth and political opinions of the people who lived in each home, and that you were given a detailed aerial photograph of the area. It would not take long to see where the most and least affluent areas are and what the

[1] Contact Author: Department of Geography, University of Newcastle, Newcastle-upon-Tyne NE1 7RU, UK

people living there tended to do and how they vote. If, however, you were interested in the whole country or rather than just part of a town, this method would no longer work. The roof signs would not longer be visible... a more subtle picture needs to be created, a picture which is not necessarily directly related to the physical geography of the country" (p. xiv). The question is really what should this picture show and how should it be designed? Indeed this is the most basic and fundamental of all the questions that could be asked.

A related concern involves questioning the traditional choropleth mapping process. Conventional cartography is quite happy with the gross generalisation of the socio-economic content of a map to five or six categories whilst seeking to portrait the *ad hoc*, essentially arbitrary, line boundaries used to represent the data at an accuracy better than the eye can resolve. Is this right? Surely both forms of data need to be in some kind of geographic scale and generalisation equilibrium, else the most detailed but irrelevant (i.e., arbitrary lines on maps) dominates the most important (i.e., the socio-economic content). Currently the balance is completely wrong when viewed from a socio-economic perspective.

These problems are least serious when there is a close correspondence between the display used to present data and the nature of the data being viewed; for example, surface representations of continuous data; although this still ignores problems such as how to deal with generalisation and noise effects. In general, when continuous data are displayed as a surface, what you see is more or less broadly what exists. Change the resolution and the surface changes in a smooth manner without too many major surprises. The problem is greatest when the data relate to discrete zones as, indeed, most socio-economic data do. What you now get is a conflated mixture of zone design, scale, aggregation, generalisation and data pattern effects. The nature of the zones used to represent the data now strongly influences, sometimes even dominates, the spatial visualisation. What you see now partly depends on the underlying microdata (prior to aggregation to the display's zoning system) and partly on the nature of the output zoning system; as well as cartographic design aspects (class intervals, colour and symbolism). If the data and the cartography are held constant, there is still almost an infinity of alternative maps that could be generated; some, maybe many of which, would probably be declared to be substantially different in the displays and stories they tell or support.

19.1.2 Modifiable Areal Unit Problems

This spatial representation problem is another instance of the modifiable areal unit problem (MAUP); (see Openshaw, 1984). The MAUP has recently been rediscovered as a significant artefact affecting statistical modelling (Fotheringham and Wong, 1991; Fotheringham *et al.*, 1995), although some still dispute whether it exists or can be made to go away (Arbia, 1989). The MAUP is unfortunately endemic to all spatially aggregated data and will affect all analysis and modelling methods that are sensitive to spatial data, only those methods which are very insensitive to spatial patterns (hence the poorest from a geographical point of view) have any chance of being robust! However, the MAUP also affects cartographic displays. The effect of scale (or size of zones) on the nature of the mappings that are produced is well known. Unfortunately, there has been a tendency to view this as either a generalisation issue or as something that does not make much difference. As you increase the scale you lose detail and see less. However, the MAUP is mainly driven by the aggregation process and this factor often tends to be conveniently overlooked because it is far more difficult to handle. Maybe an example will help. Imagine a data set consisting of 1000 fine zones. If you map these data, then there are 999 different changes of scale (and resolution) you could map, provided you assume that there is only one aggregation to each scale. This is the usual assumption, especially when hierarchically organised census or administrative area data are being processed. In fact, if the 1000 zones were UK census enumeration districts (EDs), then you could map them as 1000 EDs, as about 60 wards, and 2 districts. However, if you relax this assumption of their being only one fixed aggregation to each scale, then you can begin to appreciate the enormity of the problem. For instance, there are about 10^{1240} different 10 zone aggregations of the 1000 small zones! Which one do you use? It surely does matter! Currently this very significant source of variation is ignored because it greatly complicates most analysis procedures.

Once it mattered much less because the users had no real choice, and were constrained to use a small number of fixed zone-based aggregations because of the problems of creating spatial data. GIS removes this restriction and as the provision of digital map data improves we now have access to the full range of alternative zoning systems and the same microdata can now be given a very large number of broadly equivalent but different aggregate spatial representations. As a consequence, it is no longer possible to 'trust' any display of zone-based spatially aggregated data. There is nothing objective about a map display of a spatial variable that has been subject to aggregation! The zoning system represents a major uncontrolled source of variation in the cartographic display that can easily interact with other map design parameters; e.g., class intervals, colour and scale, and the patterns and relationships in the data to produce all kinds of potentially weird effects. If any progress is to be made, then it is important that this very significant source of variation is brought under the user's control.

19.1.3 Flexible Geographies and Multiple Representations

Suddenly users have to start seriously worrying about the nature of the spatial representation they use to report, analyse and visualise geographic information. It is argued that this problem is more likely to affect socio-economic information where because of confidentiality constraints and lingering data restrictions, attention is very often limited to the display and analysis of data that have been spatially aggregated one or more times. Openshaw (1996) explains the dilemma as follows: "Unfortunately, allowing users to choose their own zonal representations, a task that GIS trivialises, merely emphasises the importance of the MAUP. The user modifiable areal unit problem (UMAUP) has many more degrees of freedom than the classical MAUP and thus an even greater propensity to generate an even wider range of results than before" (p. 68). The challenge is to discover how to turn this seemingly insoluble problem into a useful tool for GIS applications rather than leave it as a fundamental criticism of socio-economic GIS.

There are four possible solutions: use only individual data, avoid any re-aggregation of already aggregated data, explicitly engineer subsequent re-aggregations, and use frame free-methods or continuous surface representations. The use of frame-free methods is not discussed on the grounds that an essential characteristic of socio-economic data is that they are by their nature not frame-free nor can they be sensibly converted into surfaces. Equally, it is unrealistic to anticipate that much individual socio-economic data will be available because of data protection and confidentiality constraints. The second option is briefly reviewed in Section 19.2, and the third is examined in more detail in Section 19.3. Section 19.4 gives a number of empirical illustrations to demonstrate some of the benefits of the new approach and its implications for GIS are discussed in Section 19.5.

19.2 AVOIDING RE-AGGREGATION

It is fairly obvious that the problem of not knowing which aggregation to use can be avoided by studying data at the finest level of available spatial resolution. Dorling (1992) is one of the first to realise that modern visual display and plotting technology can in fact show unprecedented levels of spatial detail. On a laser printer with 800 dpi resolution you could theoretically show on an A4 page the location of every household address in the UK to an accuracy of about 100m. If you wished and possessed a sufficiently powerful microscope, you might just be able to see individual addresses. Such a map display may even look magnificent as an artistic wonder and would probably make a superb calendar, a kind of social geographers' equivalent to a satellite image of Britain or Europe. There is still a degree of generalisation being provided but this is now mainly of a natural kind; viz. that which occurs when 'you' the observer simply cannot see or handle all the detail (Li and Openshaw, 1992). However, as geographic analysis tools this 'show all' approach has much less value. Indeed the novelty of seeing everything soon wears off! Rather more than this is expected from an information processing technology. Some form of map generalisation is needed to show the data at some higher degree of abstraction. Spatial aggregation serves a useful generalisation and pattern detection purpose, provided it can be controlled in some way and converted into a focused tool rather than left as a stochastic effect or bias.

If zonal data cannot sensibly be displayed as points, then another cartographic possibility is the cartogram. Dorling (1995) has demonstrated how useful this neglected tool can become once a good algorithm is developed for it. A cartogram is a distortion of a zoning system so that equal areas of the map are used to represent equal numbers of people (or some other variable of interest). Instead of seeing map distributions that reflect the physical land area of the data polygons, the size of the polygons is adjusted to reflect the quantity of the variable of interest located there. Dorling's principal contribution was to demonstrate in a very colourful way how cartographic visualisation could be used to support plausible and very interesting stories about socio-economic conditions in the UK. In this way the atlas of maps becomes alive and provides a useful public communication tool.

The basic Dorling algorithm may be defined as follows:

Step 1 obtain some data at some detailed geographic scale;
Step 2 map using a fine set of zones that constitute a population cartogram for some small but arbitrary areal units;
Step 3 applying various colourful symbolism use the map to support a story about the data being displayed.

It is a nice idea to correct the distortion effect that the use of arbitrary polygons of arbitrary area has on cartographic displays of variables for which area size is not a relevant factor. Why should a large but empty area grab your attention compared with much smaller but heavily populated areas that can hardly be seen due to their small physical size? It is surely only a matter of time before all GIS offer cartogram mapping capabilities. However, there are a number of weaknesses that still need to be addressed:

- there needs to be many rather than a single population cartogram;
- there are MAUP effects present in the base data that are used to construct the cartogram;
- there is no attempt to use scale as a generalisation tool; Dorling's cartogram is based on not quite the finest available data (wards) and is dependent on the number and average size of these arbitrary base areas;

- it is still a 'show everything as is' technology but what is shown is a type of gross spatial data distortion of zonal data that have already been damaged by spatial aggregation effects; and
- it is based on the Tufte (1990) assumption that "...the human eye and brain excel in their ability to see patterns, and the more detailed a picture is, the more visible is the pattern" (p. xv), however, this works best when the patterns are simple and the data not too complicated. Reality tends to need a more sophisticated technology.

Dorling (1995) argues that "if some discernible pattern is seen on a ward cartogram, it is very unlikely to have arisen out of chance, and there is almost certainly an explanation for it and a process behind it" (p. xv). However, this is a very simplistic view since the ward level data aggregation may have already greatly damaged the data and generated all kinds of false patterns about which the cartogram can do nothing helpful. There is no argument that cartograms are not potentially useful, merely a complaint that something much more powerful than this is needed to resolve the problems of socio-economic data representation in GIS. The cartogramic distortion of map space is a useful idea but it needs to be combined with some appreciation of the UMAUP. As the spatial resolution of the finest areal units continues to diminish then the need to combine the two continues to increase. Additionally, other forms of spatial data map distortion may be even more useful and are certainly worth investigating.

19.3 RE-AGGREGATION AS A DELIBERATE DESIGN TOOL

The only alternative to 'as is' spatial representation is to develop zone design as a spatial engineering tool to provide a platform for controlled visualisation, visual spatial analysis, and even deliberate spatial distortion to serve a particular purpose. GIS provides the user with the flexibility to design their own zoning systems based on their own re-aggregations of the available spatial data. The underlying hope is that by re-aggregating data in a controlled and purposeful way some of the problems associated with arbitrary aggregation may be reduced and new approaches to visualisation and map analysis identified.

The basic questions are:

1. does the choice of aggregation really matter;
2. if it does, then how to do the re-engineering efficiently;
3. is the choice of zone design criteria critical; and
4. how to use zone design as a geographic data display and analysis tool.

Previous work has demonstrated that the answer to questions (1) and (3) is yes, question (2) has an algorithmic solution, but question (4) is still to be resolved and it is here where new research is urgently needed.

19.3.1 MAUP Effects

It is very easy to demonstrate the effects of using modifiable areal units. The problem arises in two ways:

1. when individual spatial point data relating to non-modifiable entities are aggregated to a zoning system; and
2. when already spatially aggregated data are re-aggregated one or more times.

The former is often beyond the user's control. In both cases aggregation is seen as being beneficial as it reduces the volume of data, protects the confidentiality of personal data and creates geographical patterns. The problems arise because it also changes the data, can alter measurement scales and loses information; it generalises, adds noise, generates false patterns and fundamentally changes the entities (or objects) that are available for subsequent study. In a human geographic context, data about people become data about inanimate spatial objects such as places, zones, etc. With zonal entities there is also a strong implicit assumption that they are comparable entities (i.e., zone 27 can be compared with zone 59 because they are both zones of the same type), but in fact (from a socio-economic perspective based on the nature of the containing areas) they need not be comparable entities at all. For example, census zone 27 could represent a rural village whilst census zone 59 is part of an urban housing estate. This comparison of essentially different and not comparable objects can occur even with the most finely zoned census data (Openshaw, 1996). It is a major problem because in many studies these zones are compared; e.g., ranked; or treated as cases in some analysis. Yet from a socio-economic geographic point of view this basic assumption is usually incorrect.

One solution is to explicitly design zonal objects as meaningful entities related to a particular purpose so that comparison is possible and meaningful. The best example is that of local labour markets (Coombes *et al.*, 1986). Here the re-aggregation is still arbitrary but the resulting zones are declared to have an explicit validity of a substantive kind relevant to one type of study at a particular spatial scale. The problem is deciding what sorts of areal objects might be considered most useful for studying disease or housing characteristics, or even unemployment rates. The difficulty with designing set piece multi-purpose regionalisations is that there are, in theory, an almost infinite number of them. This zone design task is best done by the user to meet a specific purpose although currently there is a regrettable tendency to use the few that exist for virtually any and every data set that is available in an unthinking way (e.g., Champion, 1989). This was more acceptable when zone design was hard and a lengthy process; it is much less relevant when the technology exists to permit as a matter of course almost continuous redefinition that is tuned by the user for a particular application.

19.3.2 Why Might Re-Aggregation Be Useful?

In essence, all spatially aggregated data have been damaged by being spatially aggregated. It is further damaged by the potential that exists for endless re-aggregation. The question is, therefore, under what circumstances can controlled or purposeful spatial aggregation actually help? The answers include:

1. if it can improve consistency of spatial representation by aggregating out the anomalous areas and reducing spatially lumpiness;
2. if it can help display hidden geographic patterns by removing or reducing aggregational distortion whilst amplifying interesting patterns;
3. if it can improve the quality of the data by removing small number and unreliable data effects;
4. if it can be used as a visualisation tool able to graphically represent the data in particular purposeful and new ways;
5. if it can help simplify subsequent analysis and modelling tasks by removing those aspects of spatial data that cause the greatest statistical problems; and
6. if it can help as a visual spatial analysis tool designed to bring out some of the hidden patterns or less obvious structure.

19.3.3 Zone Design as a Visual Geographic Analysis Tool

It is argued elsewhere that far from being an insuperable problem the modifiable nature of zonal data offers the geographer an immensely flexible and powerful visualisation and analysis tool (Openshaw, 1996). The zoning system can be viewed as a pattern detector that can be visualised (because it can be mapped) and which provides a visual representation of the interaction between the spatial data being re-aggregated and the function (and constraints) being optimised by the aggregation process. The hope was expressed that viewing the zoning system created by the aggregation process might itself have some interest as a visual geographical analysis tool (Openshaw, 1984). However, this needed the development of GIS to reach a minimal level of maturity and the availability of good digital boundaries of zonal data sets before these dreams can become a practical reality. It also needs both a major shift in attitudes (Openshaw, 1996) and the development of practicable zone design algorithms that can cope with the computational complexity of the resulting zone design problems (Openshaw and Schmidt, 1996; Openshaw and Alvanides, 1999).

Some progress has been made. An Arc/Info based Zone DEsign System (ZDES) exists (Openshaw and Rao, 1995; Alvanides, 2000) that seeks to routinise zone design and offers a number of generic zone design functions. The remainder of this paper presents some empirical examples of ZDES being used on 1991 census data for enumeration districts for parts of the UK. ZDES is designed as a portable add-on to Arc/Info and a beta test version (ZDES, 1997) can be obtained from the Centre for Computational Geography, School of Geography, University of Leeds (see WWW site in references).

19.4 EMPIRICAL EXAMPLES

It is useful to consider a few illustrations of some of the issues that are being discussed here. Consider 1991 census data for the Leeds–Bradford area of the UK. The finest resolution census data are those provided for 2,315 small areas known as Census Enumeration Districts (EDs). A typical ED would contain about 200 households. Consider a simple variable such as percentage unemployment (viz. percentage seeking work divided by economically active population). This variable is mapped using Arc/Info at the ED level (Figure 19.1). The same data are available for 63 wards (EDs nest into wards). Wards are the smallest official legally recognised spatial units in the UK. The result is shown in Figure 19.2. A much simplified picture is obtained with at least some of the high unemployment areas being 'smoothed' out and thus removed by the aggregation process whilst other areas have seemingly been made more unemployed! This is due to the interaction between the class intervals and the aggregation process. Aggregation is essentially an averaging down process but the effects are not spatially consistent. Figure 19.3 shows the results of subtracting Figure 19.1 from Figure 19.2 and gives an illustration of the differences caused by the aggregation of 2315 zones into 63 regions. However, this is an illustration of scale effects as only one aggregation is considered.

There is another problem with spatial data that is often overlooked. The precision of the data is not constant. Whilst the accuracy of the census is more or less uniform, the level of precision in the cartographic displays is variable and reflects the size of the zone denominators. A 15% unemployment rate for an area with 10,000 economically active persons is much more precise than a 15% rate for an area with a denominator of only 100 persons. Likewise there are other small number effects that may matter; for instance, it is much easier to obtain an extreme result in an area with a small denominator than a large one. Bayesian mapping methods attempt to handle this problem but often generate other difficulties; for example, the proper specification of the prior distribution. A much simpler geographical solution is to aggregate the 2,315 EDs to create 63 zones of approximately equal economically active population size.

Figure 19.4 shows a much-improved representation of unemployment patterns to the extent that the effects of zone size and varying data precision have been removed. There is still, however, the assumption that 63 ward-like areas is a sensible scale for studying unemployment in this region. Some areas have lost unemployment, whilst one or two new black spots have appeared that previously had been averaged out in the ward data. Additionally the data can now be displayed as counts, since the base areas are of almost identical size (about 8,700 economically active persons). Consequently, the <5.1% category roughly translates to less than 451 unemployed, the 5.1–10.0% to 451–900, the 10.1–15% to 901–1351, and the over 15% as above 1351 unemployed persons. This could also be shown as a density surface.

Other zone design functions that may be of interest are cartograms and equal physical size zonations. The cartogram aggregation function seeks an aggregation of the data such that physical size of the zones matches the size of economically active population. Unemployment rates are then displayed against this base. As Figure 19.5 shows the result is in this instance not helpful. Maybe more zones might be better. Similarly, there is no reason why the 63 output zones cannot be re-engineered to have a similar physical size without regard to the population living there (for example, as equivalent to grid-squares). As Figure 19.6 shows these results are interesting but not again particularly useful in this instance, as they do not tell a story much different from that provided by the standard map. However, they do illustrate different zone design functions which when applied to different data, or the same data at different scales, could well have a more dramatic effect. They would, for example, offer a potentially very useful means of visualising change over time. If a historical data series were available for 10 or so time periods, then the changing zone shapes would provide a measure of the interaction between data, scale, aggregation and the zone design function being applied. This aspect is currently being investigated using 100 years of census data.

Another experiment with the Leeds–Bradford unemployment data concerns what happens if an attempt is made to split up the study region into zones of maximum accessibility to unemployed people. This is a kind of location–allocation problem except that here the aggregation process is being used to optimise both the placement and membership of these maximal accessibility regions. The hope is that viewing the results will provide some further insights into the patterns of unemployment. Small regions indicate a dense, closely-knit cluster of unemployed; whilst larger ones indicate a more spread out pattern; see Figure 19.7. Different exponents on the distance function might further help clarify the patterns and this too is currently being investigated.

A final illustration involves the use of a constrained zone design process. So far, all the zoning systems have been unconstrained, other than the constraints inherent in the zone design process (viz. coverage and internal contiguity). This can produce extremely unusual zones. There are arguments that this does not matter, either because the shape of the zones is informative about the interactions between data and zone design functions, or because the aesthetics of zones is a totally undeveloped field, or because there is already a *de facto* practice of using peculiarly shaped zones; for example, in electoral re-districting (Monmonier, 1995). Nevertheless, it is often used to seeking zoning systems that optimise a function subject to additional equality and, or, inequality, constraints on the zonal data being produced. This type of application will become much more prevalent as attempts are made to use zone design as a spatial analysis and modelling tool.

Consider the task of detecting clustering in disease data for a particular cancer in Sheffield. The available data can be aggregated to 1,057 census EDs and mapped. Figure 19.8 shows these age–sex adjusted O/E rates; values of 100 are average. Note the large apparent 'cluster' of high values in the top left of the map. Spatial epidemiology is extremely difficult; there are extreme small number problems, considerable noise due to latency, and an almost complete lack of prior knowledge about where to look or what to look for. One possible exploratory approach is to seek a zoning system that maximises the variance of (O/E) subject to having zones with similar numbers of expected cases. This standardises for geographical unevenness in populations at risk, removes small number effects, and permits maximum flexibility for multiple high and low cancer regions to emerge. Figure 19.9 provides some results for a 40 zone aggregation. Note how quite different patterns are produced. Seemingly this method has some potential as geoGRAPHICAL tool.

19.5 CONCLUSIONS

The widespread adoption of fixed sets of boundaries used in all kinds of spatial modelling, analysis and representation of data can be considered a form of 'mapism'; Monmonier (1995) explains this phenomenon as follows "to describe the ill-founded but unshakeable belief that a specific world-map projection is vastly superior to all others" (p. 3). The equivalent of a map projection in socio-economic spatial representation could be a particular zonal aggregation, but the distortion of reality created by the MAUP in the latter is not as yet known. Yet, geographers experience 'zonism' every time they have to analyse data provided for different zoning systems. Apparently, the task of converting and merging different datasets is left to the individuals; but changing zoning systems is far harder than changing map projections because there is no transformation formula and the distorted patterns are almost impossible to discover by mere observation of the shape of the zones. Additionally, there is no real knowledge of what the 'true' result is; only a glowing awareness of how easy it is to lie with maps of aggregate data.

What we argue for here is a new way of approaching the aggregating, modelling, analysing and displaying of spatial socio-economic data. This is important because most of the cartographic, mapping and GIS advances of spatial representation during the last two decades have neglected the special needs imposed by this type of data. It is apparent from this study that the range of different possible representations of the same dataset is only limited by the imagination of the user and, or, the power of their zone design algorithm. Attention is drawn here to the possible misuse of zone design as a geographical representation tool, an aspect that can be used by the naive user to further discredit spatial data. Analysing data for virtually any zoning system without any concern about the nature and limitations of the units being studied, inevitably leads to zoning anarchy (Openshaw and Rao, 1995). The way forward is to identify suitable zone-design functions and constraints, thought to be best for particular purposes and then to be able to compare and evaluate alternative zonations of the same dataset. It is left to the user to justify his or her choice and to deconstruct his or her underlying motives. However, despite problems it is quite clear that the era of flexible user-defined geographies is here and it is unlikely to go away. Indeed zone design in a GIS environment should now be routinely and easily applied in all applications where it is relevant.

The chapter has concentrated on the design of globally consistent zoning systems at one point in time. It should be noted that a related problem concerns how to design zoning systems that are consistent and comparable for different time periods and, or, as aids for studying and visualising spatial change in a variable of interest over multiple time periods. The problem at present is the lack of suitable spatial data series; viz. a time series of the same data for the same spatial reporting areas. This data problem will disappear with further developments in geoinformation, and when it does there are two alternatives: (1) apply the same zone design functions to each time period and examine how the zoning system boundaries change; and (2) apply zone design to data from all time periods simultaneously to yield a historic common set of zones. Both are possible.

A final aspect is that zone design is also a potentially important spatial data management tool. As the resolution of socio-economic databases continues to improve so it becomes important to consider the design of the very first spatial aggregations. Currently these are arbitrary and clumsily performed in safe settings on secure machines locked away in the offices of national statistical agencies. Typically they use *ad hoc* rules, often of considerable antiquity, that are designed to preserve data confidentiality, but without any clear notion of what that means in a spatial context. It is another type of zone design problem, one with confidentiality risk constraints on zone configurations combined with a desire to minimise the loss or damage done to the information being provided.

Zone design is essentially a problem in socio-economic GIS. Previously it has been largely ignored. It is a critical and most significant problem and technology designed to aid its resolution needs to be as widely diffused as possible, emphasising the positive rather than the negative aspects. The challenge now is threefold:

1. to raise awareness of what is now possible;
2. to demonstrate utility in the widest possible set of applications; and
3. to improve access to zone design technologies in the broadest range of GIS.

Hopefully this chapter will contribute to this ongoing process.

REFERENCES

Alvanides, S., 2000, Zone Design System Methods for Application in Human Geography. Unpublished Ph.D thesis, School of Geography, University of Leeds.

Arbia, G., 1989, *Spatial Data Configuration in Statistical Analysis of Regional Economic and Related Problems*, (Dordrecht, Netherlands: Kluwer).

Champion, A.G., 1989, *Counterurbanisation: the Changing Pace and Nature of Population Decentralisation*, (London: Arnold).

Coombes, M.G., Green, A.E. and Openshaw, S., 1986, An efficient algorithm to generate official statistical reporting areas: the case of the 1984 Travel-to-Work Areas revision in Britain. *Journal Operational Research Society*, **10**, pp. 943–953.

Dorling, D., 1992, Visualising people in space and time. *Environment and Planning B*, **19**, pp. 613–637.

Dorling, D., 1995, *A New Social Atlas of Britain*, (Chichester: Wiley).

Dorling, D. and Openshaw, S., 1992, Using computer animation to visualise space–time patterns. *Environment and Planning B*, **19**, pp. 639–650.

Fotheringham, A.S. and Wong, D.W.S., 1991, The modifiable areal unit problem in multivariate statistical analysis. *Environment and Planning A*, **23**, pp. 1025–1044.

Fotheringham, A.S., Densham, P.J. and Curtis, A., 1995, The zone definition problem in location–allocation modelling. *Geographical Analysis*, **27**, pp. 60–77.

Li, Z. and Openshaw, S., 1992, Algorithms for automated line generalisation based on a natural principle of objective generalisation. *International Journal of Geographical Information Systems*, **6**, pp. 373–389.

Monmonier, M., 1995, *Drawing the Line: Tales of Maps and Cartocontroversy*, (New York: Henry Holt).

Openshaw, S., 1984, *The modifiable areal unit problem*. Concepts and Techniques in Modern Geography, Vol. 38, (Norwich, UK: Geo Abstracts).

Openshaw, S., 1996, Developing GIS relevant zone based spatial analysis methods. In *Spatial Analysis: Modelling in a GIS Environment*, edited by Longley, P. and Batty, M. (Cambridge: GeoInformation International).

Openshaw, S. and Alvanides, S., 1999, Applying geocomputation to the analysis of spatial distributions. In *Geographical Information Systems: Principles and Technical Issues*, edited by Longley, P.A., Goodchild, M.F., Maguire, D.J. and Rhind, D.W. (Chichester: Wiley).

Openshaw, S. and Rao, L., 1995, Algorithms for re-engineering 1991 census geography. *Environment and Planning A*, **27**, pp. 425–446.

Openshaw, S. and Schmidt, J., 1996, Parallel simulated annealing and genetic algorithms for re-engineering zoning systems. *Geographical Systems*, **3**, pp. 201–220.

Openshaw, S., Waugh, D. and Cross, A., 1994, Some ideas about the use of map animation as a spatial analysis tool. In *Visualisation in GIS*, edited by Hearnshaw, H.M. and Unwin, D. (Wiley and Sons).

Tufte, E.R., 1990, *Envisioning Information*, (Cheshire, Connecticut: Graphics Press).

ZDES, 1997, htttp://www.geog.leeds.ac.uk/research/ccg/zdes3.html for further details.

Fig. 1 Unemployment by 2,315 EDs

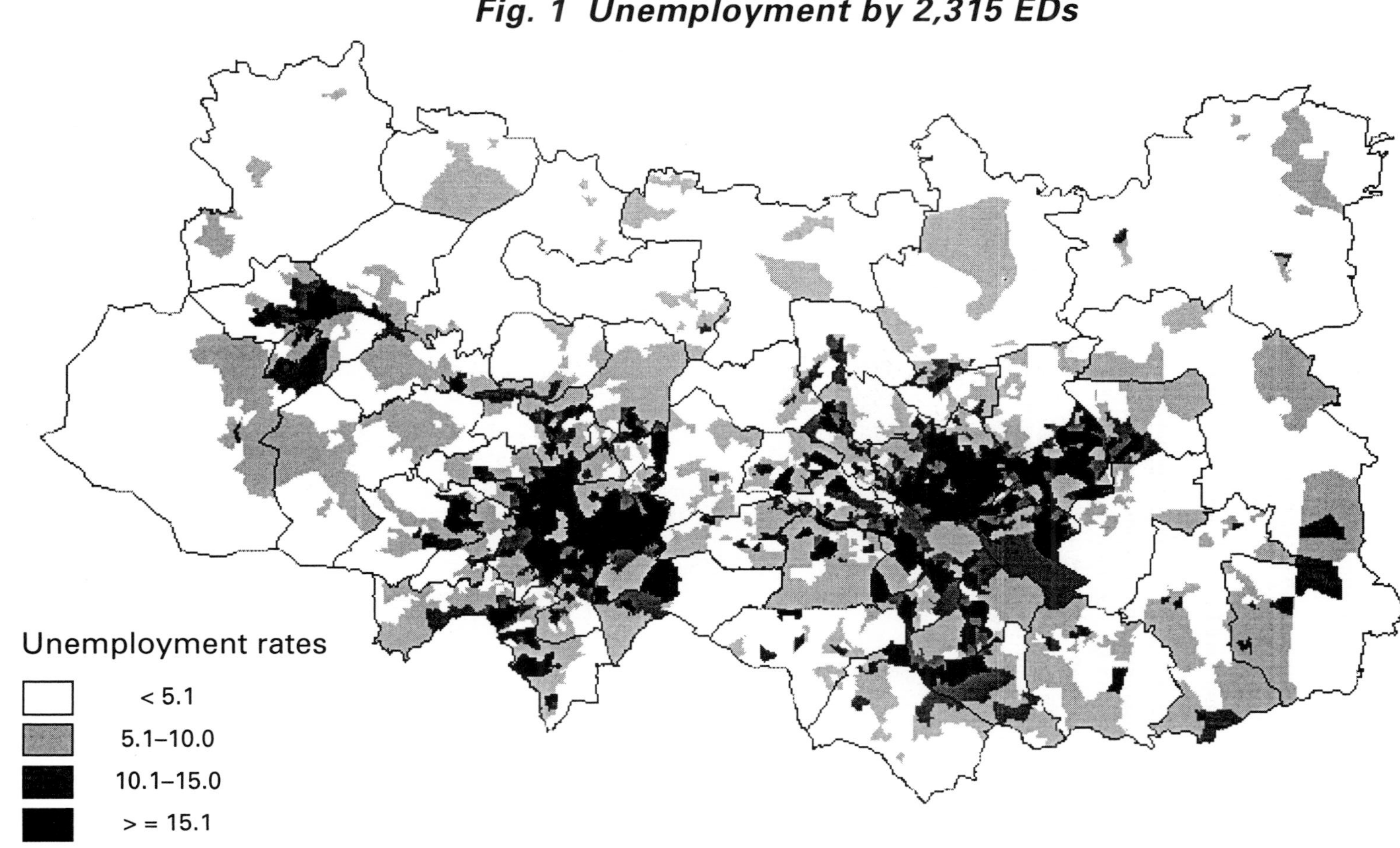

Fig. 2 Unemployment by 63 Wards

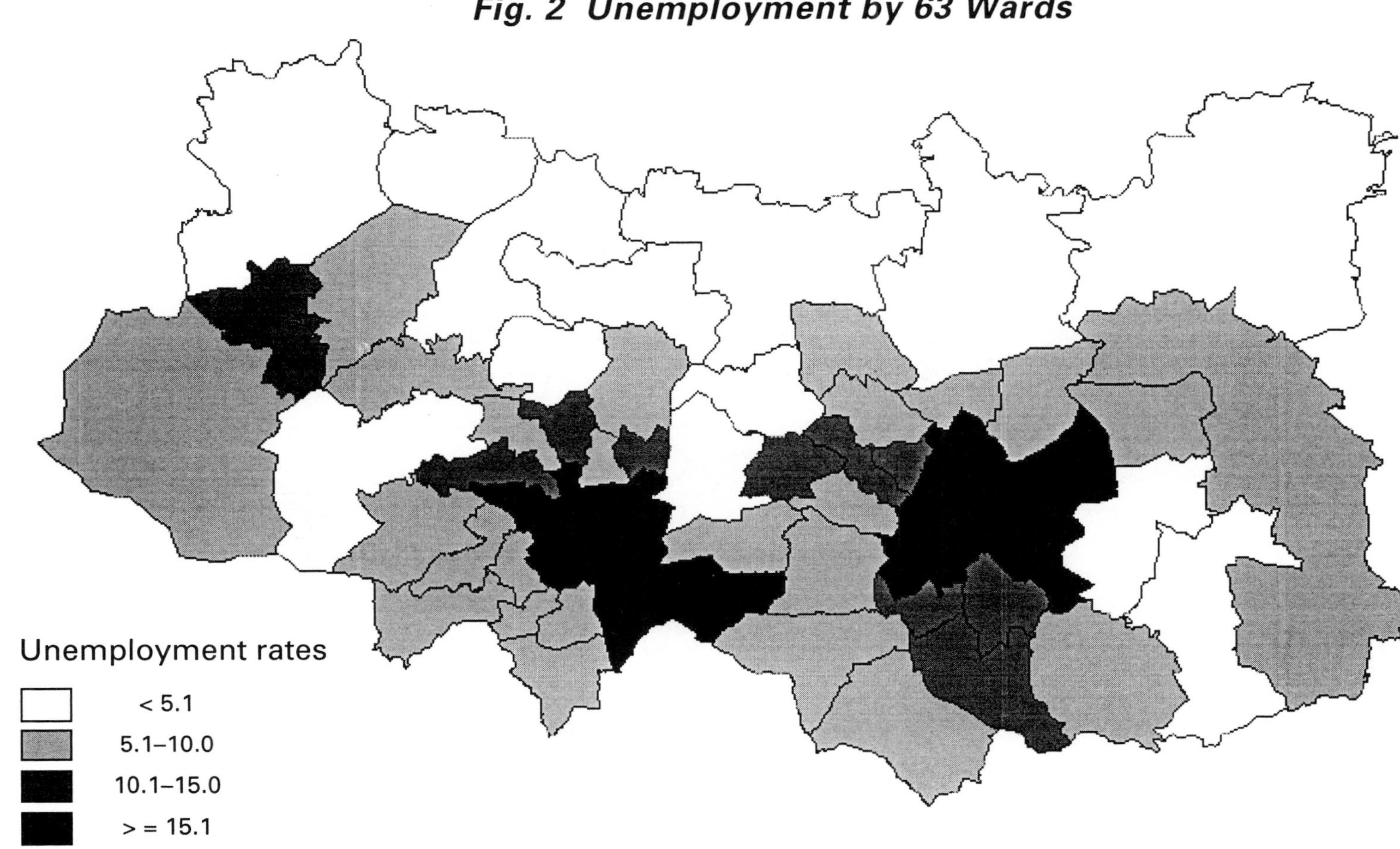

Fig. 3 Unemployment fallacies for Wards

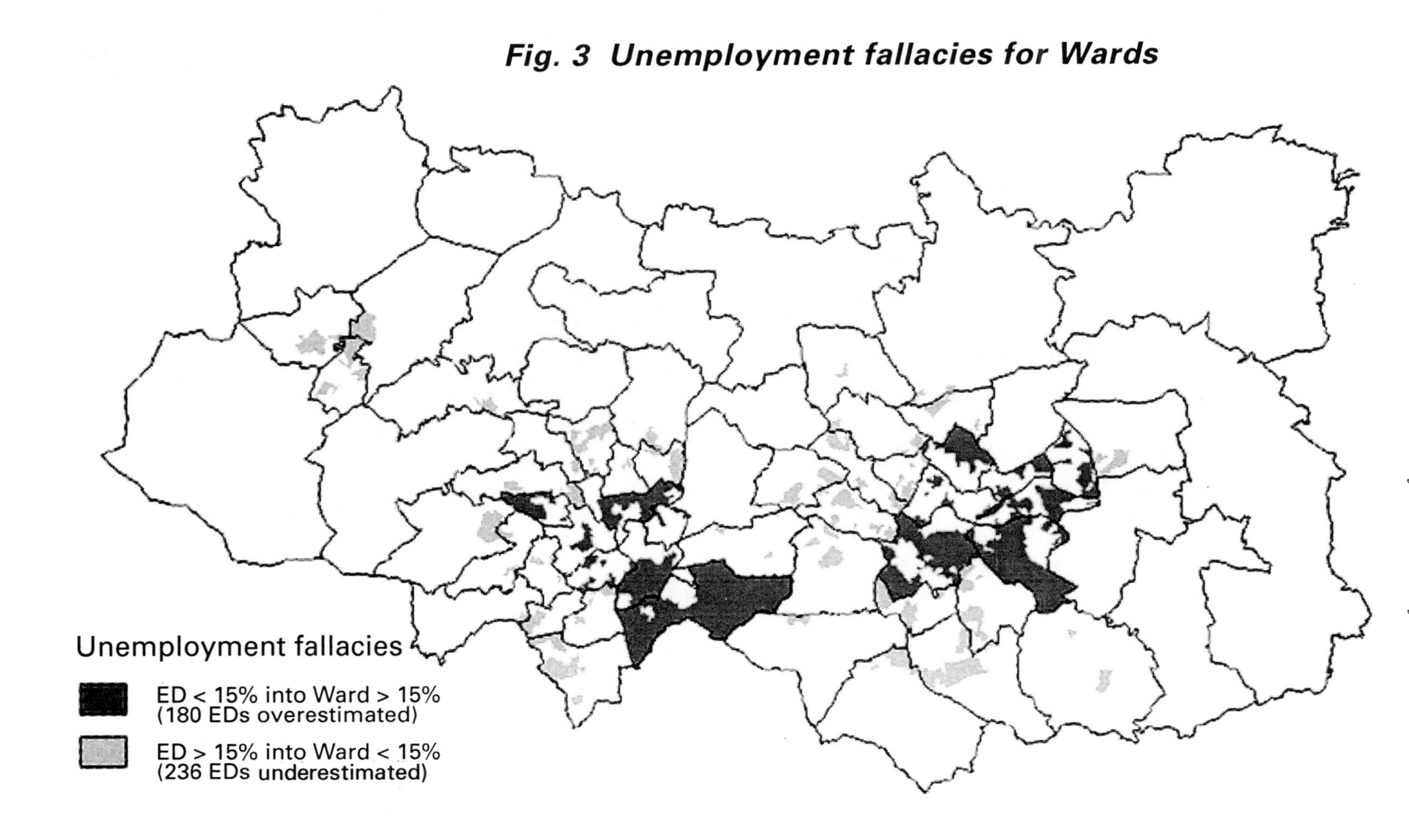

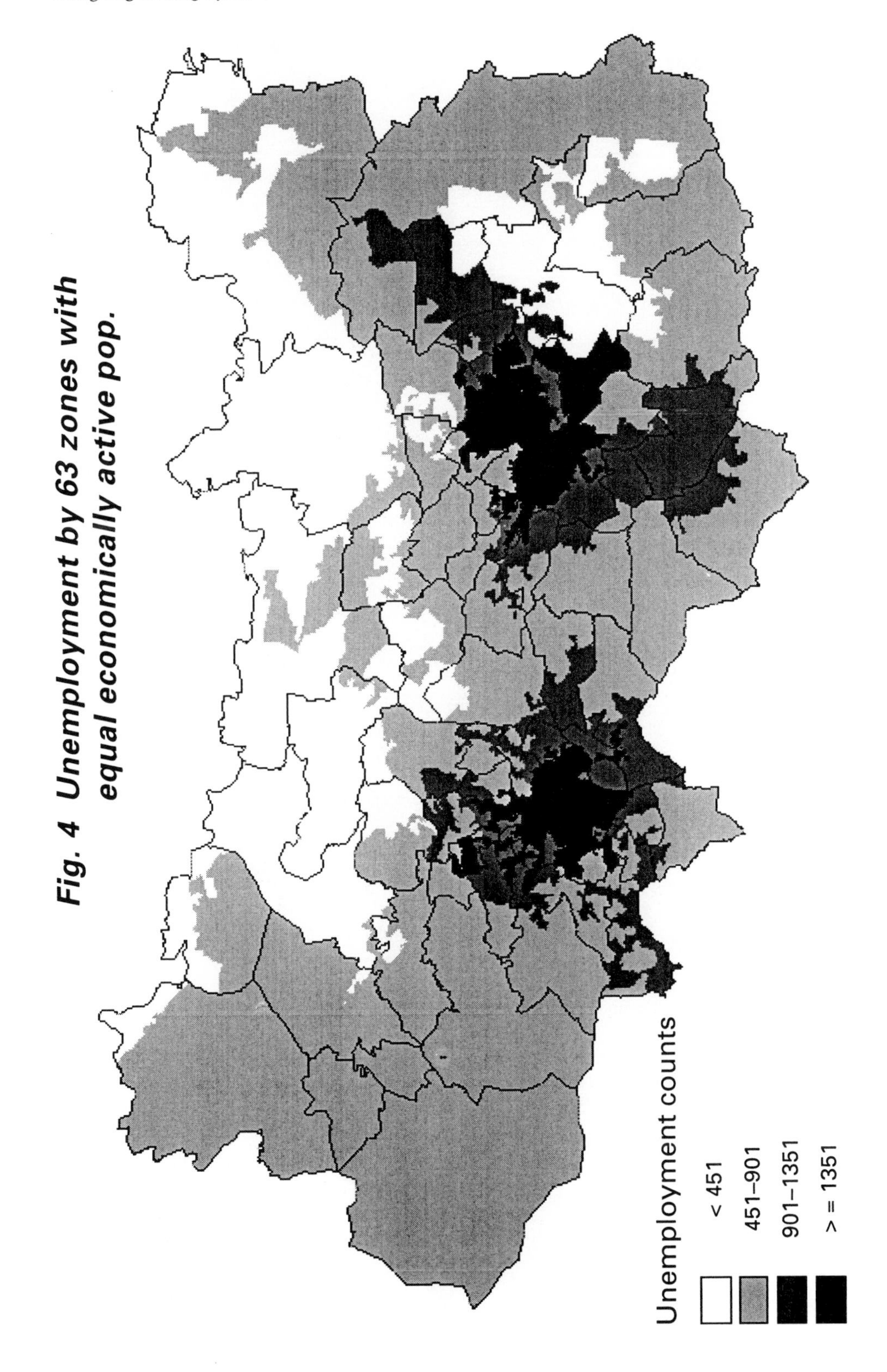

Fig. 4 Unemployment by 63 zones with equal economically active pop.

Fig. 5 Unemployment by a 63-zone cartogram of economically active pop.

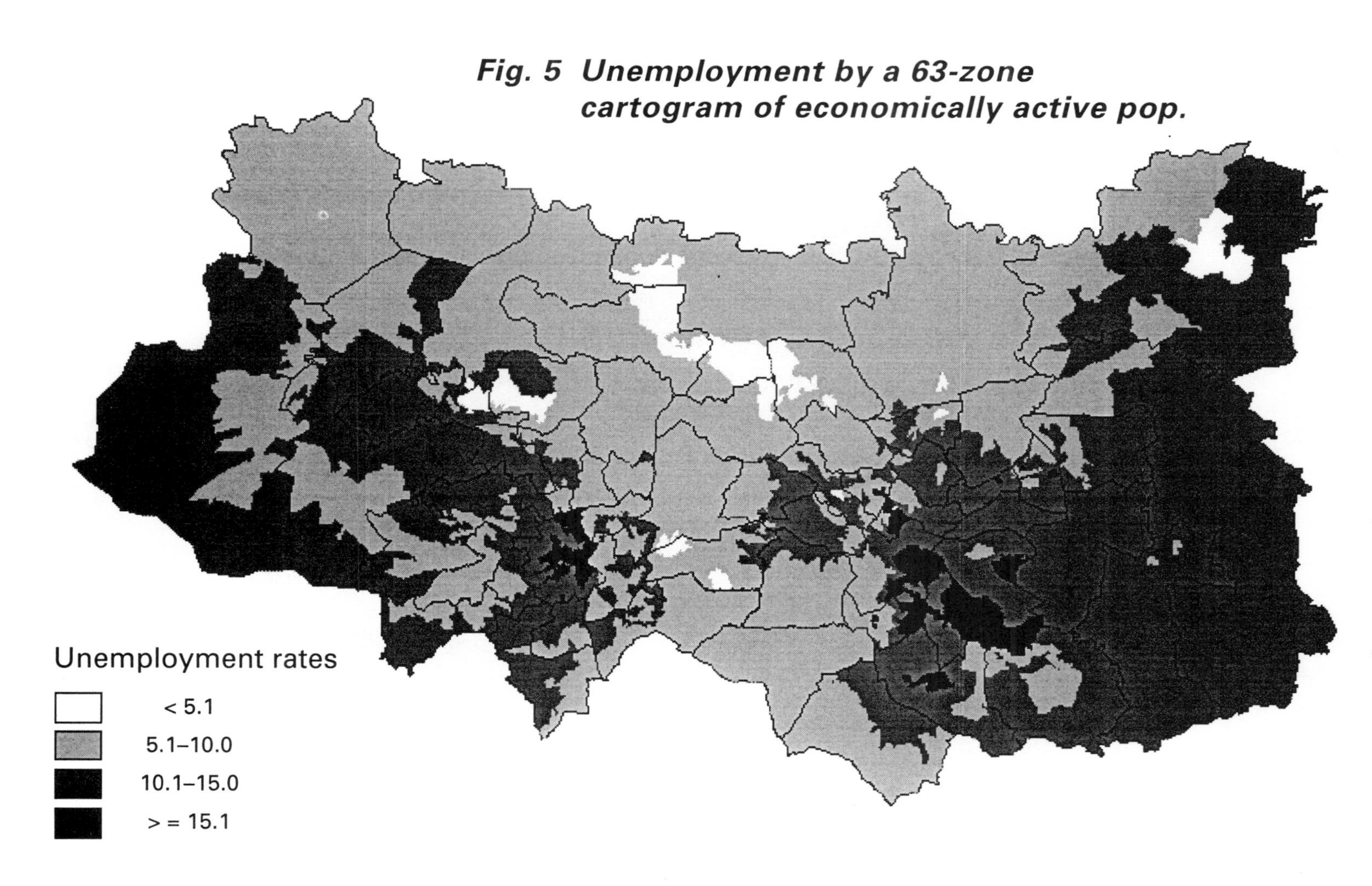

Fig. 6 Unemployment by 63 zones of similar physical size (area)

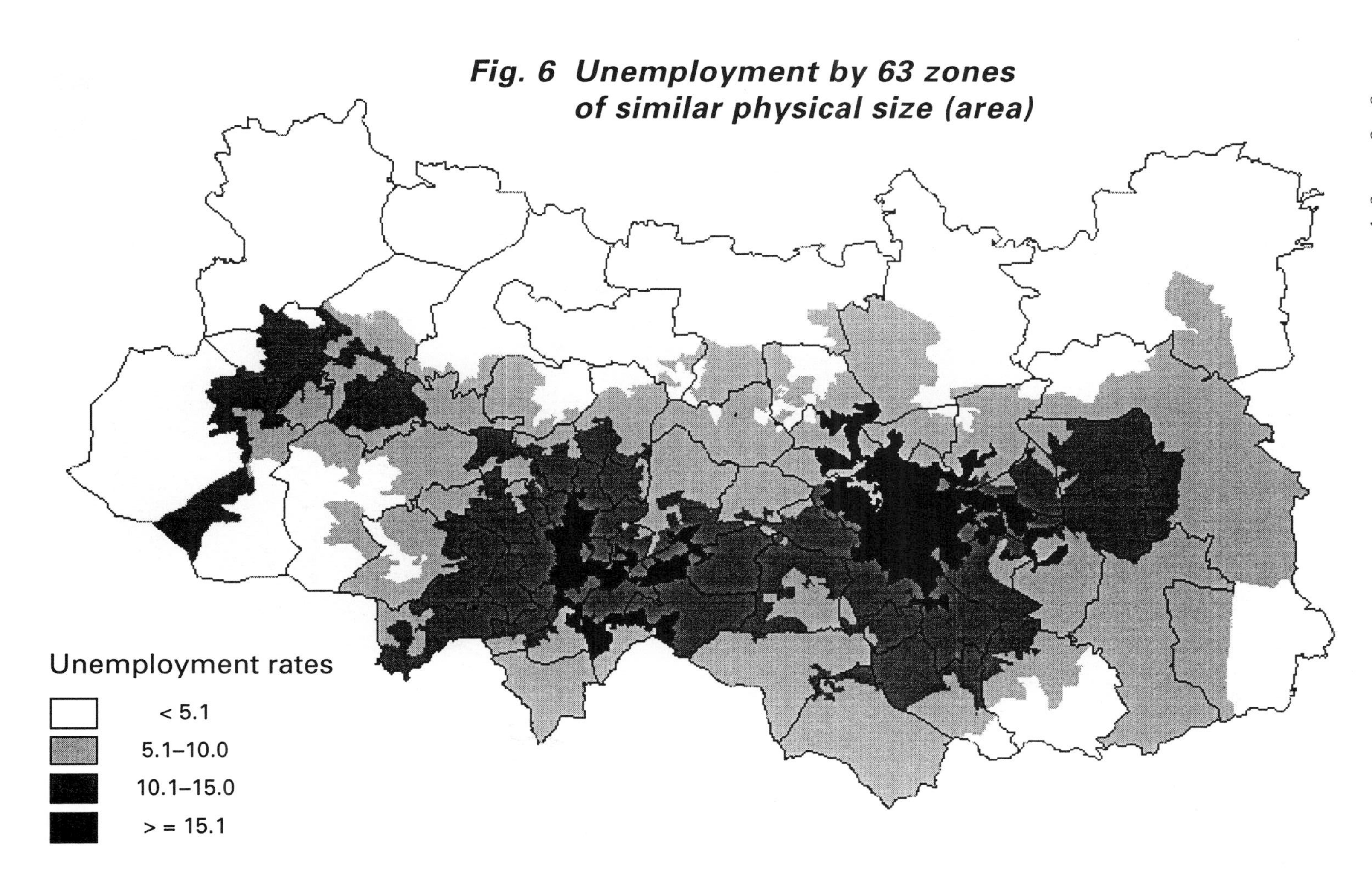

Fig. 7 Unemployment by 63 zones with accessibility on economically active pop.

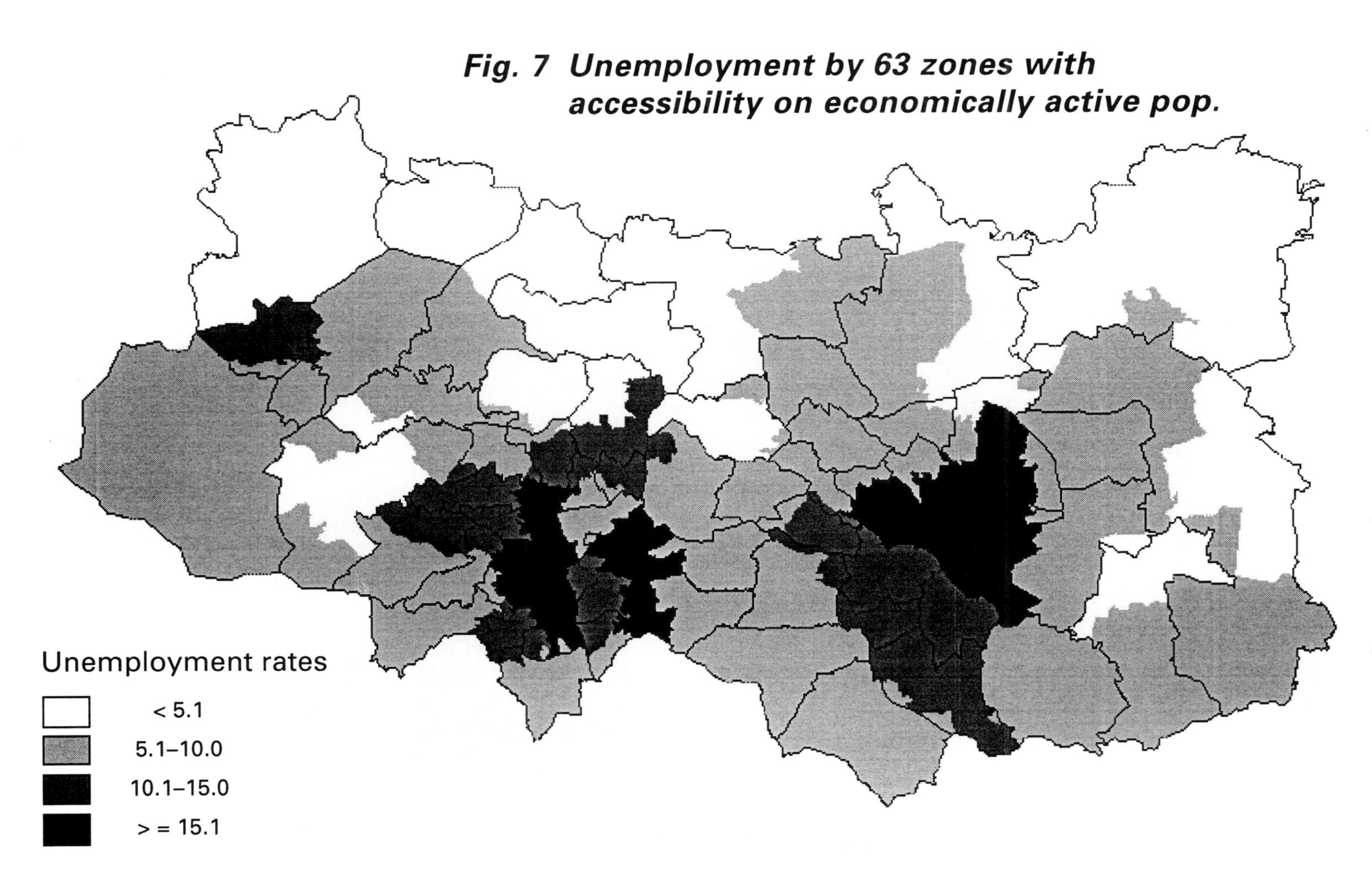

Fig. 8 Adjusted cancer rates by 1,057 EDs

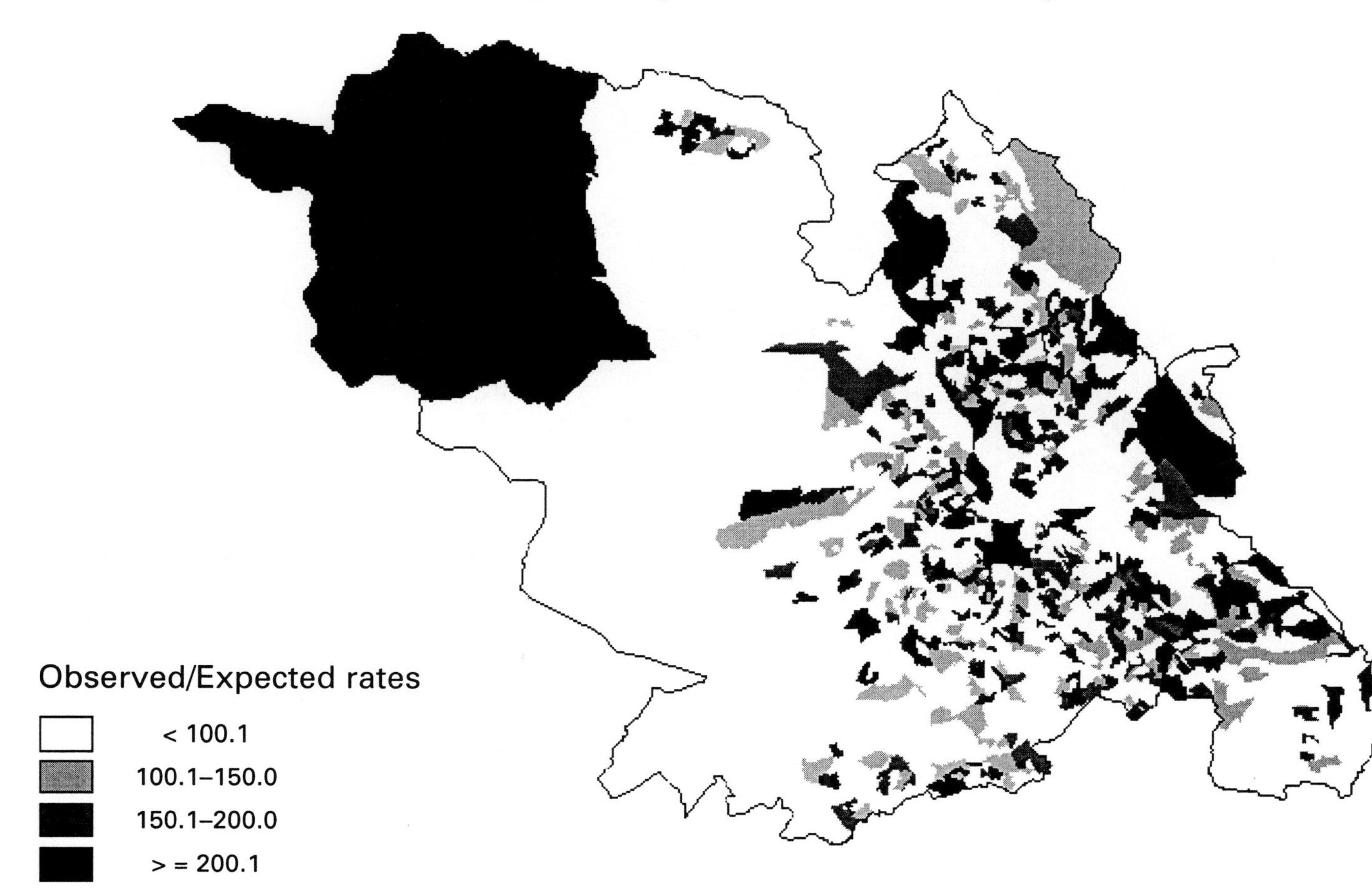

Fig. 9 Rates by 40 cluster detecting zones

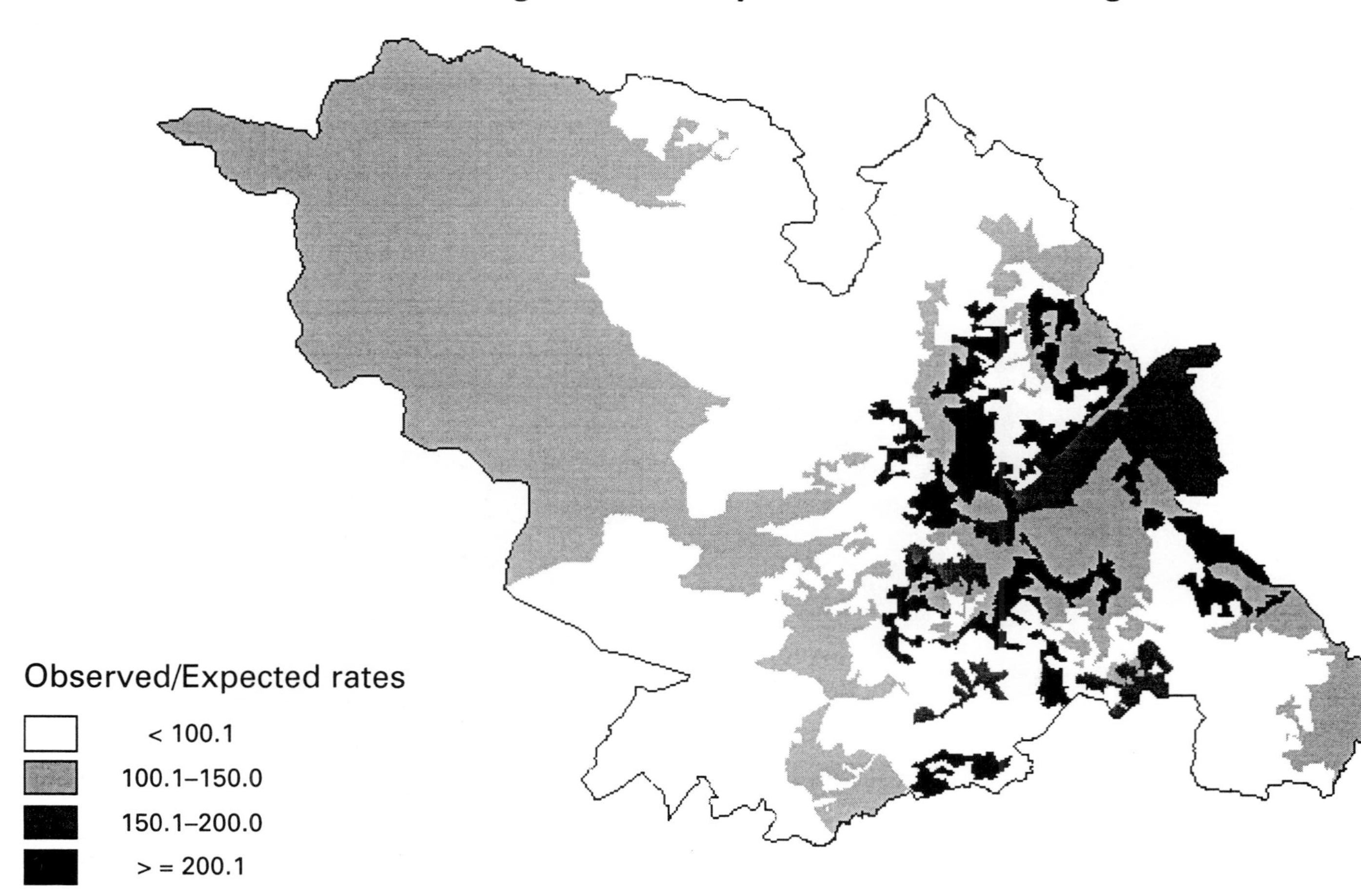

CHAPTER TWENTY

Contrasting Approaches to Identifying 'Localities' for Research and Public Administration

Mike Coombes and Stan Openshaw

20.1 INTRODUCTION

All boundaries drawn on maps are artificial. The surface of the Earth when viewed from space has few boundaries other than those between land and water, and perhaps the starker divisions within vegetation and climate colourings. Yet the human world on the surface of the Earth is ridden with boundary lines drawn or drawable on maps. They are artificial and both culture- and society-dependent. Even those boundaries which do follow coastlines or mountain ridges are the product of people's choices: for example, the way the boundary of the United Kingdom cuts through Ireland shows that physical geography does not simply determine how boundaries are drawn. Most boundaries are invisible to a person crossing them, yet boundaries can mean so much to people that they are fought over at scales varying from the nation state to the neighbourhoods of urban gangs.

This chapter is concerned with those man-made boundaries which are socio-economic rather than political. The latter type reflects the political organisation of the world as the surface of the earth is organised into countries and other units. The former type, which is the principal subject here, not only tends to be far less emotive but also to be rather more changeable. Many boundaries of socio-economic units (SEUs) have been explicitly **defined** by individuals or organisations so that their particular view of the organisation of space becomes visible on a map. They have crisp, rather than fuzzy, boundaries. As a result, changes to these boundaries are readily identifiable and tend to occur intermittently—as with, for example, changes in the areas administered by local government. These changes can then be classified (as set out in the chapter by Frank) to indicate either life or motion of the SEUs concerned.

By way of contrast, the term socio-economic unit can also be applied to a different type of spatial object which has a much more indeterminate boundary (Burrough, 1996); for example, a town's hinterland, or the area where a language is spoken. In most cases, these SEUs with fuzzy boundaries will have come into existence more or less unbidden. Their existence may be essentially neither more nor less than the clustering of spatial behaviour which has meaning but which is not very readily identified (Miller, 1991). These fuzzy SEUs are thus **organic** rather than externally defined: they are also prone to changing in a near continuous process of evolution. This level of volatility, combined with the inherent fuzziness of organic SEUs' boundaries, makes it much less clear whether it is possible to identify an individual boundary change and to classify it as an example of life or of motion. Even so, these organic SEUs are extremely important because they usually are the basis upon which crisp SEUs are defined for spatial management purposes. It is also becoming increasingly possible to visualise these previously unseen patterns in datasets covering these patterns of behaviour.

This chapter is concerned with the representation of these organic SEUs in a crisply defined way; a process of simplification which helps them to be visualised, and which prevents their complex overlapping structure and fuzziness overwhelming the user with too much complexity. The objective is for the fuzziness to be handled in a way which recognises the organic SEUs' fuzzy identities, and then to clarify these to provide a crisp representation based on the best available evidence at a particular time. A familiar example is provided by many countries' periodic implementation of the UN guidelines for the definition of urban areas on the basis of their physically built-up limits at a single point in time within the process of their constant evolution. In this chapter the focus is on settlement patterns and, in particular, on a type of fuzzy SEUs which is of increasing interest to social science viz. 'localities' (Coombes *et al.*, 1988). The crucial challenge is to deal with the way in which the organic SEUs are changing—in practice or potentially—more or less continuously whilst (as stressed in the chapter by Salvemini) the crisp SEUs' updating will always lag behind.

The most familiar approach to this problem is to define the crisp boundary on the basis of the most recent available evidence that is taken to characterise the organic SEUs of interest. For example, commercial organisations interested in the area served by a local newspaper or radio station will want to use the most up-to-date available map of these areal boundaries. As a result, such maps tend to be updated fairly frequently if it is thought that the behaviour concerned—in this case, the take-up of the newspaper or radio station by the people of different areas—is changing noticeably. It is notable that because the emphasis is on the most up-to-date information, there is not often much direct interest in the **change** between the two sets of SEUs of this kind. Most users will only pay attention to differences between the two maps as a way of making a mental note in order to replace those parts of the old map that they had memorised with the equivalent new boundaries. In short, the priority in these cases is on having the most up-to-date information on SEU boundaries so that changes to the SEUs are not of interest in their own right but instead are sought as a means of reducing the out-dazedness of the boundaries.

A much more ambitious approach to representing constantly changing fuzzy SEUs is to shift the emphasis to those structural features of the organic SEUs that in fact show considerable inertia. The turbulence of the shifting boundaries of local newspapers' service areas, in the earlier example, may well be obscuring the fact that most newspapers have been in existence for long periods during which their primary service areas may have altered very little. Thus the aim becomes to analyse not only the most up-to-date information on behaviour but also that of previous times, so as to emphasise common features throughout the time period covered. An example might be an analysis of economic data for several years which identifies consistently distinctive regions such as a group of textile towns or a mining district—even though in the later years such industries may no longer exist in some parts of their former regions of distinctiveness. This identification problem becomes even more important as developments in information technology fuel processes that are changing the entire basis of economic activity. For example, the emerging Internet-based economy has the capacity to totally ignore existing political and economic structures. Yet the evidence to date is that its actual development is preserving and reflecting geographic structures based on economic systems of the past, and is not extensively replacing physical with virtual journey to work activity.

The contrast then is between, on the one hand, changing SEU definitions in pursuit of up-to-date boundaries and, on the other hand, focusing the definitions on the underlying stability that lies behind the changes around the edges. This chapter takes these two approaches and illustrates each in turn with examples of analysis, which have been undertaken on British data. For the emphasis on updated boundaries, the example used is the set of Travel-to-Work Areas (TTWAs) which seeks to represent local labour market areas by computer analysis of the latest available data on commuting flows. The example of the contrasting approach is a set of Localities defined by analyses of various datasets (including the information on commuting) which covered earlier as well as more recent years. This approach depends on an innovative step of collating numerous prior boundary datasets and then using this multi-sourced and multi-time data in a spatial classification process.

A final section of the chapter summarises the differing advantages to these two approaches. In particular, an empirical element is introduced to illustrate the effectiveness of the TTWAs and the Localities at representing on-going changes in the types of underlying organic SEUs that both these sets of boundaries seek to portray. The example chosen is urbanisation, a clear and familiar phenomenon, which has the fuzzy characteristics of being difficult to demarcate precisely and of being in a continual process of change. A new measure of urbanisation is presented, and its results compared for two years (1981 and 1991). This comparison is undertaken using both the TTWAs and the Localities definitions to illustrate the impact of the choice of one approach versus the other. The conclusions are set within a brief wider discussion of the issues raised here that are of more general interest and applicability.

20.2 TRAVEL-TO-WORK AREAS AS UPDATED SEU BOUNDARIES

The British government has a need for a set of accurate and consistent definitions of labour market areas for statistical and policy purposes. For several decades this need has been met by defining Travel-To-Work Areas (TTWAs) based on the most up-to-date data available (Smart, 1974). For the last two decades the definition of TTWAs has attracted widespread interest, due to the importance of their boundaries in determining which areas gained access to certain policies which brought financial assistance. The need for objectively-defined boundaries culminated in the 1981-based definitions becoming unique among official boundaries in Britain by being derived directly from research undertaken by academics (Coombes *et al.*, 1986). One reason for this innovation was that the TTWAs' credibility depends on them being seen as based on statistical and scientific criteria, and the fact that independent academics devised and implemented the analysis provided a rigorous 'quality control' aspect to the definitions.

20.2.1 TTWA Definitions

The main design objective is that there should be as many TTWAs as possible—so as to maximise the detail of the unemployment statistics that are published for these areas (Dept. of Employment, 1984)—but this is balanced against statistical constraints, which are determined by the way the unemployment rate data are compiled. These constraints ensure that TTWAs are not so small as to make them statistically volatile, and that all TTWAs satisfy a self-containment minimum (that is, a relatively low proportion of a TTWA's resident workforce cross its border to their workplace **and** a similarly low proportion of local jobs are taken by those living elsewhere). The technical challenge which arose—especially when working with the computer hardware of 1983—can be seen from the fact that the analysis has to flexibly process a matrix of commuting flows between over 10,000 wards (the sub-local authority areas, which are the smallest standard area for Census commuting data).

There have been many alternative approaches to the definition of labour market areas, and to flow data regionalisation more generally (see by way of contrast: Smart,

1974 and Dawson *et al.*, 1986; or Slater, 1984 and Boatti, 1988). It is clear that sub-setting the analysis leads to sub-optimal results in the border regions, so new software was needed so that each run could process the full matrix of over 100 million ward-to-ward flows; whilst still allowing multiple stages of disaggregation and re-grouping of wards in each iteration towards a preferred set of TTWA boundaries. The basic algorithm and data handling software development was devised for the 1981-based TTWA definitions (Openshaw *et al.*, 1988). A key feature of the algorithm is that it is a multi-step procedure. Extensive empirical testing of alternative one-step statistical and Markov chain based clustering methods suggested that none could handle the complexity of the data. The method had to be able to cope with geographical conditions as different as London and the Outer Hebrides whilst, in all areas, producing results which were broadly consistent with widely accepted local knowledge. The fundamental features of the algorithm can be summarised as follows:

1. identify all plausible foci for TTWAs (viz. concentrations of in-commuting flows, or highly self-contained areas);
2. group together these foci with notable commuting flows between themselves;
3. gradually allocate each of the remaining wards to the proto-TTWA with which it is most strongly linked; and
4. discard proto-TTWAs failing to meet the prescribed criteria for TTWAs (starting with the least satisfactory one, re-assigning its areas individually using the above steps, then re-testing the remaining proto-TTWAs).

Eurostat (1992) evaluated the TTWA algorithm against the definitions of local labour market areas in other European Union countries and singled out the TTWA method as a model for more general adoption.

20.2.2 Updating the Definitions

What was not then known was how this form of analysis would perform when, as an example of the phenomenon discussed by Stubkjaer (this volume), the same method was applied to an updated version of the same dataset. As in previous decades, the period between the 1981 and 1991 Censuses witnessed lengthening average commuting distances. A number of factors contributed to this trend: loss of traditional industrial jobs, which were often staffed locally; continuing dispersal of population; increasing affluence of the population and further growth in car usage. Of course, there were also some countervailing pressures, most notably the strong growth in part-time working and the wider increasing in the female workforce—both categories of workers who tend to have shorter commuting distances (Simpson, 1992). The suburbanisation of some types of jobs could reduce commuting distances too, although experience in the North America suggests that this is unlikely.

The single most influential factor in the TTWA definitions is the self-containment of commuting. As commuting distances between 1981 and 1991 have lengthened, more commuters tend to be crossing the TTWA boundaries and so self-containment values fall. A broad estimate across all TTWAs is that self-containment values have declined between 1981 and 1991 by an average of about five percentage points (a slightly faster decline than between 1971 and 1981). Of course, this trend is likely to have continued since 1991 (indeed the National Travel survey data on commuting trip lengths suggests that this trend may have accelerated further). Retaining the same threshold on the self-containment criterion from one set of definitions to the next makes it inevitable that there **must** be change in the TTWA boundaries. This is to say no more than that an exclusive emphasis on using the most up-to-date data must involve sacrificing stability through time. The more general risk is that an emphasis only on change tends to focus attention on large numbers of minor adjustments, whilst over-looking the deeper geographical structure which may well have changed very little.

This dilemma has been emphasised by the fact that the 1984 TTWA definitions maximised the number of separate areas which could be defined whilst still ensuring that every area met the self-containment and size criteria. This success was largely due to partitioning many parts of the country into numerous local labour market areas, which only marginally surpassed the statistical criteria. As a result, even a slight generalised decline in self-containment levels can cause a surprisingly high proportion of all the existing TTWAs to fall below the critical threshold. In fact, over a fifth of the 322 existing TTWAs were defined with self-containment values which, on 1981 Census data, were so close to this threshold that the 1991 data finds them falling below that critical level. Most of these then became merged with a larger neighbouring area—thereby causing the instability of boundaries to also affect these areas, whose self-containment had **not** been marginal. The instability can then ripple out still further because the TTWA algorithm has an optimisation element to it and often finds advantage in adjusting the boundary between two areas once a third area has been joined to one of them.

To summarise, the optimality of a set of boundary definitions such as the TTWAs is inevitably rather short-lived and this in turn emphasises their essential instability. Having said that, it is clear that the trends in commuting between time-points are not random, and short-term predictions of changing levels of self-containment could be made with a fair degree of confidence. The relevance of this final point is partly that the delays in Census data availability mean that the TTWA definitions at best represent a 'snap shot', which is already several years out-of-date when they are newly produced. After this they remain unchanged for a further ten years as if changes to commuting—and resultant change in self-containment levels—could not be anticipated.

Coping with the inevitability of change has not been a strong feature of government statistical systems in many countries. This underlying reluctance to take on board the implications of change was recently given an ironic emphasis in Britain once the level of instability in the TTWA definition was revealed by applying the criteria to the 1991 commuting data. The substantial adjustment to TTWA boundaries which was clearly required was deemed problematic by the then government and the 1991-based TTWA re-definition procedure was out on hold for almost four years. Yet a significant reduction in TTWA numbers remains essential to maintain the self-containment levels, and the up-to-date basis of the definitions, which are the *raison d'etre* of TTWAs as statistical units. In this way, the definition of TTWAs can be seen to provide an illustration of both the strengths and weaknesses of boundary definitions which are constrained to being based on the most up-to-date data available.

20.3 LOCALITY DEFINITIONS BASED ON DATA FROM SEVERAL YEARS

The research interest in localities often shares the TTWAs' focus on labour market issues (Coombes *et al.*, 1988), although this can also broaden into more social and environmental concerns such as the impact of the area's local industries on pollution levels or the workforce's class structure. For example, an area once dominated by textile manufacturing is likely to have a legacy of certain types of long-term illness, and also a workforce in which there is a long tradition of high levels of economic activity among women, while wage levels will probably have traditionally been low. A suitable form of locality definition that will be appropriate for analyses of these types of issue will need to group together all the contrasting social neighbourhoods within a single town or city **and** its surrounding rural areas. It is widely recognised that local authority areas are far too arbitrary and inconsistent in their definitions to be useful in this research-driven concept of localities.

An example may illustrate the issue here. The 1981-based set of TTWAs included both Rotherham and Sheffield as separate areas. The trends in local employment levels and commuting patterns has led to the two towns being combined in the 1991-based TTWA boundaries, due mainly to increasing commuting from Rotherham to Sheffield. Yet it remains true that Rotherham is a locality which has been dominated by coal mining, whereas Sheffield is traditionally a steel-making city which is now also a major service centre. If the two are combined, then the two areas' differing characteristics will be mixed and this will not particularly well represent either locality in studies of longer-term processes such as the effect of local industrial structure on health or migration behaviour.

For a method of defining boundaries to provide a more broadly based view than that which is derived exclusively from analysing the latest data on commuting patterns, it is essential to be able to synthesise the essential characteristics of patterns which lie within a number of different datasets. Once it becomes possible to bring together information from different datasets, it will also be possible to analyse data for a number of different time-periods. To summarise, the objective becomes to split the whole regionalisation procedure into two phases:

1. compile a range of relevant analyses from numerous datasets, and then
2. collate the results from these analyses within a single synthesising identification process.

It is the second phase for which there were no known precedents in the regionalisation literature.

20.3.1 Synthetic Data for Locality Definitions

The new method that has been devised involves creating what are termed *synthetic data*, which provides the basis for phase two of the regionalisation—using as input the initial, phase one, analyses. Each of these phases one analyses has produced a classification of all parts of the country that can be expressed in terms of the 10,529 1991 wards which are the building blocks for this analysis. Such a classification identifies which of these wards are grouped together within the same local labour market area (or migration region, or TV area, or whatever it is that the classification represents). Thus the key information in each of the multiple classifications can be reduced to binary data by taking each pair of areas and identifying whether they are (= 1) or are not (= 0) classified into the same region. In this way, a classification list—which had assigned a region number to each of the 10,529 wards—is re-expressed as a binary matrix of 10,529*10,529 zero-one cells (N.B. the matrix is of course symmetrical, so only half of it is in fact needed). For example, if area B was in the same region as area C but in a different region to area D, then the cell BC would take the value 1, while cell BD would be 0 (NB. cell CD would also take the value 0).

The crucial benefit from re-expressing each separate classification in this binary form is that these matrices can then be aggregated to produce what is in effect a synthetic flow data matrix based on multiple input classifications. For example, if three input analyses were collated in this way, the value in each cell of the synthetic flow data matrix would vary from 0 (for any pair of wards which were not in the same region according to any of the three input sources) up to 3 (i.e., for any pair of wards which all three analyses had put in the same region). In GIS terms, it is analogous to layering the sets of boundaries on top of each other and counting the number of layers in which there is no boundary between that pair of wards. Thus, for example, the original travel-to-work flow data can be partially represented by a TTWA classification, which can then be combined, as input to the synthetic data, with other relevant information (such as TV regions, previous TTWA classifications, cultural area definitions, etc.) which can be expressed as classifications of wards. They are then processed to yield a synthetic 'flow' data matrix, which in fact contains no physical flows at all, but represents the strength of the evidence for the togetherness of each possible pair of wards.

It can be seen that this approach provides an assessment of the strength of evidence that those two wards should be grouped together. The methodological innovation of creating synthetic flow data has provided this advantage by removing the technical limitations that have been inherent in all the preceding methods relying upon a single analysis of a single dataset. For example, it becomes possible for something of a fuzzy approach to then be adopted within the analysis itself by initially carrying out more than one form of analysis of the same dataset and accepting that each of these may provide a valid insight into the patterns lying within that dataset. Each of these analyses can then provide a separate input classification to the synthetic dataset, which will automatically identify those findings that these analyses have in common and also those on which they differ. In addition, of course, the synthetic data can draw upon the evidence of many different datasets—and not only the most up-to-date datasets but also, for example, the results of analyses of previous Censuses datasets. Reducing the time-dependence of the synthetic dataset in this way has been taken further by including as inputs a range of existing sets of boundaries—such as local authority areas—because these are also indicative of which areas might be better kept together and which kept separate. Some of the boundaries considered to be relevant for the definition of localities in Britain were those of Counties which ceased to exist over 20 years ago, but which are often cited as still shaping people's sense of identity today.

In this way, the two-phase regionalisation method removes the limitations that are inherent in all regionalisations based on a single analysis of a single dataset. These limitations were partly technical, but the more important limitations were in terms of the coverage of the inputs—and in particular the time period covered by the relevant evidence such as existing boundary sets—which could be drawn upon for the analysis. The synthetic data matrix thus represents a summary of multiple sets of information for a range of time periods. A set of Localities can be defined by applying a customised version of the TTWA algorithm to this synthetic 'flow' dataset. No methodological problems arise from analysing the synthetic data matrix as if it were a dataset of commuting or migration flows. Initial reaction to the results is that the method has indeed been able to identify Localities which will be appropriate for researchers with a wide range of interests, and which could well be far more enduring than a highly data specific set of TTWAs could ever be.

20.3.2 Automated Locality Definitions

The continued increase in computer workstation speeds creates the prospect that end-users will soon be able to design their own customised spatial frameworks, optimised for their particular form of SEU application. In the 1970s the TTWAs were designed using a mix of manual methods (including use of a slide rule for calculations), which took at least 6 months and several person-years of effort to complete. In the 1980s the TTWA identification process was computerised with mainframe run times of an hour or more, allied to expert knowledge. In the 1990s typical workstation runs are now reduced to a small number of minutes and doubtless it will soon be a few seconds. If the software were available and if the users knew enough about the TTWA algorithm, they are now much more likely to have available the computer power needed to create their own statistical reporting geographies, which are based on commuting or other flow data, or indeed synthetic data of their own creation.

The development of a synthetic data approach which can handle multiple types and sources of data dramatically increases the potential practical usefulness of the resulting definitions. The boundaries have become simultaneously more useful—because they are more broadly based—and also more robust, because they are not reliant on only one type of data collected at one point in time. The remaining difficulty in diffusing this technology is that it is not easy to understand or to apply. The synthetic data were analysed by a variant of the TTWA algorithm to produce Locality definitions in the way briefly outlined above. Unfortunately, this methodology may well be restricted to research applications because the optimality of the results depends on expert knowledge about various technical parameters and thresholds used by the algorithm. Although the TTWA algorithm is a fully automated procedure, its application has always involved a degree of manual tuning of the parameters in order to obtain good results.

As an alternative approach, an unsupervised artificial neural network regionaliser was also investigated based on Kohonen's self-organising map. A full description of the basic method and Fortran code is given in Kohonen (1984) and Openshaw (1994). Openshaw and Openshaw (1997) provide a basic introduction to the underlying idea, which is very simple. The input in this case is not the synthetic flow data, but all the different classifications of the wards that were used to create the synthetic data. The 10,529 wards are classified using 45 sets of variables, which are the various classifications of the wards that went into the synthetic flow data creation process previously described. For example, if a set of TV regions has 10 codes, this is represented as a set of 10 variable zero/one values. Every ward in the same TV region will have an identical set of 10 zero/one values. When applied to all 45 data sources, the result is a data matrix of 10,529 wards and 17,200 zero/one variables. It is this dataset which is classified using a Kohonen self-organising map classifier, because a more conventional cluster analysis procedure would not only not be sufficiently flexible, it would also not be capable of handling this number of variables. The self-organising properties of this neural network allow the maximum opportunity for the implicit structure to emerge from the data. The artificial intelligence in the classifier also takes advantage of distinctive features of the dataset: for example, the fact that some of the variables are fine scaled and others are very coarse (for example, TV regions). The relative importance of the different variables to the final results will depend on the scale of the outputs (i.e., the number of localities which are required). A variable that is macroscopic in spatial extent will only impact on the boundary definitions in those areas where the other datasets have produced a very marginal outcome.

An important feature of the self-organising map regionaliser is that the neurons compete to define localities using those variables which happen to be most relevant as boundary discriminators at the scale at which they are being asked to work. The only user controlled variable is the number of localities. The early results indicate that this neural network regionaliser, when run with the synthetic data offers the user the opportunity to define their own sets of areas in a few hours of workstation compute time. If this form of analysis can be made widely available, then the simplicity of the user's task—just choosing the number of output areas required—seems likely to greatly increase its value to non-expert users. Using this approach, the expert knowledge used to tune the parameters in the TTWA algorithm is side-stepped and attention turns instead to the selection of input datasets.

20.4 AN ASSESSMENT OF THE TWO BASIC APPROACHES

This section of the chapter moves on to consider the relative effectiveness, in representing organic SEU change, of the two basic approaches which have been outlined in the two previous sections. The empirical analysis here focuses on the settlement pattern, which is the most basic, feature underlying the distribution of socio-economic units. The nature of an area's settlements, and in particular the level of urbanisation of that part of the country, is an ever-present influence on the patterning of people's behaviour, which in turn is reflected in commuting flows and other key indicators relevant to the definition of SEU boundaries (cf. Wilson, 1970). It is clear that settlements are indeed organic SEUs in that their boundaries tend to be subject to a constant process of adjustment. The officially recognised definitions of British urban areas, for example, are updated once every 10 years (for example, Office for Population Censuses and Surveys, 1984).

The interest here is in a more synoptic view of population distribution than can be obtained from a binary classification of settlements as being either urban or rural. A more detailed analysis of population distribution is needed to represent the incremental nature of most changes to the settlement patterns of a highly developed country. For this purpose, an urbanisation index provides a useful starting point. The index presented here has been developed to provide measures at the finest resolution for which the relevant data is available—and also then to be aggregatable so that summary statistics can be produced for larger areas. The index is conceptually simple, but has a number of parameters within it so that it can be customised to emphasise those issues that are of most interest for any particular application. The question of interest is how well the TTWAs and the Localities represent the process of urbanisation over this 10-year period.

20.4.1 An Urbanisation Index

The basic interpretation of the term urbanisation which is adopted here is that the more urbanised the area then the higher the levels of visible activity and of physical development. This definition is then put in a spatial framework **not** by analysing the circumstances within the pre-defined boundaries of a socio-economic unit, but by considering the nature of the areas around the central point of interest. To be specific: the index value for a small market town will partly reflect the low level of urbanisation in the areas adjacent to the town, it will **not** simply reflect the circumstances within the town's own boundaries. The method adopted is to identify a centroid for each area of interest, and then to analyse the available data using a two-dimensional moving window which is centred on each centroid in turn. For each area of interest the index value is built up from the relevant data on all the areas whose centroids fall within that window. As in (Martin and Bracken, 1991), each of these areas' contributions to the index value is weighted inversely by the centroid-to-centroid distance by which it is displaced from the middle of the moving window:

$$U_i = \Sigma \, (s_j/d_{ij}{}^{\alpha}) \qquad (20.1)$$

Where:

U is the urbanisation index

i is the area of interest

j is an area whose centroid falls within the moving window

d is the distance between the centroids of i and j

s is a measure of scale or intensity of settlement.

This approach provides the following relevant parameters by which the index can be customised for each particular application:

Window: The window size is a critical parameter to reflect the wider or narrower focus of the analysis; the window also need not be circular, and could censor areas such as those which are not readily accessible (e.g., the mainland from an island, or *vice versa*);

Distance: Straight line distances can be replaced with measures more sensitive to actual travel time (if this is the relevant issue), whilst the power function applied can be adjusted as required (for example, it can be set at a high level to strongly emphasise information on the immediate neighbourhood); and

Settlement: The information on settlements might be simply population size or perhaps the number of buildings, whilst it might also be a combination of datasets (e.g., the resident population could be measured by one dataset whilst at the same time an effort could be made to follow the lead of Goodchild and Janelle (1984) by also examining the number of jobs in the area as a daytime population measure).

An important technical flexibility to this form of analysis is that there is no requirement to bring all the data and analyses into a single set of areal units. For example, if a combination of input datasets is drawn upon, then they need not use the same set of areas—each unit in each dataset is a distinct j for the purpose of the analysis. More importantly, there is no need at all for the input data areas to be the same as the output areas for which the index values are required. The index values are especially robust for areas that are larger than the input data areas. The ideal method may be to obtain output for a higher level units by deriving population-weighted averages of the results from analyses which have been undertaken on lower level units (i.e., at a finer level of spatial resolution).

20.4.2 Results for TTWAs and Localities

In the example data presented here, the basic analysis is undertaken using the 113,062 Enumeration Districts into which England and Wales were subdivided for the 1991 Census. The moving window around each area's centroid searched for the centroids of the 9,289 wards which were used for the 1981 Census. In general, it will be preferable if this ratio between numbers of input and output areas was reversed (i.e., there were **at least** as many input as output areas), but the input areas are restricted to those for which both 1981 and 1991 population data is available. The results would not be reliable in detail at the level of analysis (Enumeration Districts), but these resolution errors will not be significant after the large-scale aggregation to the Locality or TTWA scale.

The results from the index are scaled to the highest value for any Enumeration District in either year, with values expressed as percentages of this maximum. The highest value was for an area in central London in 1981—as expected; the highest value in 1991 was a little lower due to the 1980s' net population movement away from the areas of most intense urbanisation. The average index value for the set of 262 Localities in England and Wales was 10.676 in 1981 (i.e., just over 10% of the value for that small area in central London that was the most urbanised point in Britain). The mean value for Localities in 1991 had risen to 10.736 (an increase of 0.060 percentage points). There was a slight population growth nationally, but the main influence is the drift away from the more urbanised parts of the country. This analysis is a mean value across Localities and numerically this set of areas is dominated by the smaller towns that have mostly gained from the pattern of population distribution. Thus the most frequent movement by

people has been to less urbanised settings, but the **overall effect** has been that more Localities have become more urbanised, whilst the areas which are losing population are relatively few in number because they are large.

The mean 1981 urbanisation index value (8.473) for the 1981-based TTWAs in England and Wales was lower than the equivalent Localities' value. This is because there are even fewer and larger TTWAs in comparison to Localities in the main metropolitan areas, and more very small TTWAs than Localities in the remote rural areas. Thus the difference between the two sets of areas' mean values—based on the same analysis at Enumeration District level—is as would be expected. Of course, the particular interest here is in the fact that for 1991 there is a new set of TTWAs for which to present 1991 urbanisation index values. Due to the lengthening of commuting trips in the more urbanised parts of the country in particular, the fact that there are 20 fewer 1991-based TTWAs than 1981-based TTWAs in England and Wales reflects a process in which there are declining numbers of separable TTWAs **especially** in the metropolitan areas. The result is that the 1991 TTWAs are even more numerically dominated by smaller towns than were the 1981 TTWAs—and so this contrast with Localities has grown stronger. The consequence is that the mean 1991 urbanisation index value for 1991-based TTWAs is only 7.961 (whereas the equivalent 1991 value for 1981-based TTWAs is 8.594).

20.4.3 Implications for the two Approaches

It is important to summarise the separate points that have emerged from these statistical analyses. Population redistribution in Britain in the 1980s has been mainly characterised by movement away from the largest cities. The overall effect is that there are many parts of the country which have seen slowly increasing urbanisation, whilst the main metropolitan areas are dominated by population decline. To the extent that it is the smaller towns and cities which predominate **numerically** in almost all sets of areas, then the urbanisation index analyses outlined above emphasise these less urbanised areas' low but growing values. Of the two approaches outlined in this chapter, the TTWAs show this effect to a more extreme degree and so have lower mean values than the Localities have.

The flexibility of the urbanisation index **does** allow values for 1991 to be calculated for the 1981-based TTWAs, but for many analyses this would **not** be the case. For most updated sets of areas, the only available measure of change over time would be a comparison of the time t_1 values for the t_1-based areas against the t_2 values for the t_2-based areas. Such a comparison in the TTWA case would apparently show a sharp **decline** from 8.473 for 1981 to 7.961 for the later year. Interpretation of this change would then be dependent on a judgement as to how much of an effect was due to the change in the units of analysis—but there would be no data to inform that judgement. In fact, the other data presented here shows that this effect of the change in boundaries is actually greater than the change between those two values, but that over the same time there has been a widespread slow **increase** in urbanisation levels in Britain's towns and cities.

The conclusions from this exemplar empirical analysis are perhaps unsurprising: emphasis on updating the units of observation can readily confuse attempts to observe the actual pattern of change in the organic SEUs themselves. The example here used TTWAs—a set of crisp SEUs defined using a commended algorithm (Frey and Speare, 1995) which reflects Britain's settlement structure—and examined urbanisation, which is perhaps the most basic feature of any settlement pattern. Yet the results of such an analysis will often be dependent on informed guesswork as to the relative influence on the results obtained of the SEUs themselves having changed. For this analysis it **was** possible to compute data for both time periods for the 1981-based TTWA boundaries, but in most equivalent cases this will **not** be the case because the data available for updated

SEUs will most often only cover the period from the time of their definition to the time when a new and more up-to-date set is produced.

By way of contrast, maintaining a fixed set of crisp SEU boundaries does allow the data on change in the organic SEUs to be shown without other confounding effects. The fact that the Localities included both 1981 and 1991 data as input to their definitions allows them to represent change over that period sympathetically. Figure 20.1 shows that the underlying pattern of urbanisation in England and Wales during the 1980s is quite readily identified using this method. The most and least urbanised Localities (with index values of over 40 or under 5 respectively) have changed little, with the former group of Localities' index values falling very slightly whilst most of the latter were slowly urbanising. In between these two extreme groups, there has been a remarkably clear inverse relationship between Localities' 1981 levels of urbanisation and the changes in those levels during the 1980s. The most important point which arises for this chapter is that this pattern could only be revealed with a set of areas such as Localities whose definitions explicitly set aside most of the change in organic SEUs in order to focus on the underlying stability within the settlement system.

20.5 DISCUSSION

The contrast made here between the definition of TTWAs and of Localities has centred on the formers' emphasis on analysing the single most up-to-date dataset, and the latters' concern to be more broadly-based and hence to reflect longer-term patterns. The implication of this difference in approach is that the TTWA definitions are inherently sensitive to secular changes in commuting behaviour—indeed, their very objective requires them to reflect the latest evidence and to discard any boundary which portrays a superseded pattern of commuting—whereas the Localities seek to represent underlying patterns which recur through time and across datasets, and which are less volatile as a result. Neither approach can be appropriate for every different context: the important prior step when selecting a boundary set will always be to reflect on which characteristics the areas need to make them relevant to that application. This requirement to think through the implications of the boundaries used should be increasingly stressed as GIS and related techniques make a wider range of options available.

The innovation reported here of generating boundaries from 'synthetic flow data' is a case in point. As it becomes more generally feasible to manipulate different boundary sets in order to create new ones it will be essential to be aware of the way in which boundaries reflect the circumstances of different periods being combined, so that the temporal dimension becomes one more parameter which can then be manipulated explicitly. At the same time, the end-users may be gaining access to technology with which they can customise locality definitions to the purpose for which they are needed, as well as to revise them to keep abreast of subsequent changes which are important to that particular application. There should be less need for heroic assumptions such as that a set of definitions are sufficiently general purpose for them to meet all needs equally well, or that a set of definitions will continue to be valid even though everything which they represent has changed. As the world moves away from physical place-based economic activities so the flexible nature of the synthetic data approach may well become increasingly relevant. What is left as a question for the future is whether or not the multivariate data patterns that characterise the nature of localities might be better handled via computer vision technology. Regardless of how this identification process is engineered, there can be little doubt that the identification of SEU boundaries will remain an essential tool for organising and making visible the spatial patterning of human behaviour which continues to shape human society into the Internet era.

ACKNOWLEDGEMENTS

CURDS is the Centre for Urban & Regional Development Studies: This chapter draws on research by Mike Coombes (assisted by Simon Raybould) sponsored by ESRC as a Senior Fellowship on "New GIS based measures of population distribution" (reference no. H52427501095), and also on two CURDS projects—"Localities & City Regions" sponsored by ESRC (reference no. H507255129), and the TTWA research sponsored by the Dept. of Employment and the Office for National Statistics—which were undertaken by the CURDS NorthEast Regional Research Laboratory (NE.RRL) team in conjunction with Colin Wymer (Planning Dept.) and Stan Openshaw (Leeds University).

REFERENCES

Boatti, G., 1988, I sistemi urbani, struttura del territorio lombardo. *Edilizia Popolare*, **25**, pp. 2–27.

Burrough, P.A., 1996, Natural objects with indeterminate boundaries. In *Geographic Objects with Indeterminate Boundaries*, GISDATA Series Vol. 2, edited by Burrough, P.A. and Frank, A.U. (London: Taylor & Francis), pp. 3–28.

Coombes, M.G., Green, A.E. and Openshaw, S., 1986, An efficient algorithm to generate official statistical reporting areas: the case of the 1984 Travel-to-Work Areas. *Journal of the Operational Research Society*, **37**, pp. 943–953.

Coombes, M.G., Green, A.E. and Owen, D.W., 1988, Substantive issues in the definition of localities: evidence from sub-group local labour market areas in the W. Midlands, *Regional Studies*, **22**, pp. 303–318.

Dawson, J.A., Findlay, A.M. and Sparks, L., 1986, Defining the local labour market: an application of log linear modelling to the analysis of labour catchment areas. *Environment & Planning A*, **18**, pp. 1237–1248.

Dept. of Employment, 1984, Revised travel-to-work areas. *Employment Gazette,* September, Occasional Supplement 3.

Eurostat, 1992, *Study on Employment Zones*. Luxembourg: Eurostat E/LOC/20.

Frey, W.H. and Speare, A., 1995, Metropolitan areas as functional communities. In *Metropolitan and Nonmetropolitan Areas: New Approaches to Geographical Definition*, edited by Dahmann, D.C. and Fitzsimmons, J.C. (Washington, DC: US Bureau of the Census).

Goodchild, M.F., and Janelle, D., 1984, The city around the clock: space–time patterns of urban ecological structure. *Environment and Planning A*, **10**, pp. 1273–1285.

Kohonen, T., 1984, *Self-organization and Associative Memory*, (New York: Springer-Verlag).

Martin, D. and Bracken, I. 1991, Techniques for modelling population-related raster databases. *Environment and Planning A*, **23**, pp. 1069–1075.

Miller, H., 1991, Modelling accessibility using space-time prism concepts within geographical information systems. *International Journal of Geographical Information Systems*, **5**, pp. 287–301.

Office for Population Censuses and Surveys, 1984, *Key Statistics for Urban Areas*, (London: HMSO).

Openshaw, S., 1994, The neuroclassification of spatial data. In *Neural Nets: Applications in Geography*, edited by Hewitson, B C. and Crane, R.G. (Dordrecht: Kluwer).

Openshaw, S. and Openshaw, C.A., 1997, *Artificial Intelligence in Geography*, (Chichester: Wiley).

Openshaw, S., Wymer, C. and Coombes, M.G., 1988, Making sense of large flow datasets for marketing and other purposes. In *NorthEast Regional Research Report 88/3*, Newcastle upon Tyne: CURDS, Newcastle University.

Simpson, W., 1992, *Urban Structure and the Labour Market*, (Oxford: Clarendon).

Slater, P.B., 1984, *Migration Regions of the United States*, (Santa Barbara: University of California Press).

Smart, M.W., 1974, Labour market areas: uses and definitions. *Progress in Planning*, **2**, pp. 239–353.

Wilson, A.G., 1970, *Entropy in Urban and Regional Modelling*, (London: Pion).

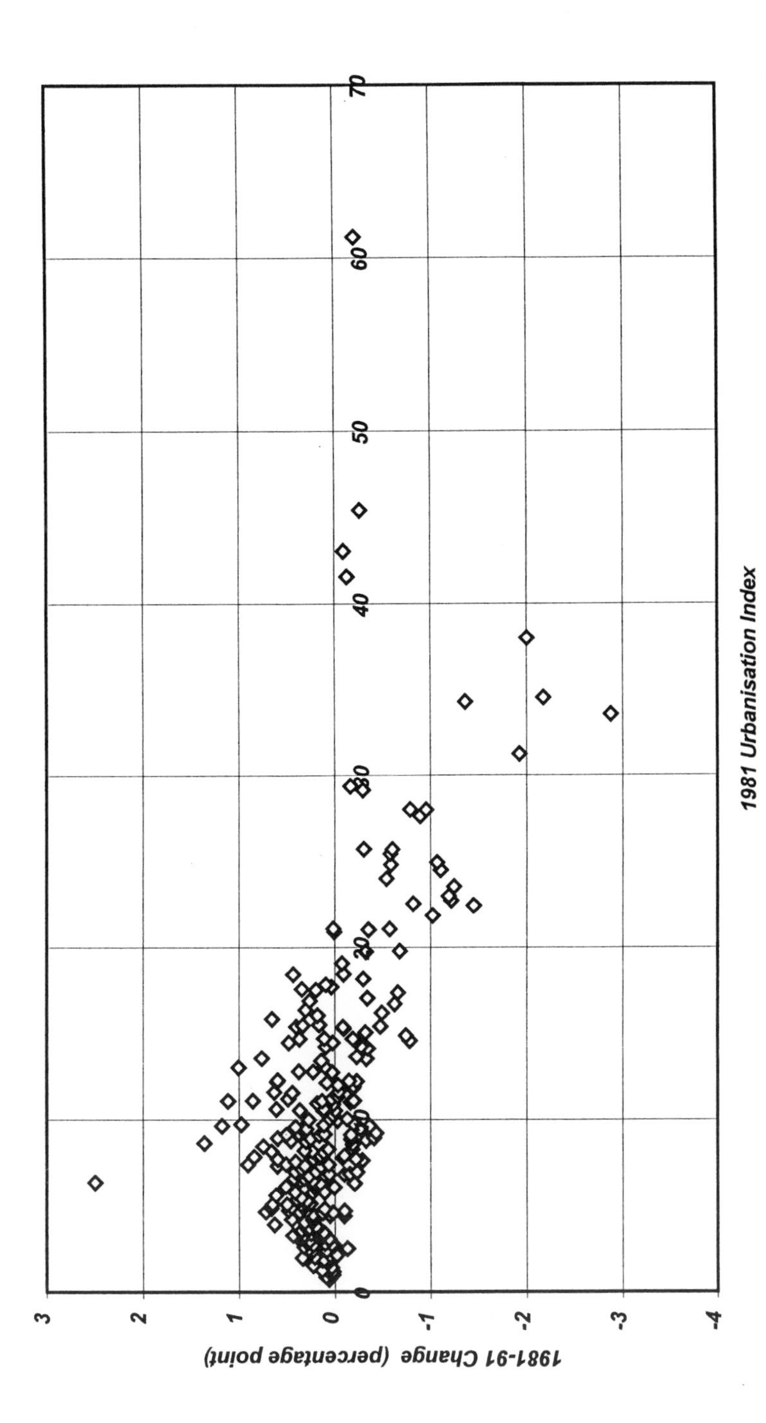

Figure 20.1 Urbanisation Index 1981 and 1981–91 change (localities in England and Wales)

CHAPTER TWENTY-ONE

On the Creation of Small-Area Postcodes from Linear Digital Spatial Data

Rui Manuel Pereira Reis

21.1 INTRODUCTION

The importance of spatial socio-economic units (SSEUs) for geographical analysis derives, to a great extent, from their potential to locate socio-economic data relating to individuals. This characteristic of SSEUs allows, among others, the study of spatial patterns, the temporal evolution of socio-economic phenomena and the overlay of different kinds of data. One of the main data sources for socio-economic applications is the decennial census of population and it was in this context that the first studies on geo-referencing population-related data appeared. The US Census Bureau pioneered such studies from which GBF/DIME was the first result and, more recently, by creating TIGER-Line. Both systems use linear geo-referencing, that is, the geo-referenced SSEUs are described by associating them to line entities whose geographical position is known.

This study was developed in connection with the revision of the Portuguese postcode system to create smaller postal units, but the theoretical background and methodologies developed are applicable to most socio-economic applications that share the need to build SSEUs from basic geometry. For example, another application area that could apply the methods described here would be the census to build new basic collection or reporting areas. Due to the practical advantages and wide availability of geo-referenced linear geometry we will give special attention to the line-to-area transformation processes, which are of special interest when, in addition to SSEU creation, one of the intended purposes is the optimisation of delivery routes.

This chapter will first summarise the background context to this study followed by a brief characterisation of the data available. The desirable geo-referencing characteristics of the units to be built are then assessed and more specific objectives for the study are defined. Next, due to the need to have areal units on which to base GIS analysis, we review some related methodologies in areas meaningful for this study and a tentative general framework for the creation of new areal units is drawn up in the context of spatial data transformations. General strategies for the generation of socio-economic units from areal, point and linear data are then discussed, followed by the description of some methodologies for the creation of such units from linear data. Finally, some examples of application are presented for a test area in Lisbon and a tentative evaluation of the performance and validity of the methods tested is presented.

21.2 BACKGROUND

The background to this study was covered in more detail in (Reis and Raper, 1994). In summary, due to the stated requirement of the Portuguese Post Office (Correios de Portugal—CTT) to optimise mail delivery and rationalise existing resources, it was judged that GIS techniques could improve the existing analysis capabilities and provide mail process automation by the development of smaller areas. Preliminary studies on the constraints imposed by the data suggested that the optimum spatial resolution for the geometry from which the new unit was to be built was at the street-segment-side level (corresponding to the block frontage in urban areas) to which all other necessary information can be linked. Using this data not only allows the implementation of the intended analysis capabilities on existing commercial GIS, but also provides other advantages, among which is the capability to use this data as the basis for a wide range of other application areas and thus, to finance the expansion of the system by developing derived products.

Methods for generating new SSEU systems have been developed in the USA to review the boundaries of electoral districts after the decennial census, a process known as redistricting. For early texts on this subject see, for instance, Hess *et al.* (1965), Garfinkel and Nemhauser (1970), Sammons (1978) or more recent treatments by Johnson *et al.* (1989) or by Macmillan and Pierce (1994). Openshaw (1977a, 1977b, 1978a, 1978b) and Openshaw and Rao (1994a, 1994b) use several exploratory data analysis techniques to find alternative aggregations of a basic set of SSEUs and to find the optimal arrangement in order to achieve a minimal within zone variation and to maximise variation between contiguous units.

In contrast with the situation in the USA, in Europe there are fewer examples of applications of this kind. However, the Enumeration Districts for the 1991 Censuses in Scotland were built from postcodes and, more recently, a study appeared (Coombes *et al.*, 1993) dealing with the assessment of the required data and GIS functionalities for the Local Government Commission (LGC), an organisation whose objectives include reviewing local government boundaries and electoral arrangements in non-urban counties of England.

Elsewhere in Europe, changes to electoral constituency boundaries, are probably rare but do occur from time to time. For example, in Portugal a debate has been going on since 1994 about a proposal to reduce the size of the electoral constituencies. At present, Portuguese electoral constituencies coincide with the administrative districts, a unit whose size is between that of the European NUTS II and NUTS III. Each administrative district elects a variable number of Members of Parliament (*Deputados*) dictated by its voting population. In order to hold proportionality within bounds the National Electoral Committee (CNE-Comissão Nacional de Eleições) makes adjustments to the number of elected MPs by administrative district on a regular basis. It could, however, alter the size of the electoral units.

The need to generate new SSEU systems also appears in other contexts. Work has been carried out on the creation of new units for the census (Witiuk, 1990), in location–allocation problems (Goodchild, 1979 and Fotheringham *et al.*, 1995), in transportation studies (Masser and Brown, 1978 and Masser *et al.*, 1978) or related to postal services such as the creation of new postcodes (Raper *et al.*, 1992 and Reis and Raper, 1994).

Other areas sharing the need to control the SSEU building process include geo-demography and planning applications where the geographic objects of interest are settlements and neighbourhoods, when using socio-economic information to perform spatial analysis operations in a GIS context (Martin, 1991). Often, the original patterns are obscured because the data is already aggregated. Another possible use is for understanding the processes underlying the detection of regions of significant change in order to assess the validity of existing boundaries (Oden *et al.*, 1993). The need for areal interpolation arises then and, more generally, whenever data is available for sets of areal units that are not coincident with the units for which the analysis must be performed. The units of analysis may be altogether different in nature from the reporting units, or the boundaries may just have changed over time. See, for instance, Lam (1983) for a review on spatial interpolation methods, and, for studies on the problems of transferring data between SSEU systems Openshaw (1976, 1984) and Flowerdew and Green (1989, 1991).

21.3 OBJECTIVES

Most of the reviewed applications dealing with the creation of SSEU systems assume that a set of units already exist and then new systems are to be created either by aggregating smaller units into larger ones, by partitioning large units into smaller ones, or by combined methods involving alternative aggregation and partitioning. Thus, common approaches for the creation of new zonal systems assume, as a starting point, that a set of SSEUs already exist. Accordingly, the emphasis usually goes to the optimisation methods used, thus overlooking the process that gave origin to the original set of SSEUs. However, certain questions come to mind:

a. What if there are no SSEUs to start with? and
b. What are the methods available for generating SSEU systems in this case?

Socio-economic data might be associated with any of the basic geometry types, e.g., point, line or area. Thus, an obvious approach for creating new sets of units, when no SSEU system already exists is to use basic geometry. For example, population counts might be associated with addresses that can be geo-referenced by point coordinates. Alternatively addresses could be associated with linear entities like street segments, or to areal entities, like census units. The problem can then be studied as a subset of the spatial data transformations, specifically, those resulting in new areal entities.

The overall objective of this chapter is, then, to discuss methods for the automated generation of SSEU systems from digital spatial data in its various geometric forms. Among these, we will give special attention to the point-to-area and line-to-area transformation processes.

21.3.1 Data Issues

Postal delivery is among the most important activities from which a regionalisation at the national level can be derived in any modern state. The wealth of information used and maintained for mail delivery purposes includes data on mail volumes, travel times, delivery points, administrative units and address ranges. This information is important, not only to those directly involved, but also to a broad class of other socio-economic activities. The creation of a new set of SSEUs can also be seen as a form of spatio-temporal evolution/progression of the regionalisations used in a state; once created they are only rarely altered.

In order to start with a manageable size, to allow for a natural growth path and to be able to accomplish all the intended objectives special care had to be taken in the choice of data structure. Typical network analysis functions like the optimisation of delivery and collecting routes were among the main functionalities required. Thus, the main options were to use the address as the minimum building block, or to use the street-segment. It was also clear that, in order to guarantee consistency and versatility, the data should be organised in terms of the smallest part on which a 'postman's walk' can be subdivided. A street-segment can be defined as the part of a street between two junctions or between one junction and the end of a street with no exit (a 'dangling segment'). In urban areas the sides of street-segments can be associated to the frontage of city blocks and these are commonly used to subdivide delivery routes. Thus, the street-segment-side was chosen as the unit to which mail-related data, address-ranges and administrative units and other data were linked. Road centrelines provide the geo-referencing information needed.

For the purpose of testing the different methodologies for SSEU creation a test dataset was created from alphanumeric data provided by CTT covering an area in Lisbon. This dataset contains information on administrative units, address ranges, mail delivery and collection data (delivery routes, categorised delivery point counts and mean mail volumes). The original data was organised by postman's walks. Each data item could represent an entire street, just one side, both sides or, more often, an address range (12–28). This data could not be used directly to provide the linkage with the street centrelines due to the lack of internal correspondence between them and because it was not organised by street-segment-side. For example, in the original alphanumeric dataset, a street can be referred to as a single item of data or can be subdivided in various possible ways:

a. one data item might refer to the even-numbered side and another to the odd-numbered side, or
b. an item could hold data from the beginning until certain even and odd numbers, or
c. data item could represent the even side, the odd side from the beginning until No. 23 and the odd side from No. 25 until end.

In order to guarantee the consistency between alphanumeric and positional information the alphanumeric data had to be disaggregated into street-segment-sides.

Due to the lack of street centreline digital products covering the test area in Lisbon at the time, street centrelines were manually digitised from paper maps. Then the street centrelines were topologically structured and the disaggregated alphanumeric data was linked to it. Thus, in the dataset used each street centreline segment is linked to data on its left and right sides.

Figure 21.1 Extract of road centrelines from the Lisbon dataset

During the course of this study, CTT began development of an internal GIS and has now adopted a national level road-centreline data creation program using internal resources and external digitising services. At present, road-centreline data covers Lisbon, Oporto and several other municipalities. Medium-sized towns and rural areas are intended to be covered later, due to their lowest relative weight in terms of mail volumes. The main workload, however, is the reorganisation of postmen's walks descriptions into left and right sides of street segments.

The availability of street-centreline digital data covering all the urban areas in any country is a valuable dataset in itself. When associated with fine resolution population related information it would surely represent a major step towards the creation of the basic conditions for the development of advanced spatial analysis dealing with population characteristics. The versatility of road centreline data as well as its importance in the development of GIS applications is indicated by the EU initiative led by CEN/TC278 which gave origin to the Geographic Data Format (GDF).

21.3.2 Geo-Referencing Issues

A unique characteristic of this new data among other mail-related datasets is that it is organised by street-segment-side and associated with street centrelines. Usually mail-related datasets have very poor geo-referencing for GIS usage and *a posteriori* enhancements have proved difficult to implement. The preferred approach to the geo-referencing of postal codes in most previous work (Raper *et al.*, 1992), often dictated more by financial than technical constraints, has been to use seed points to represent the addresses covered by a unit. Digitising the boundaries for these postal units when they are small in geographic area is a process that few postal services can fund on a large scale. In contrast to this, with data structured by street-segment-side geo-referencing is inherent and follows hand in hand with the creation of the areal units. In fact, the approach followed here has more to do with approaches found in the census field then with traditional postal applications.

A unique characteristic of this new data among other mail-related datasets is that it is organised by street-segment-side and associated with street centrelines. Usually mail-related datasets have very poor geo-referencing for GIS usage and *a posteriori* enhancements have proved difficult to implement. The preferred approach to the geo-referencing of postal codes in most previous work (Raper *et al.*, 1992), often dictated more by financial than technical constraints, has been to use seed points to represent the addresses covered by a unit. Digitising the boundaries for these postal units when they are small in geographic area is a process that few postal services can fund on a large scale. In contrast to this, with data structured by street-segment-side geo-referencing is inherent and follows hand in hand with the creation of the areal units. In fact, the approach followed here has more to do with approaches found in the census field then with traditional postal applications.

The representation of areal entities by means of basic linear entities can be achieved by assigning to each base line information about the left and right areal entities. If the left and right areas are the same, then the line is inside the area; if they are different, then the line is the boundary of some areal entity. The representation of point entities by means of linear entities can be achieved in a number of ways, and with varying degree of precision. The position of one point entity might be referenced by the distance from the start of the line or just by its reference (that is, no position along the line is given).

Desirable characteristics for new areas based on street geometry, as general purpose basic geo-referencing units (see Visvalingam, 1991), at least for population-related use in urban areas, include, among others:

- Size: they should be smaller than comparable, already existing, spatial units (usually the census Enumeration District-ED);
- Exact fit: in the census and administrative units (once, in urban areas, they also follow the geometry of streets);
- Stability: in consolidated urban areas they are stable to a large extent (only the disappearance or creation of roads makes them change);
- Coverage: they should cover all the streets, even those without addresses.

Despite these advantages there are some drawbacks to street segments as building blocks for SSEUs. Linear geo-referencing is an indirect representation and thus data has to be pre-processed upon input and queries involving other data types might become elaborated. Special care has to be taken in order to assure the logical consistency of the database. Moreover, once address data is involved, inconsistencies in addressing schemes make it difficult to implement automatic processing techniques. For example, there are rules indicating where street numbers should begin but sometimes these are not followed. Another example of an inconsistency of the addressing scheme is the assignment of the left and right inside the stairs of an apartment building. Despite an existing rule that states that the door to the left, when going upstairs and reaching the first floor should be assigned to the left category, quite often, it is not.

21.3.3 Space Coverage

Postal SSEUs, *per se*, do not exhaust space. In fact, for postal operations the main concern is with the path followed, not the area covered. If space exhaustion is to be achieved, it must be imposed by a process involving the creation of zones from the basic linear entities. One way to interpret the zones to be created is as measures of workload and thus as those resulting from postman's walks. An actual postman's walk, however, not only depends on the number of delivery points but also on mail volumes and on a number of other factors. For instance, the distance from the delivery centre and the transportation means used, irregularity of the terrain and, even, expected weather conditions in the area. All these and other factors influence, to a variable extent, the final number of mail objects delivered and the population covered by an individual walk.

It is generally accepted that the wider usefulness of the areas to be generated depends to a great extent on their characteristics as measures of population. Thus, they must be uniform and cover a small population. The number of residential delivery points can serve as an indicator of the amount of population covered. On the other hand, if their role in data integration is to be fully exploited they must exhaust space.

If the geo-referencing capabilities are intended to be versatile and use of the system widespread, not just urban areas, but full national coverage has to be considered. Postal Service priorities, however, are likely to be dictated first of all by operational factors (units with large mail volumes) and full national coverage will certainly take some time to achieve. On the other hand, it has to be assessed whether geo-referencing population-related data using road centrelines still holds its comparative advantages when applied to rural areas. For example, their size, in terms of the covered population, is expected to vary widely but to decrease sharply in rural areas. In urban areas the problem with the variability in size is the inverse: how to treat 'vertical streets' (for example, large residential blocks). This question becomes pertinent if the intended size is to be comparable with the best systems of this kind already implemented. Assuming a target of 15 'delivery points' as the target for the average number of dwellings covered by a unit, if a block has 4 dwellings in each floor, this represents more than a medium-sized block even if it has only 4 floors.

For application areas where confidentiality constraints have to be considered the disaggregated use of SSEUs might not be possible in a large percentage of cases. If the aggregation process is to be performed on an application-by-application basis, incompatible systems will inevitably result. To overcome this problem the aggregation must be made beforehand. The simplicity of the proposed geo-referencing scheme would be somehow lost: the correspondence between a (single) street-segment centreline and two data records, corresponding to the odd and even sides, would not hold anymore in cases involving aggregation.

21.3.4 Types of Spatial Partitions

The intrinsic characteristics of the units will depend on the data available, intended use, and methodology used for their creation. Once created a set of units will partition space in some way. One way to classify different space partitions is according to space exhaustion. We can distinguish at least two types:

1. The complete filling of space: the more commonly used units for spatial analysis in GIS fall in this category—space is completely occupied by the units and thus they share boundaries.
2. Allowing empty sub-spaces: in this case, space is not completely covered by the components of the partition, the units concentrate where the objects of interest are. They might share boundaries, or not.

Another way to classify space partitions is according to the relations between components (units) themselves. We can distinguish (Kriegel *et al.*, 1991):

1. pure space partitions: when the components are pairwise disjoint. That is, they do not overlap with other units of the same partition; or
2. coverings: in this case the components might overlap and thus some portions of space might belong to more than one unit.

21.4 GENERAL FRAMEWORK FOR THE CREATION OF NEW SSEUS

One way to put the problem of creating new areal units into context is to treat it as a member of a class of transformation problems. To treat the problem in a general sense we cannot assume that only one type of data is at hand. More often new zones need to be created from data which:

- is characterised by having different geometry: points, lines, areas, and surfaces;
- is observed in various forms: areal data, for instance, may be reduced to point form, or;
- that have different intrinsic characteristics (e.g., continuous or discontinuous), and
- have different levels of measurement (e.g., nominal, ordinal, interval or ratio).

From the nine possible spatial data transformations only four result in the creation of zones (areal entities), namely: Point to Area (P–A), Line to Area (L–A), Area to Area (A–A) and Surface to Area (S–A). In the next section we will take a look at existing general methods for generating new zones, classified according to the geometric type of the data being processed.

21.4.1 Zones Derived From Point Data

The most natural representation for addresses, mail counts, etc. is by using points. Thus, in this context, point data might represent houses, aggregations of people or an estimate of these. Averak and Goodchild (1984) describe methods for boundary definition from point data. They divide existing methods into statistical, including, probability surface mapping, density contouring, probability contouring and discriminant analysis and geometrical methods based on convex hulls, spanning circles, etc. In general terms the creation of zones from point data can be achieved either by processing the points locally or by applying a global process to them.

When zone generation is based on local point processing we can distinguish between:

- direct Point to Area transformation processes, including those resulting from the direct connection of points with their neighbours (Delaunay triangulation, for instance);
- composite Point to Area transformation processes, those involving intermediate transformation(s) to other geometric data types, for instance, Point to Line followed by Line to Area.

The methods for zone creation based on the global processing of points might be classified in statistical and geometrical methods:

- statistical methods: characterised by using the value of attributes associated with each point (e.g., number of dwellings, height). Lam (1983) gives an extensive review of this kind of methods, namely, quadrate count methods, probability surface mapping, analysis of variance, fourier and spectral analysis, space–time spectral analysis, density contouring, probability contouring and discriminant analysis;
- geometric methods are those based on computational geometry. Among these the more important in this context are methods for computing the convex hull, spanning circles and Voronoï regions (simple or weighted).

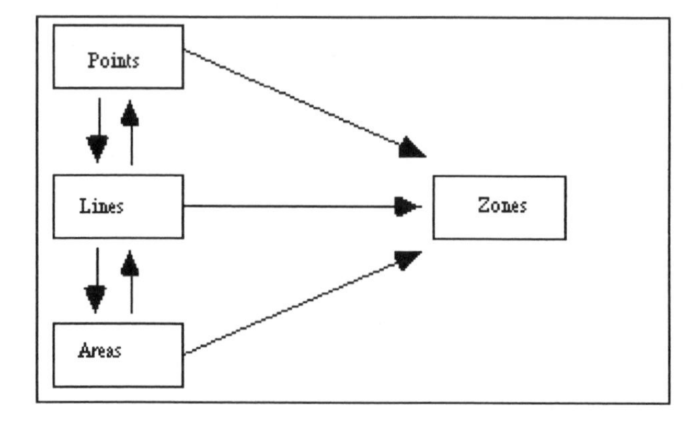

Figure 21.2 Spatial data transformations and zone building

21.4.2 Zones Derived from Line Data

For the purpose of constructing zones, lines might be considered as edges and we can distinguish between:

a. seed-based methods which form areas from lines by connecting the lines surrounding a point (seed);
b. path-based methods (using connectivity relations), see Visvalingam *et al.* (1986) for the description of a process for building areas from line geometry.

Another way to look at lines is to consider them as skeletons of an original area. The representation of areas by their skeletons is a well-known computational geometry technique introduced by Pfaltz and Rosenfeld (1967). We can either reconstruct the original area from width information stored with the skeleton, if available, or to use the line Voronoï diagram, either simple, or weighted, or possibly constrained to create a set of areas surrounding the lines and completely filling space. See (Okabe *et al.* 1992) for a review of the methods to build Voronoï tesselations. Further elaboration on these and other approaches will be discussed below.

21.4.3 Zones Derived from Areal Data

If the original areas exhaust space then existing approaches for the optimisation of zonal systems might be used. When the original areas do not exhaust space the area Voronoï diagram might be used to generate a tesselation.

21.4.4 Zones Derived from Surface Data

Usually surface data is collected as a set of sample points. Based on these observations a mathematical surface is then fitted to the data points in order to minimise the discrepancies. Units can be derived from this fitted surface, either by interpolating boundaries by classification techniques (Bracken, 1994), or to extract boundaries by using the landscape metaphor developed by the ARGUS project at the Department of Geography, University of Leicester.

21.5 STRATEGIES FOR THE CREATION OF SOCIO-ECONOMIC UNITS

Most of the approaches found in the literature assume as a starting point the existence of areas covering the space under study. Different manipulations are then applied to these units in order to generate new zones by one of the following methods:

- by subdivision of the zones with the help of ancillary data, resulting in finer resolution areal classifications, or
- by a process involving the aggregation of the basic areal entities, ruled by a number of restrictions and objectives, or
- by mixed processing usually involving aggregations of the basic areas, followed by the evaluation of certain criteria and, if necessary, swapping of base areas between neighbouring zones in order to achieve a better solution both locally and globally.

Despite this brief overview of the problems associated with the generation of SSEU systems, our aim here, in common with these techniques, is to generate an optimised set of areas with contiguity constraints. It differs from them because we do not assume as the starting point that an existing contiguous space partition exists, but that the process can be initiated by other geometric data types. We can then include the problem at hand—following the classification of spatial analysis operations proposed by Goodchild (1987)—in the more general class of problems dealing with the creation of new objects from one or more existing classes of objects.

The general objective of this section is to describe methods using the strategies outlined above for generating small-area systems from linear basic spatial data. Usually optimisation processes work assuming some initial conditions and, given the objective(s) expressed by means of a function or parameter(s), a transformation process is applied resulting in a final state achieved upon the objective(s) fulfilment. Of course, the validity of the results obtained by any method will depend on correctly choosing the parameter(s) to be optimised and these will dictate the data needed for the implementation. Our approach consists of, first, deriving a basic set of units from base geometry only. After this step the usual optimisation processes can then be applied in order to reach the objectives of the application field.

If we consider only linear entities the process of generating zones can be tackled according to one of two main strategies. The normal case is when polygons are formed by connecting linear entities that eventually recurve to enclose a set of segments (dangling segments). Another case occurs when it is not possible to generate one area by this process—the unconnected case. The connected (normal) case can be solved by applying algorithms that look for the connectivity of lines by propagating the search according to a predefined rule (always turn right, for instance) and to stop when the original segment is found again.

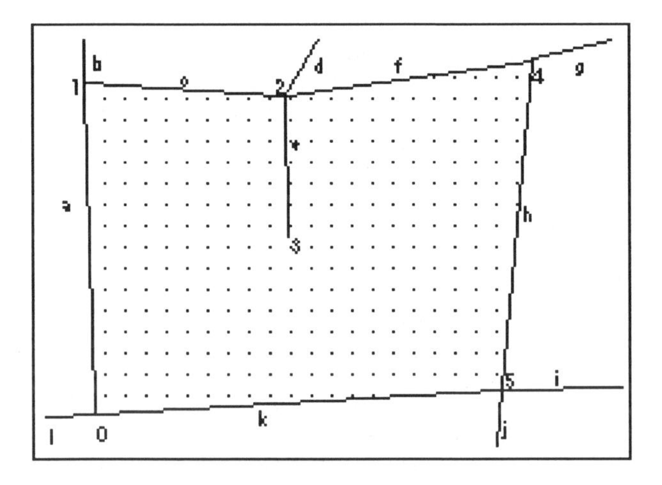

Figure 21.3 The connected case

A common method for the connected case consists of following all the alternatives, based on continuity information associated with lines, and might be described as: beginning with one edge (*a* in the example) or node (*0*), search for the first edge in the counter clockwise/clockwise direction—for example, the edge for which the angle made by it and the previous edge is minimum/maximum—(*c* in the example). When a dangling arc is found, the strategy is to return to a node where there are more than two segments (*2* in the example) and to restart the search with the second smallest angle (*f* in the figure). In a tree-like situation this procedure can be repeated as many times as needed. The unconnected case can be reduced to the connected case by generating the convex hull around the study area and then proceeding as in the previous case or by generating the generalised Voronoï diagram with lines as the generators (Okabe *et al.*, 1992). However simple, when applied to real data, this method soon reveals its weaknesses. In fact, complex network configurations, which are the norm in real situations, result in huge processing times due to the unguided exhaustive search performed.

These general methods are not of much use in our case because of the structure and contents of our data. In fact we are not interested in the lines themselves but on the information associated with their sides. That is, if we start the area generation process by using lines it corresponds to performing an initial aggregation of the information on the left and right, what can be undesirable. What is intended is to maintain line sides (left and right) apart and to use them as generators for the areas. A partial solution consists of performing a triangulation constrained to maintain the original lines as edges of the final triangles. In order to avoid triangles with points belonging to different lines the medial axis of the empty space is calculated beforehand and used in the constrained triangulation.

Preparata and Shamos (1985) show that there is a mathematical equivalence between the Delaunay triangulation and the Voronoï diagram. Moreover the skeleton of a polygon is the loci of the inscribed circles tangent to at least two points in the polygon boundary. The same authors also point out that the skeleton might be calculated during the Voronoï diagram determination. Two assumptions are made in the proof of the mathematical equivalence between Delaunay triangulation and the Voronoï diagram. One is that no four points in the dataset are co-circular (for example, it is not possible to build a circle passing through any four points in the dataset). The other assumption is a consequence of the definition of Delaunay triangulation, that is, the triangulation must not be constrained. More often than not, the triangulation wanted in this application is of a constrained nature and thus if the Voronoï diagram of such a dataset is constructed it will be plagued with degeneracies. The most straightforward way is to use a GIS to calculate the polygons and put seeds in them; another way is to use the process described by Hoult and Parker (1994) and also implemented in some commercially available GIS software packages. This process assumes that polygon seeds already exist, resulting from an automated process or by manual input. Starting with a bounding line segment, the process continues by searching in a clockwise direction for connecting line segments ordered by internal angle. The process is then repeated until a circuit is formed or until searching all the line segments connected to the initial one. The process uses a doubly linked list to store the sequence of lines, together with a flag indicating if a line defines the boundary of the polygon, or if it is interior.

A completely different approach to the problem is based on the Space Syntax theory developed at the Bartlett School of Architecture and Planning, University College London and described by Hillier and Hanson (1984) and Hillier *et al.* (1987). The methodologies associated with this theory aim to classify the global structure of urban settlements by subdividing the street centrelines in the study area according to their axiality, convexity, depth and shadowness. This involves the so-called alpha analysis, which is a model for the syntactic representation, analysis and interpretation of settlements. The alpha analysis involves building an axial map of the open space structure of the settlement, which corresponds to the least set of straight lines which passes through each convex space and makes all axial links, and the convex map, which is the least set of 'fattest' spaces that covers the system. Thus the Space Syntax involves a Top-Down approach to the problem operating from the whole dataset, by splitting space with several axes, until a certain resolution is achieved. Existing computerised implementations of Space Syntax methods of analysis rely on the manual input of the axial lines for a map and thus on human judgement.

The author is working on the automation of this task in order to keep uncontrolled factors to a minimum. One option, depicted in the next figure, is to work from an empty space triangulation by aggregating triangles into long and thin polygons (axis). Another option is to implement direct representations of empty space and, thus, to find the axial lines directly. These direct empty space representations have been used in robotics, namely to find the best paths for robot motion in the presence of obstacles.

Figure 21.4 Triangulation of empty space

21.6 EXPERIMENTS

In order to avoid the Voronoï diagram degeneracies which are introduced when the triangulation is constrained we had to abandon the concept of Voronoï diagram and to adopt another definition for the proximal regions around each line. Voronoï vertices (for example, the vertices of Voronoï polygons) are, by definition, the intersection of the three bisector lines of the Delaunay triangles. Degenerate cases appear whenever the intersection of the bisector lines (half-spaces in the generalised case) falls outside the triangle. The change to the definition of the proximal region was made in such a way that the vertices of the proximal regions will fall inside the triangles in all cases.

The adopted solution consists of (after performing a constrained triangulation of the dataset) of changing the method for determining the vertices of the proximal regions. That is, to define these vertices as the intersection of the three lines, passing through the triangle vertices and bisecting the angles at the triangle vertices. In turn, the edges of the proximal region are defined as the straight lines passing through the proximal region vertices and the mid-point of the triangle edges around a triangulation vertex.

Figure 21.5 shows the constrained triangulation of street-centrelines, which is at the core of the tested methodology.

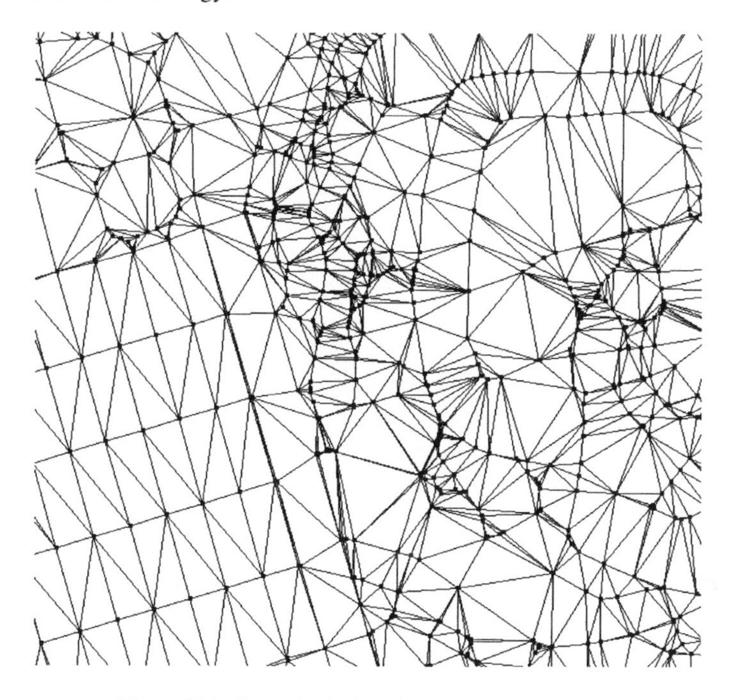

Figure 21.5 Constrained triangulation of street centrelines

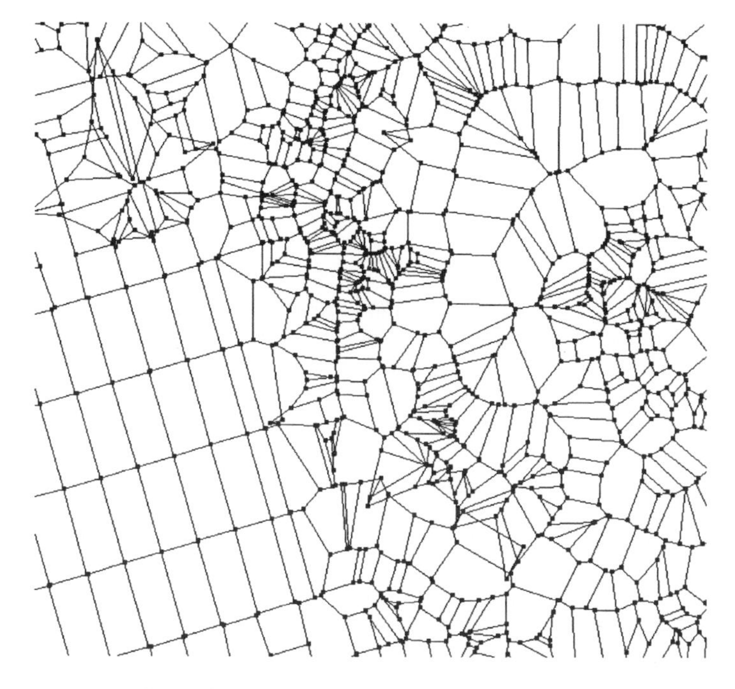

Figure 21.6 Voronoï polygons showing degeneracies

Figure 21.6 illustrates, for the same area as above, the Voronoï polygons obtained from the constrained triangulation, affected with degeneracies, for example, Voronoï vertices falling outside the base triangle.

Finally, Figure 21.7 depicts the proximal regions around street centrelines obtained when we change the definition of the proximal regions.

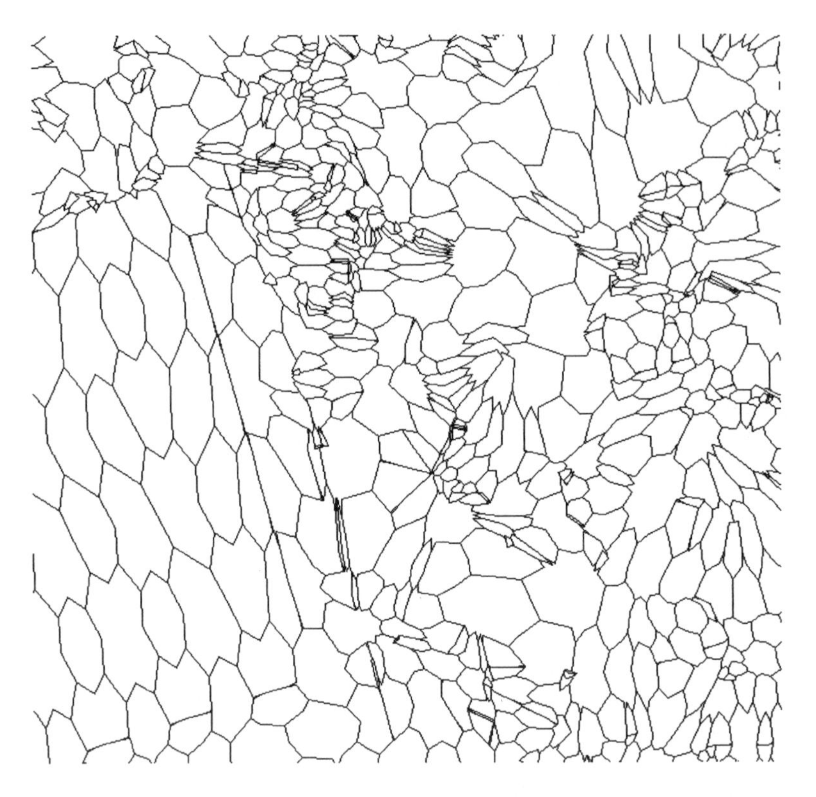

Figure 21.7 Proximal regions around street-centreline

21.7 CONCLUSIONS

The results so far show some promise in the line of thought followed which encourages the further exploration of this approach. Directions for further testing include the use of different proximal region definitions, the implementation of a triangulation of the original dataset of street-centrelines allowing for the discrimination between left and right sides, along the lines described above, and the use of Space Syntax methods or some other.

We reviewed some general strategies for the creation of socio-economic units from areal, point and linear data followed by the description of some methodologies for the creation of such units from linear data. The generation of zone systems from point-based data is undoubtedly the better-studied problem of those treated here. At the other extreme there are few references to the use of linear data types for this purpose. Thus, the line-to-area transformations deserve some further study and testing. Another topic deserving attention is the combined study of the line-to-area methods with already existing zone optimisation methods.

REFERENCES

Averack, R. and Goodchild, M., 1984, Methods and algorithms for boundary definition. In *Proceedings of the International Symposium on Spatial Data Handling, SDH '84*, Zurich, Switzerland, edited by Marble, D.F., Brassel, K.E., Peuquet, D.J. and Kishimoto, H., August 20–24, Vol. 1, pp. 238–250.

Bracken, I., 1994, A surface model approach to the representation of population-related social indicators. In *Spatial Analysis and GIS*, edited by Fotheringham, S. and Rogerson, P. (London: Taylor & Francis).

Coombes, M., Openshaw, S., Wong, C. and Raybould, S., 1993, GIS in community boundary definition. *Mapping Awareness & GIS in Europe*, **7** (4).

Flowerdew, R. and Green, M., 1989, Statistical methods for inference between incompatible zonal systems. In *The Accuracy of Spatial Databases*, edited by Goodchild, M. and Gopal, S. (London: Taylor & Francis), pp. 239–247.

Flowerdew, R. and Green, M., 1991, Data integration: Statistical methods for transferring data between zonal systems. In *Handling Geographic Information: Methodology and Potential Applications*, edited by Masser, I. and Blakemore, M. (Harlow: Longman), pp. 38–54.

Fotheringham, A.S., Densham, P.J. and Curtis, A., 1995, The zone definition problem in location–allocation modeling. *Geographical Analysis*, **27** (1), pp. 60–77.

Garfinkel, R.S. and Nemhauser, G.L., 1970, Optimal political redistricting by implicit enumeration techniques. *Management Science*, Series B, **16** (8), pp. 495–508.

Goodchild, M.F., 1979, The aggregation problem in location–allocation. *Geographical Analysis*, **11** (3), pp. 240–255.

Goodchild, M.F., 1987, A spatial analytical perspective on geographical information systems. *International Journal of Geographical Information Systems*, **1** (4), pp. 327–334.

Hess, S.W., Weaver, J.B., Siegfeldt, H.J., Whelan, J.N. and Zitlau, D.A., 1965, Non-partisan political redistricting by computer. *Operations Research*, **13**, pp. 998–1006.

Hillier, B. and Hanson, J., 1984, *The Social Logic of Space*, (Cambridge: Cambridge University Press).

Hillier, B., Hanson, J. and Peponis, J., 1987, Syntactic analysis of settlements. *Arch. & Comport./Arch. Behav.*, **3** (3), pp. 217–231.

Hoult, C., and Parker, D., 1994, The F in GIS: The use of face-based data in large scale GIS. In *Proceedings of Conference 'GIS Research in the UK' (GISRUK '94)*, Leicester, UK, April 11–13, 1994.

Johnson, D., Aragon, C., McGeoch, L. and Schevon, C. 1989, Optimization by simulated annealing: an experimental evaluation; Part 1: Graph partitioning. *Operations Research*, **37**, pp. 865–892.

Kriegel, H.-P., Horn, H. and Schiwietz, M., 1991, The performance of object decomposition techniques for spatial query processing. In *Advances in Spatial Databases, Proceedings of 2nd Symposium SSD '91*, Zurich, Switzerland, August 28–30, 1991, Lecture Notes in Computer Science 525, (Berlin: Springer-Verlag), pp. 257–276.

Lam, N.S.-N., 1983, Spatial interpolation methods: A review. *The American Cartographer*, **10** (2), pp. 129–149.

Macmillan, W. and Pierce, T., 1994, Optimization modelling in a GIS framework: The problem of political redistricting. In *Spatial Analysis and GIS*, edited by Fotheringham, S. and Rogerson, P. (London: Taylor & Francis), pp. 221–246.

Martin, D., 1991, Understanding socioeconomic geography from the analysis of surface form. In *Proceedings of the European GIS Conference (EGIS '91)*, Utrecht, pp. 691–699.

Masser, I., Batey, W.J.P. and Brown, P.J.B., 1978, Sequential treatment of the multi-criteria aggregation problem: a case study of zoning system design. In *Spatial Representation and Spatial Interaction*, edited by Masser, I. and Brown, P.J.B. (Leiden: Martinus Nijhoff Social Sciences Division), pp. 27–48.

Masser, I. and Brown, P.J.B., 1978, An empirical investigation of the use of Broadbent's rule in spatial system design. In *Spatial Representation and Spatial Interaction*, edited by Masser, I. and Brown, P.J.B. (Leiden: Martinus Nijhoff Social Sciences Division), pp. 51–69.

Oden, N.L., Sokal, R.R., Fortin, M.-J. and Goebl, H., 1993, Categorical wombling: Detecting regions of significant change in spatially located categorical variables. *Geographical Analysis*, **25** (4), pp. 315–336.

Okabe, A., Boots, B. and Sugihara, K., 1992, *Spatial Tesselations: Concepts and Applications of Voronoï Diagrams*, (Chichester: John Wiley & Sons).

Openshaw, S., 1976, A geographical solution to scale and aggregation problems in region-building, partitioning and spatial modelling. *Transactions of the Institute of British Geographers*, New Series, **2**, pp. 459–472.

Openshaw, S., 1977a, Optimal zoning systems for spatial interaction models. *Environment and Planning A*, **9**, pp. 169–184.

Openshaw, S., 1977b, Algorithm3: a procedure to generate pseudo random aggregations of N zones into M zones where M is less than N. *Environment and Planning A*, **9**, pp. 1423–1428.

Openshaw, S., 1978a, An empirical study of some zone design criteria. *Environment and Planning A*, **10**, pp. 781–794.

Openshaw, S., 1978b, An optimal zoning approach to the study of spatially aggregated data. In *Spatial Representation and Spatial Interaction*, edited by Masser, I. and Brown, P. (Leiden: Martinus Nijhoff Social Sciences Division), pp. 95–113.

Openshaw, S., 1984, The Modifiable Areal Unit Problem. Concepts and Techniques in Modern Geography No. 38, Environmental Publications, University of East Anglia.

Openshaw, S. and Rao, L., 1994a, A zone design system for ARC-INFO: Methods for re-engineering Census Geography. In Proceedings of *Conference 'GIS Research in the UK' (GISRUK '94)*, Leicester, April 11–13, p. 21.

Openshaw, S. and Rao, L., 1994b, Re-engineering 1991 Census Geography: serial and parallel algorithms for unconstrained zone design. Working Paper, School of Geography, Leeds University.

Pfaltz, J.L. and Rosenfeld, A., 1967, Computer representation of planar regions by their skeletons. *Communications of the ACM*, **10** (2), pp. 119–125.

Preparata, F.P. and Shamos, M.I., 1985, *Computational Geometry: An Introduction*, (Springer-Verlag).

Raper, J.F., Rhind, D.W. and Shephard, J.W., 1992, *Postcodes: The New Geography*, (London: Longman).

Reis, R.M.P. and Raper, J.F., 1994, Methodologies for the design and automated generation of postcodes from digital spatial data. In *Proceedings of the Fifth European Conf. and Exhibition on Geographical Information Systems—EGIS/MARI '94*, Paris, March 29th to April 1st, 1994, edited by Harts, J.J., Ottens, H.F.L. and Scholten, H.J., Vol. 1, pp. 844–851.

Sammons, R., 1978, A simplistic approach to the redistricting problem. In *Spatial Representation and Spatial Interaction*, edited by Masser, I. and Brown, P. (Leiden: Martinus Nijhoff Social Sciences Division), pp. 71–94.

Visvalingam, M., Wade, P. and Kirby, G.H., 1986, Extraction of area topology from line geometry. In *Proceedings of the Auto-Carto Conference*, London, edited by Blakemore, M., Vol. 1, pp. 156–165.

Visvalingam, M, 1991, Areal units and the linking of data: Some conceptual issues. In *GIS for Spatial Analysis and Spatial Policy: Developments and Directions*, edited by Worral, L. (London: Belhaven Press), pp. 12–37.

Witiuk, S.W., 1990, A spatial decision support system for autodistricting collection units for the taking of the Canadian census. Unpublished Ph.D. Thesis, University of Edinburgh.

CHAPTER TWENTY-TWO

Dealing with Boundary Changes when Analysing Long-Term Relationships on Aggregate Data

Jostein Ryssevik

22.1 INTRODUCTION

Regional or aggregate data are of great importance to social scientists studying social, cultural or economic change. This is especially true if the phenomena under investigation belong to a period where no microdata or sample survey data are available. But even for periods where such data are on hand, data describing spatial units (like census tracts, communes, etc.) might for various reasons be a valuable compliment or even a substitute.

One of the major problems facing the users of regional data, is the relative instability of the data-carrying units. The sets of units used for data collection and data distribution are normally administrative or political constructions object to frequent changes. When studying long-term relationships, this relative instability my have severe consequences for the cross-time comparability of data.

The aim of the chapter is to present and evaluate a solution to this problem, originally developed in connection with a database containing a major part of the official statistics ever published about Norwegian communes. The database covers a period of more than 200 years and is to a large extent used for cross-time comparisons and study of change. The obstacles caused by shifting borders between data-carrying units are as a consequence quite challenging.

The solution to the problem is based on an algorithm that whenever a change in the system of areal units occurs, recalculates data values to match the new units. The recalculations are based on information about population transfers and the type of data under consideration. As a consequence, the users of the database are able to retrieve data from various points in time and standardise them to a freely chosen set of units.

The logic of the procedure is based on several assumptions which are not always fully met. The accuracy of the recalculations might therefore vary. However, the tests and evaluations that have been carried out so far, are rather promising. Recalculation of data based on population transfer does not seriously effect conclusions based on statistical analysis of all or a large subset of regional units. When studying longer time-series for a few cases, the effects on data quality might in some cases be of a more severe character.

22.2 THE PROBLEM

Until very recently, the amount of *available* data about regional units has to a large extent been equal to the amount of *published* (for example, printed) regional data. Although the majority of data describing regional units is produced from information about individuals, only the aggregates have been published or taken care of for further use. The microdata has either been destroyed or preserved in a form that did not easily lend itself to re-aggregation. As a consequence, the major part of available quantitative information describing the geographical distribution of various social and economic phenomena is found in the printed publications of the statistical agencies.

For researchers studying long-term change this situation is far from ideal. On the one hand the types of regional units used for publishing official statistics might vary across time as well as across subject areas. On the other hand, even if the same *type of units* is used (i.e., communes, counties, etc.), the actual partitioning of the geographical space into this set of units will normally differ from one point in time to the other. Researchers trying to reconstruct past processes from the snapshots that can be extracted from publications of official statistics are therefore faced with the problem that the snapshots describe incomparable sets of areal units. The magnitude of the problem will vary across several dimensions. One obvious factor is the length of the time-period under investigation. The number of changes in the sets of units will normally be a function of the length of the research period.

It might be added that this problem is not only a product of the rigid nature of historical data. Even today, large parts of the official statistics are either published for fixed sets of areal units or kept in databases that do not allow for flexible re-aggregation to whatever set of units the researchers are demanding. Within a large number of subject areas the perfect world of geocoded microdata has not yet arrived.

22.2 GENERAL BACKGROUND

For almost 160 years the communes have been the smallest political and administrative units of Norway. While these units were originally based on the parishes (the subdivisions of the church), their recorded history can be stretched even further back in time. As viable political and administrative units the communes have throughout this period been the main data-carrying units for publication of official statistics, a fact which has given us 200 years of good and diverse aggregated information at this level.

In order to encourage the use of these valuable sources of information, the Norwegian Social Science Data Services (NSD) has built a database containing the major part of this statistics. The project started 25 years ago mainly as a vehicle for studying regional contrasts and long-term changes in electoral behaviour. Over the years, the scope of the database has gradually been broadened. The current Commune Data Base (CDB) covers the period from 1769 to present and contains close to 150 000 variables taken from a broad range of sources (demographic data, economic data, communal accounts, election data, data about the production of public services, etc.).

The main reason for integrating data from different sources in the same database is analytical. Firstly, if one is to move beyond mere description, some measure of covariance is needed, meaning that we need our measures as variables in the same data matrix. For obvious reasons, measures of the various phenomena we want to correlate are unlikely to come from the same source. Secondly, when studying phenomena over time multiple sources are a necessity. Integrating information from various sources may be anything from trivial to almost impossible. If every researcher or research team has to go through more or less the same procedure, there is an enormous effort, which in some instances may result in unreliable data and questionable findings.

CDB has thus become an important part of the research infrastructure in Norway serving students and researchers from a variety of disciplines. Most frequently the CDB is used for retrieving data for statistical analysis. For this purpose, the system has been supplied with procedures to output data formatted for standard statistical packages like

SPSS. A parallel coordinate database for the corresponding units and time periods has also been developed, allowing the users to produce the necessary input to standard software for map drawing. In addition, the database includes vast amounts of meta-data. As a general rule, all relevant information available in written form about variables, groups of variables or sources are included in the database and can be retrieved along with the data.

During the time span covered by the CDB, the communal subdivisions have changed dramatically. In 1837 the country was divided into less than 400 units. Due to numerous splits, the number increased gradually reaching its peak level of 750 units in the 1950s. After a thorough reconstruction of the communal subdivision in the 1960s the number was again reduced to about 450. Over the last three decades some major reconstructions of urban areas have gradually reduced the number even further. Moreover, according to a proposal from the government the current number of 435 communes ought to be reduced to about 250. Whether this reform will be implemented or not, awaits to be seen.

As a consequence of this process, there are only a handful of communes whose territory has been stable during the entire period. Accordingly, there are only a few years without any changes in communal borders. The types of changes that have taken place are also numerous, ranging from simple splits or merges to complicated rearrangements involving several partners. For researchers studying long-term change, this rather 'messy' situation represents quite a challenge. The development of a method that could provide a general and feasible solution to the problem of changing borders, has consequently been a high priority task within the CDB-project.

22.4 THE SOLUTION

At the most general level, the problem could have been approached in three different ways:

1. The database could simply store the data as they were and leave the problem of standardisation to the users.
2. The necessary standardisation could be done once and for all before the data were loaded into the database. Such a strategy would produce a harmonised database where both the initial problems and their solutions were concealed for the users.
3. The database could store the data as they appear in the sources and then be supplied with the necessary tools, rules and knowledge base to do various types of standardisation.

While the first strategy would leave the entire problem to the researcher, the second one would take away all of their control. Consequently, for a database serving a variety of researchers and research purposes only the third strategy would do the job.

Our solution has therefore been to maintain the original information as detailed and identical to the source as possible. Accordingly, data from 1900 are stored on the communal units that were in existence in 1900, while data from 1995 are stored on the units from 1995. As a second step, a database system has been developed which:

- stores and gives access to relevant information for standardisation;
- includes tools that simplify standardisation while utilising all available knowledge;
- provides the user with all relevant information about the data and how it has been modified by the standardisation process.

Standardisation of areal units over time (when no microdata is available) may be achieved in at least three different ways:

1. All areal units involved in changes during the period of study may be deleted from the data matrix, leaving only the set of unchanged units for the analysis. (This method could, of course, be modified by ignoring relatively small and insignificant changes). When there are few changes, this can be a workable solution. But it is far from satisfactory since areas affected by boundary changes tend to be non-typical areas. Moreover, when studying long-term change, the probability of losing most of the data matrix is rather high.
2. Another slightly more advanced possibility is to aggregate data to the largest possible set of common units. When units split, data are aggregated backwards in time. Accordingly, when units merge, data are aggregated forward. At least in our Norwegian case, this has proved to be a rather unsatisfactory solution. As changes of borders tend to be rather complex, involving several communes and a combination of splits and merges, the resulting units will very often be relatively large and consequently of little value to the analyst studying variation. A reduction in the number of units will always conceal information.
3. A third possibility is to measure the amount of resources redistributed when the borders between units are changed. If such measures are available, they can function as standardisation coefficients, in principle making it possible to calculate the effect of every change among the units. For obvious reasons this is a rather information intensive solution. While a particular change of borders between two units will effect different variables in different ways, the system would require a unique coefficient for every variable and every change in order to produce accurate results. The amount of information that is necessary to build coefficients at this level of accuracy is simply not available, and if it were, the amount of time needed to construct them would be forbidding. A more feasible solution would therefore be to base the coefficient matrix on estimates rather than exact values. The most likely candidate for such estimates, is the number of people transferred from one commune to another when borders change. Firstly, information about population transfer is in most cases rather easy to acquire. Secondly, the major part of variables in CDB are attributes of the population (age groups, people with higher education, number of votes for any given party, etc.). Assuming that these attributes approach a homogeneous distribution across the population within each commune, coefficients based upon population transfer will in most cases provide us with fairly accurate estimates of redistribution.

Users of CDB can choose between all of these three solutions when retrieving data for their analysis. The retrieval system will also supply them with hints and rules as to which method to use. If basic assumptions are violated, the system might even refuse to execute a query.

The third solution in the above list is clearly the best if the goal is to study all or at least a large subset of the communes by statistical analysis. However, the assumptions that this solution is based on, are not always met.

One type of problematic variables might be called *global attributes*. This is data describing the commune as an entity—like communal accounts, administrative apparatus, public services, etc. For variables like these, there are few other options than to assign missing data values whenever there is a change in borders. Alternatively a threshold can be defined to allow the system to ignore insignificant changes.

Another type of problem occurs for aggregated variables that measure characteristics not related to the population—for example, areal data, farm production, etc. For these types of data one must either assign missing values or be willing to use more drastic assumptions.

In order to prevent unjustified use of data, each variable in CDB is assigned a computability-indicator, which informs the retrieval program if and to what extent the coefficients can be used to manipulate values. This indicator can take on four different values:

Type 1: Standardisations based on the coefficients are allowed. This value is normally assigned to variables measuring aggregates of individuals.
Type 2: Variables in this class will only be recalculated when whole units are merged and then by adding values. This applies to situations where it is unlikely that information about population transfers renders valid estimates for the transfer of the resource under consideration. Examples might be farm production or the number of factories within a particular industry. In these cases it is possible to calculate the correct value merely by adding, otherwise missing data will be assigned.
Type 3: The crucial question concerning this class of variables is whether units involved in a change have the same value or not. Examples might be different types of typologies, indices, etc. If two communes that are being merged have identical values, the new unit will receive this value; otherwise missing data will be assigned.
Type 4: No recalculation whatsoever is possible. This class consists of global variables where a specific value is an inherent part of the communal unit. Every change of borders will in this case produce missing data.

The most valuable property of the system is to allow the user to retrieve data from various points in time and standardise them to a freely chosen set of units (year). As the system is able to recalculate data forward as well as backward in time, there are, in fact, no restrictions on the choice of year. However, one of our general recommendations is to recalculate data towards fewer units. As a rule of thumb, aggregations will normally render more accurate data than disaggregations. The time-span of the retrieved data (i.e., the number and types of changes involved) is therefore an important parameter, which should be taken into consideration when deciding which set of units to use.

The users are allowed to enter a missing data threshold, which comes into use whenever the computability indicator recommends to assign a missing value to a variable. Changes that affect a smaller portion of the population than indicated by the threshold will be ignored, leaving the involved units with their original data values.

To be able to evaluate the end result, the users can order a standardisation report, which allows them to follow a variable through the entire process of recalculations. This is important information when deciding whether to accept the produced data matrix or to change the parameters in a second run

22.5 EVALUATION

In order for the system to render accurate data, two important assumptions must be met:

1. The attribute being measured must approach a homogeneous distribution across the population within each commune. If we, as an example, are measuring the number of persons with higher education, the proportion of the population that belongs to this educational group should be constant across the territory of each commune.
2. The number of people living in the transferred part of a commune must be a constant proportion of the total population of that commune during the time-span of standardisation.

For obvious reasons there are numerous situations where these assumptions are not met. If the transferred part of a commune differs from the rest when it comes to social or economic structure, there is a great chance that the first assumption is violated. If, as an example, the transferred part of a commune forms an industrial enclave in an otherwise agricultural territory, any recalculation of variables measuring phenomena that are related to the industrial/agricultural dimension will be distorted. A comparable situation occurs when a suburban area is transferred from an otherwise rural commune to an expanding city—a type of change that is rather frequent in our material. Any recalculations of variables related to the rural/urban split will in this case be affected.

The second assumption is in some cases even harder to meet. If we, as an example, standardise data from the 1900-Census to the 1970-units, and a commune is involved in a change in 1965, the recalculations of the 1900-data will be based on a coefficient derived from the settlement structure of 1965. If the population growth has been higher in the transferred part of the commune than in the rest (in the period from 1900 to 1965), too much of the 1900-resources of that commune will be transferred to the receiving commune(s). The errors introduced by the violation of this assumption will normally be larger, the longer the time-span is between the collection of the data and the year of the applied set of units.

Different types of tests have been carried out to measure the amounts of errors introduced by the violation of these assumptions. A more detailed documentation of the tests can be found in (Alvheim, Olaussen and Sande).

In one group of tests recalculated data have been compared with correct data aggregated from information about census tracts. The data used for this group of tests was taken from the Census of 1960 and 1970—the start- and endpoint of a decade where the number of communes was reduced from 750 to 450. As the census tracts normally formed the building blocks of the reconstruction of the communal borders in this period, it is possible to use aggregated census tracts data to produce real 1960-data on 1970-units and vice versa. By comparing these real data with the corresponding recalculated data produced by CDB, we were able to measure the quality of our estimates. Two different variables were used in the tests: *the number of people employed in industry* and *the number of women 30–49 years of age*. The first of these variables is deemed specially sensitive to violations of the first assumption mentioned earlier, the second is not.

Bearing in mind that the reconstruction of the communal borders that took place in the 1960s was more dramatic than in any other period, the results were rather promising. Three important conclusions can be drawn from the experiment:

1. When studying just a few cases, the method should be used with caution.
2. Recalculations towards bigger units (from 1960 to 1970) render more accurate results than recalculations towards smaller units (from 1970 to 1960). In the first case correlation between actual and recalculated data was in the range from .992 to .999; in the second case the correlation was in the range from .702 to .925.
3. The differences in errors between the two variables were insignificant. This might indicate that violation of the second assumption is more hazardous than violation of the first.

In a second group of tests, factor-analyses were run on a set of variables describing the political, occupational and residential structure of the communes around 1970. The set of variables were recalculated to match the communal subdivisions from seven different points in time (from 1835 to 1980), and the otherwise identical analyses were run on all of these sets. Although the number of units ranged from 362 (in 1935) to 732 (in 1960), the various coefficients produced by the different runs were almost identical. All seven runs resulted in a 3-factor solution accounting for 77.8 % to 82.4 % of the variance. The various communalities, eigenvalues and factor-loadings were only minimally altered across the different sets of units.

It might therefore be argued that a factor-analysis run on these variables would lead to the same substantive conclusion whatever set of units it were applied on. Actual coefficients might be slightly modified, but that would not lead the researchers to different conclusions about the underlying structure of the data.

In a final group of tests, variables from separate points in time were correlated using different sets of units. The variables in question were taken from two referendums: votes for prohibition in 1919 and votes against Norway's entry into the European Common Market in 1972. As the outcome of both of these referendums has been explained as a kind of counter-cultural protest of the periphery, some positive correlation between the two variables was expected. The variables were standardised to match the units from 1919 (n = 700), 1960 (n = 732) and 1972 (n = 444). Again the substantive conclusion that can be drawn from the analysis is the same irrespective of the applied set of units. The correlations are all strongly positive ranging from .46 (1960-units) to .59 (1972-units).

22.6 IMPLICATIONS

The described solution to the problem of boundary changes is clearly not a perfect one. The ambition has been to develop a 'second best solution' in a situation where the 'best' is not feasible. The relative success of the approach should therefore be measured against:

- the amount of error and bias that are introduced to the data by this method;
- the amount of error and bias that would have been introduced by alternative methods;
- the loss of accumulated knowledge that would have been the result if the data were not made available for the research community in this way.

Concerning the first of these points, the few and scattered tests that are presented in this chapter do not warrant any strong conclusions. However, given that the intention has been to test the behaviour of the system in what we regard as 'worst-case-situations', it might be argued that the standardisation method does not seriously effect conclusions based on statistical analysis of all, or at least a large subset, of the communes. When using longer time-series for just a few cases, greater caution is called for.

When it comes to the second point, no systematic comparison of methods has been carried out so far. However, there are firm reasons to believe that at least some of the more likely alternatives might have more severe effects on data quality than the method described in this chapter:

Deleting all units involved in changes during the period of investigation will in most cases produce an unrepresentative or biased sample. When studying long-term change the number of units will also be dramatically reduced.

Aggregation towards the smallest stable units might be recommended when studying data from a shorter period or from a period where the number of changes is relatively small. In other cases, this method tends to produce very large units where a substantial part of the information (variance) will be concealed.

Concerning the third point, only speculations can be made. There are, however, no doubts that the CDB have been instrumental in promoting an extensive use of regional data among various Norwegian scholars, not only geographers but also economists, sociologists, political scientists, historians, etc.

For the time being, the biggest challenge is to establish a system where the accumulated experiences of the users can be fed back into the database. With a system for amending our knowledge about the data combined with continuously enhancing tools for implementing rules and automated knowledge-based decisions, we believe that the reliability of a given set of data can be improved, thus providing more and more accurate information.

REFERENCE

Alvheim, A., Olaussen, T.G. and Sande, T., An evaluation of solutions to the problem of boundary change when analyzing long-term relationships on aggregate data. In *Historical Social Research*, **29**, (Cologne: Quantum).

Postscript

This volume points out eloquently the need for an extension of current GIS with functions that can properly handle temporal data. It is certainly true that observations of time can be stored in current databases for GIS as well as observations of other properties, but this does not make these GIS 'temporal'. A temporal GIS must have, firstly, operations to deal with a data type 'time' properly—to calculate intervals, analyse the relations between intervals (Allen, 1981)—and, secondly, must provide constructs to describe changing objects.

This book has, unlike other similar volumes or meetings on temporal GIS (e.g., Egenhofer and Golledge, 1998), concentrated on a particular subset of all the possible applications of temporal GIS. All the chapters contribute—at different levels from the philosophical base to the concrete applications—to the analysis of the issues related to geographic databases, which *manage data collected with respect to socio-economic units over periods of time*. The prime examples are demographic databases, as they result from regular census operations. The spatial units, for which data are collected, are politically-administrative units, from country size down to small enumeration districts. These districts appear as stable, but seen over longer periods, they are subject to change. As early as 1978 Joel Morrison suggested that a spatio-temporal database be constructed for the historic county boundaries of the USA in order to properly document changes. This book addresses therefore a set of very particular and real needs, culminating perhaps in a report of a practical, useful and working system in Norway, which makes historic census data collected for different areal units available for comparative analysis (Ryssevik, Chapter 22).

The concentration on a particular set of issues as they are posed by applications has allowed the editors to build a pyramid of chapters, where the foundation chapters at the beginning of the book each address some of the issues as seen from the perspective of one of the scientific disciplines that are at the foundation of temporal GIS: philosophy and mathematics. These views are then gradually filtered to review the difficulties with the implementation of databases for GIS and the current state of the art. Finally this process leads to a review of several application areas, where the changing socio-economic units pose problems, and, at the end, details of the process of defining particular socio-economic units in different countries are tackled.

With such a fixed and relatively narrow topic, an interdisciplinary dialogue between the authors was necessary to prepare for the meeting and continued through the rewriting of the contributions. We hope that this collection amounts to more than the currently fashionable multi-disciplinary approach, i.e., more than an assembly of papers from different disciplines with a somewhat common topic, where each author tackles subjects, independently and with his own terminology and understanding a possibly related topic without engaging in a dialogue with the other disciplines. We have made efforts to help to bridge the gaps in the terminology, which is necessarily different for different disciplines, and to point out similarities and common points across disciplines even where they are not immediately obvious. Our goal was truly interdisciplinary scientific work, where the viewpoints and methods of different disciplines are contributing to a joint discussion of a subject and the result is more then the sum of the individual parts.

The concentration on this single topic excludes much. An initial survey of the issues surrounding temporal GIS revealed two major types of systems: systems where data regarding small, relatively fast moving objects are reported against a static spatial situation. The prototypical application would be car navigation systems, but also air traffic control, or parcel tracking. The other type of systems is changes of areal units in their form, location and extension. The most obvious application are demographic databases where data is collected with respect to a fixed subdivision of space, but detailed observation reveals that this subdivision of space is not completely fixed if considering long enough periods of time. The results of the workshop confirm that these are the two major types of temporal GIS, as this volume arrives at the same subdivision from different points of view (for example, Stefanakis and Sellis, Chapter 12). Other researchers must now take up the challenge of the first type of temporal GIS.

On the fundamental side, Güting has recently—within the Chorochronos research project—investigated the representation of moving objects in a database as 'current point and movement vector' and thus introduced an alternative to the storage and rapid update of mere positions. He then goes on to discuss how such moving objects can be approximated with the finite representation in computer systems, extending work from the representation of geometry to this realm (Erwig *et al.*, 1997). Güting extends the concept of data types for GIS, which are implied to be always seen as static, container-like, and introduces a new data type for moving objects, which includes the dynamism in the data type, not in the frequent update! The same concept should be applied to spatial socio-economic units to provide a higher, more meaningful abstraction, which can be used in models of spatial interaction. This will at least allow to understand spatial socio-economic units as changing and to properly model their interaction with the (mostly social) environment.

There are different views for temporal GIS. The authors of this volume seem (mostly) to agree that a careful analysis of the conceptual foundations, specifically the ontology implied, is the necessary starting point to make progress. However, it is immediately noted that the ontologies assumed differ not only in the level of detail studied and the terminology used for their description—both being aspects typically related to differences between disciplines as they are always found in a multi-disciplinary volume. The differences in ontologies fundamentally are in what they focus on: abstract space populated by objects and events, or objects that exist in space and time, or—most radically—concentrating on change in itself.

The focus of a post-modern approach is on change—but this coincides with the push of some of the technical studies: temporal GIS need not only describe the state of the world at different points in time, but need to provide facilities for users to analyse change and to model change with respect to space (Gautier, Chapter 14; Frank, Chapter 2) and such needs must eventually be reflected in the proposed data type for moving objects by Erwig *et al.*, (1997).

The analysis of time in GIS is often related to an analysis of a spatial or temporal hierarchy or hierarchy-like structure. The inadequacy of our understanding is already documented by subsuming a variety of phenomena under the term hierarchy—there are generalisation hierarchies and aggregation hierarchies (these are better understood in the database conceptual modelling literature (Brodie *et al.*, 1984)). However, there are also 'filter hierarchies' (Timpf, 1998) and hierarchies of operations, again of various types (Timpf *et al.*, 1992; Voisard and Schweppe, 1996). The inadequacy of our current understanding is best documented by the term hierarchy—nearly all the phenomena thus described are not strictly hierarchical but directed acyclic graphs (DAG), but often more restricted structures, for example, lattices. It is one of the results of this research collaboration that the analysis of objects and their interaction must be expanded to include on another scale also the change in the units themselves—in a way reflecting the social theory critique outlined by Aitken. Like the spatial socio-economic units we have studied, the boundaries and objects in the research landscape have also changed.

REFERENCES

Allen, J.F., 1981, *A General Model of Action and Time*, (New York: Department of Computer Science, University of Rochester).

Brodie, M.L., Mylopoulos, J. and Schmidt, J., 1984, *On Conceptual Modelling: Perspectives from Artificial Intelligence, Databases, and Programming Languages*, (Springer-Verlag).

Erwig, M., Güting, R.H., Schneider, M. *et al.*, 1997, *Spatio-Temporal Data Types: An Approach to Modeling and Querying Moving Objects in Databases*, Report No. 224 (Hagen: FernUniversitaet).

Egenhofer, M. and Golledge, R., Eds, 1998, *Spatial and Temporal Reasoning in Geographic Systems*, (New York: Oxford University Press).

Timpf, S., 1998, Hierarchical Structures in Map Series. Ph.D. thesis, Department of Geoinformation, (Vienna: Technical University Vienna).

Timpf, S., Volta, G.S. *et al.*, 1992, A conceptual model of wayfinding using multiple levels of abstractions. In *Theories and Methods of Spatio-Temporal Reasoning in Geographic Space*, Lecture Notes in Computer Science 639, edited by Frank, A.U., Campari, I. and Formentini, U. (Berlin: Springer-Verlag), pp. 348–367.

Voisard, A. and Schweppe, H., 1996, A multilayer approach to the open GIS design problem. In *Proceedings of 2nd ACM GIS Workshop*, Gaithersburg, MD (New York: ACM Press).

Index